M
A

MICRO-ASSEMBLY TECHNOLOGIES AND APPLICATIONS

IFIP TC5 WG5.5 Fourth International Precision Assembly Seminar (IPAS'2008) Chamonix, France February 10-13, 2008

Edited by

Svetan Ratchev
*University of Nottingham
United Kingdom*

Co-edited by

Sandra Koelemeijer
*École Polytechnique Fédérale
de Lausanne
Switzerland*

Micro-Assembly Technologies and Applications

Edited by Svetan Ratchev and Sandra Koelemeijer

p. cm. (IFIP International Federation for Information Processing, a Springer Series in Computer Science)

ISSN: 1571-5736 / 1861-2288 (Internet)
ISBN: 978-0-387-77402-2
eISBN: 978-0-387-77405-3

Printed on acid-free paper

Copyright © 2008 by International Federation for Information Processing.
All rights reserved. This work may not be translated or copied in whole or in part without the written permission of the publisher (Springer Science+Business Media, LLC, 233 Spring Street, New York, NY 10013, USA), except for brief excerpts in connection with reviews or scholarly analysis. Use in connection with any form of information storage and retrieval, electronic adaptation, computer software, or by similar or dissimilar methodology now known or hereafter developed is forbidden.

The use in this publication of trade names, trademarks, service marks and similar terms, even if they are not identified as such, is not to be taken as an expression of opinion as to whether or not they are subject to proprietary rights.

Printed in the United States of America.

9 8 7 6 5 4 3 2 1

springer.com

TABLE OF CONTENTS

Preface   xi

International Advisory Committee   xiii

PART I – Design of Micro Products   1

Chapter 1. Methods for Design for Micro-Assembly   3

Application of a DFµA Methodology to Facilitate the Assembly of a Micro/Nano Measurement Device   5
Carsten Tietje, Richard Leach, Michele Turitto, Ronaldo Ronaldo, Svetan Ratchev

A DFA Framework for Hybrid Microsystems   13
Marcel Tichem, Dafina Tanase

Statistical Assemblies with Form Errors – A 2D Example   23
Pierre-Antoine Adragna, Hugues Favrelière, Serge Samper, Maurice Pillet

Chapter 2. Methods and Solutions for Micro-Product Design   35

A Classification and Coding System for Micro-Assembly   37
Amar Kumar Behera, Shiv G. Kapoor, Richard E. DeVor

A Method for Three Dimensional Tolerance Analysis and Synthesis
applied to Complex and Precise Assemblies   55
Frédéric Germain, Dimitri Denimal, Max Giordano

New Designs for Submillimetric Press-Fitting   67
Ludovic Charvier, Fabien Bourgeois, Jacques Jacot, Grégoire Genolet, Hubert Lorenz

Design and Testing of an Ortho-Planar Micro-Valve   75
O. Smal, B. Raucent, F. Ceyssens, R. Puers, M. De Volder, D. Reynaerts

Robust Design of a Lens System of Variable Refraction Power 87
with respect to the Assembly Process
Ingo Sieber, Ulrich Gengenbach, Rudolf Scharnowell

PART II – Micro-Assembly Processes and Applications 95

Chapter 3. Process Modelling for Micro-Assembly 97

Product-Process Ontology for Managing Assembly Specific 99
Knowledge between Product Design and Assembly System
Simulation
Minna Lanz, Fernando Garcia, Timo Kallela, Reijo Tuokko

Bridging the Gap – From Process Related Documentation to an In- 109
tegrated Process and Application Knowledge Management in Micro Systems Technology
Markus Dickerhof, Anna Parusel

Distributed Simulation in Manufacturing using High Level Architecture 121
J. Rodríguez Alvarado, R. Vélez Osuna, R. Tuokko

Chapter 4. High Precision Packaging and Assembly Processes 127

Adaptive Packaging Solution for a Microlens Array placed over a 129
Micro-UV-LED Array
Markus Luetzelschwab, Dominik Weiland, Marc P. Y. Desmulliez

Solder Bumping – A Flexible Joining Approach for the Precision 139
Assembly of Optoelectronical Systems
*Erik Beckert, Thomas Burkhardt, Ramona Eberhardt,
Andreas Tünnermann*

FluidAssem - A New Method of Fluidic-Based Assembly with 149
Surface Tension
*N. Boufercha, J. Sägebarth1, M. Burgard, N. Othman,
D. Schlenker, W. Schäfer, H. Sandmaier*

Concepts for Hybrid Micro Assembly using Hot Melt Joining 161
Sven Rathmann, Annika Raatz, Jürgen Hesselbach

Application of Microstereolithography Technology in Micromanufacturing *Hongyi Yang, Gregory Tsiklos, Ronaldo Ronaldo, Svetan Ratchev*	171
In Situ Micro-Assembly *Ronaldo Ronaldo, Thomas Papastathis, Hongyi Yang, Carsten Tietje, Michele Turitto, Svetan Ratchev*	177

Chapter 5. Micro-Assembly Applications — 187

Interest of the Inertial Tolerancing Method in the case of Watch Making Micro Assembly *Maurice Pillet, Dimitri Denimal, Pierre-Antoine Adragna, Serge Samper*	189
Precision Assembly of Active Microsystems with a Size-Adapted Assembly System *Kerstin Schöttler, Annika Raatz, Jurgen Hesselbach*	199
Assembly of Osseous Fragments in Orthopaedic Surgery: The Need for New Standards of Evaluation *Olivier Cartiaux, Laurent Paul, Pierre-Louis Docquier, Xavier Banse, Benoit Raucent*	207

PART III – Gripping and Feeding Solutions for Micro-Assembly — 219

Chapter 6. Micro-Gripping Methods and Applications — 221

A Critical Review of Releasing Strategies in Microparts Handling *Gualtiero Fantoni, Marcello Porta*	223
A Low Cost Coarse/Fine Piezoelectrically Actuated Microgripper with Force Measurement Adapted to the EUPASS Control Structure *Kanty Rabenorosoa, Yassine Haddab, Philippe Lutz*	235
Development of a Monolithic Shape Memory Alloy Manipulator *Kostyantyn Malukhin, Kornel Ehmann*	243

Statically Determined Gripper Construction *Ronald Plak, Roger Görtzen, Erik Puik*	251
Precision Positioning down to Single Nanometres Based on Micro Harmonic Drive Systems *Andreas Staiger, Reinhard Degen*	257
Assembly of a Micro Ball-Bearing using a Capillary Gripper and a Microcomponent Feeder *C. Lenders, J.-B. Valsamis, M. Desaedeleer, A. Delchambre, P. Lambert*	265

Chapter 7. High Precision Positioning and Feeding Techniques — 275

Pneumatic Contactless Microfeeder: Design Optimisation and Experimental Validation *Michele Turitto, Carsten Tietje, Svetan Ratchev*	277
Pneumatic Positioning System for Precision Assembly *Martin Freundt, Christian Brecher, Christian Wenzel, Nicolas Pyschny*	285
Manufacturing of Devices for the Parallel Precision Alignment of Multiple Micro Components *Christian Brecher, Martin Weinzierl*	297

Chapter 8. Micro-Metrology — 305

Towards a Traceable Infrastructure for Low Force Measurements *Richard K. Leach, Christopher W. Jones*	307
When Manufacturing Capability Exceeds Control Capability: The Paradox of High Precision Products *S. Koelemeijer Chollet, M. Braun, F. Bourgeois, J. Jacot*	315
Impact Forces Reduction for Micro-Assembly *Ronald Plak, Roger Görtzen, Erik Puik*	325

PART IV – Development of Micro-Assembly Production Systems — 333

Chapter 9. Design of Modular Reconfigurable Micro-Assembly Systems — 335

Strategies and Devices for a Modular Desktop Factory — 337
Arne Burisch, Annika Raatz, Jurgen Hesselbach

A Decision Making Tool for Reconfigurable Assembly Lines – EUPASS Project — 345
F. Wehrli, S. Dufey, S. Koelemeijer Chollet, J. Jacot

Standardised Interface and Construction Kit for Micro-Assembly — 353
Matthias Haag, Samuel Härer, Andreas Hoch, Florian Simons

Towards a Publish/Subscribe Control Architecture for Precision Assembly with the Data Distribution Standard — 359
Marco Ryll, Svetan Ratchev

Smart Assembly – Data and Model Driven — 371
Juhani Heilala, Heli Helaakoski, Irina Peltomaa

Chapter 10. Assembly System Integration — 383

Man - Robot Co-operation – New Technologies and New Solutions — 385
Timo Salmi, Ilari Marastio, Timo Malm, Esa Laine

Integration of Design and Assembly using Augmented Reality — 395
Juha Sääski, Tapio Salonen, Mika Hakkarainen, Sanni Siltanen, Charles Woodward, Juhani Lempiäinen

Concept for an Industrial Ubiquitous Assembly Robot — 405
Juhani Heilala, Mikko Sallinen

Author Index — 415

Preface

Micro-assembly is a key enabling technology for cost effective manufacture of new generations of complex micro products. It is also a critical technology for retaining industrial capabilities in high labour cost areas such as Europe since up to 80% of the production cost in some industries is attributed directly to assembly processes. With the continuous trend for product miniaturisation, the scientific and technological developments in micro-assembly are expected to have a significant long-term economic, demographic and social impact.

A distinctive feature of the process is that surface forces are often dominant over gravity forces, which determines a number of specific technical challenges. Critical areas which are currently being addressed include development of assembly systems with high positional accuracy, micro gripping methods that take into account the adhesive surface forces, high precision micro-feeding techniques and micro-joining processes.

Micro-assembly has developed rapidly over the last few years and all the predictions are that it will remain a critical technology for high value products in a number of key sectors such as healthcare, communications, defence and aerospace. The key challenge is to match the significant technological developments with a new generation of micro products that will establish firmly micro-assembly as a core manufacturing process.

The book includes contributions by leading experts in the field of micro-assembly presented at the 4th International Precision Assembly Seminar (IPAS'2008) held from 10 to 13 February 2008 in Chamonix, France. The seminar has established itself as a premier international forum for reporting and discussing research results and technical developments and charting new trends in micro assembly. The published works have been grouped into 4 parts. Part 1 is dedicated to micro-product design with specific emphasis on design for micro-assembly (DFµA) methods and solutions. Part 2 is focused on micro-assembly processes and includes contributions in process modelling, high precision packaging and assembly techniques and specific examples of micro-assembly applications. Part 3 describes the latest developments in micro-gripping, micro-feeding and micro-metrology. Part 4 provides an overview of the latest developments in the design of micro-assembly production systems with specific emphasis on reconfigurable modular micro-assembly equipment solutions.

The seminar is sponsored by the International Federation of Information Processing (IFIP) WG5.5, the International Academy of Production Research (CIRP) and the European Factory Automation Committee (EFAC). The seminar is supported by a number of ongoing research initiatives and projects including the European sub-technology platform in Micro and Nano Manufacturing MINAM, the UK

EPSRC Grand Challenge Project 3D-Mintegration, The EU funded coordinated action Micro-Sapient and the EU funded integrated project EUPASS.

The organisers should like to express their gratitude to the members of the International Advisory Committee for their support and guidance and to the authors of the papers for their original contributions. Our special thanks go to Professor Luis Camarinha-Matos, Chair of the IFIP WG5.5 and Professor Helmut Bley, Chair of the STC A of CIRP for their continuous support and encouragement. And finally our thanks go to Ruth Strickland and Rachel Watson from the Precision Manufacturing Centre at the University of Nottingham for handling the administrative aspects of the seminar, putting the proceedings together and managing the detailed liaison with the authors and the publishers.

Svetan M Ratchev

International Advisory Committee

Prof T Arai - University of Tokyo, Japan
Prof H Asfarmanesh – University of Amsterdam, The Netherlands
Prof M Björkman - Linköping Institute of Technology, Sweden
Prof H Bley - University of Saarland, Germany
Prof C R Boer – ICIMSI-SUPSI, Switzerland
Prof I Boiadjiev - TU Sofia, Bulgaria
Prof L M Camarinha-Matos - Univ Nova, Portugal
Prof D Ceglarek – Warwick University, UK
Prof A Delchambre - ULB, Belgium
Prof M Desmulliez – Heriot-Watt University, UK
Prof S Dimov – University of Cardiff, UK
Prof G Dini - Univ di Pisa, Italy
Dr S Durante – DIAD, Italy
Prof K Ehmann – Northwestern University, USA
Prof R Fearing - University of California at Berkeley, USA
Prof R W Grubbström - Linköping Institute of Technology, Sweden
Dr C Hanisch - Festo AG & Co, Germany
Prof T Hasegawa - Kyushu University, Japan
Mr J Heilala - VTT, Finland
Prof J M Henrioud - LAB, France
Prof J Jacot - EPFL, Switzerland
Dr M Krieger – CSEM, Switzerland
Dr P Lambert – ULB, Belgium
Dr R Leach – National Physical Laboratory, UK
Prof P Lutz - LAB, France
Dr H Maekawa – Nat. Inst. of Adv. Industrial Science & Technology, Japan
Prof B Nelson – ETH, Switzerland
Prof J Ni – University of Michigan, USA
Prof D Pham - Cardiff University, UK
Prof M Pillet – Polytech Savoie, France
Dr G Putnik - University of Minho, Portugal
Prof B Raucent - UCL, Belgium
Prof K Ridgway – Sheffield University, UK
Prof K Saitou – University of Michigan, USA
Prof G Seliger - TU Berlin, Germany
Dr W Shen – Nat. Research Council, Canada
Dr M Tichem - TU Delft, The Netherlands
Prof R Tuokko – TUT, Finland
Prof E Westkämpfer – Fraunhofer IPA, Germany
Prof D Williams – Loughborough University, UK

PART I

Design of Micro Products

Chapter 1

Methods of Design for Micro-Assembly

APPLICATION OF A DFμA METHODOLOGY TO FACILITATE THE ASSEMBLY OF A MICRO/NANO MEASUREMENT DEVICE

Carsten Tietje[1], Richard Leach[2], Michele Turitto[1], Ronaldo Ronaldo[1], Svetan Ratchev[1]

[1]Precision Manufacturing Group, The University of Nottingham,

University Park, Nottingham, NG7 2RD, UK.
[2]Industry & Innovation Division, National Physical Laboratory,
Hampton Road, Teddington, TW11 0LW, UK.

epxct1@nottingham.ac.uk, richard.leach@npl.co.uk, epxmt2@nottingham.ac.uk, ronaldo.ronaldo@nottingham.ac.uk, svetan.ratchev@nottingham.ac.uk

Abstract A lack of well defined Design for Microassembly (DFμA) methodologies to enable an increased transfer of prototypes from the research lab to production on industrial scale has been identified. The main benefit of such a methodology is the adaptation of the design by matching it with microassembly process characteristics. In addition there needs to be a push in metrology equipment to respond to the ongoing trend of miniaturisation, enabling quality assurance for three dimensional products with nanometer scale features. The presented paper addresses these two gaps by utilising a novel DFμA methodology to enable a state-of-the-art CMM stylus assembly, which is characterised by extremely rigid and challenging requirements. The design of the parts to be assembled is shown. Furthermore the selection of the most suitable assembly equipment is supported. Finally the actual assembly system is described and illustrated as proof of validation.

1 Introduction

The key challenges of manufacturing in a commercial environment can be summarised as delivering products with a competitive price, competitive quality and competitive delivery time. These challenges are supported by a number of production objectives, such as high flexibility, high productivity, constant and high product quality, short throughput times and low production costs [1]. Within the production chain, assembly is a critical step because it forms a substantial part of the total production cost and throughput time. This is particularly true for the microdomain, where it is widely accepted that up to 80% of the production cost of miniaturised systems or hybrid systems occur in assembly [2, 3].

The Nexus Market study 2005 predicts the market size for microsystems and microtechnologies to more than double from €16 billion in 2005 to €36 billion by the year 2009 [4]. The major market sectors to benefit from the trends of miniaturisation and automation are in the high-technology areas of medical/surgical, automotive and transport, biotechnology and consumer products [5]. To enable these developments it is imperative to develop appropriate manufacturing tools and methods. Under reflection of the above described trends the following two critical needs have been identified and singled out:

- A lack of sufficient Design for Microassembly (DFµA) methodologies to enable an increased transfer of prototypes from the research laboratory to production on industrial scales has been identified. The main benefit is the adaptation of the design by matching it with assembly process characteristics.
- Because "measurement underpins manufacturing technology"[6], there needs to be a push in metrology equipment to respond to the ongoing trend of miniaturisation to enable quality assurance for arising three dimensional products with nanometre scale features.

This paper addresses these two gaps by utilising a novel DFµA methodology [7] to enable the state-of-the-art coordinate-measuring machine (CMM) stylus assembly, which is characterised by extremely rigid and challenging requirements. A CMM is a programmable, versatile instrument that is used to measure dimensional data for many types of manufactured component. CMMs have three or more measurement axes, usually linear or rotary or a combination of the two. The measurement axes are combined in series so that a unique combination of axes positions defines a single point in space. Measuring an object using a CMM is achieved by moving a measuring probe to a number of points on the object surface in sequence and measuring the position of the probe at each point via the machine scales. The presented paper shows that the DFµA methodology influences the design of the parts to be assembled and furthermore enables the selection of appropriate assembly equipment. The final assembly system and processes and their validation are described and illustrated.

2 Microassembly and -metrology – scope definition

2.1 Critical challenges in microassembly

"Assembly is defined here as bringing together parts and / or subassemblies, so that a unit comes into being. A subassembly is a composition of parts into a product unit. The assembly process is determined by the manner and the sequence in which the product parts are put together into a complete product. The assembly process comprises a cycle of operations. These operations can be divided into: feeding, handling, composing, checking, adjusting and special processes" [1]. Thus it becomes clear that assembly is more than putting parts together. In fact, "Assembly is the capstone process in manufacturing" [8]. Furthermore, assembly is undoubtedly the "least understood process in manufacturing" [8]. Assembly in the microworld faces new challenges hence the complexity increases enormously.

- Sticking effects: Superficial tension, electrostatic forces, van der Waals forces
- Required high positional accuracy of automatic assembly machines
- Difference in the mechanics of object interactions due to scaling effects
- Loss of direct hand-eye coordination for the operator
- Cost of manipulation in the micro world (serial versus parallel assembly)
- Clean room environment
- Limited standard equipment

Fig.1. Assembly problems in the micro domain

The main difference between assembly in the macro- and in the microworld is the required positional precision and accuracy. Closed-loop strategies are needed to compensate for poor kinematic models and thermal effects. Another important difference between assembly in the macro- and the microdomain is related to the mechanics of object interactions due to scaling effects [9-11]. In the microworld forces other than gravity dominate. Surface related forces, such as *van der Waals, surface tension forces* and *electrostatic forces* become dominant over gravitational forces. Because of this scaling behaviour, manipulation in the microdomain is entirely different from manipulation in the macroworld. As these forces are very difficult to control, they are likely to disturb the assembly process. The object might jump to the gripper and lose orientation or stick to the gripper so that releasing the parts becomes difficult [9, 12]. An additional problem to sticking effects in the microworld is the loss of direct hand-eye coordination for the operator [9]. Microscopes and tools limit the ability to directly see and sense the objects to be handled.

2.2 Increasing demand for high accuracy measurement systems

Most of the devices used for microprocess examination originate from the macroworld and do not fulfil the required demands of microtechnology. Moreover, the downscaling of methods and techniques for quality control from the macroworld is problematic, because experiences and results cannot easily be transferred into the microdimension, e.g. resolution, measuring range or image quality restrict the usability [13]. However, the delivery of micro products with nanometre scale features needs to be supported by reliable metrology [6].

Figure 2 gives a schematic overview about micrometrology and identifies four areas, namely material testing, completeness check, dimension and position measurement and function test. The metrological tasks are used mainly for three different components: electronic components, optical components and mechanical components. PFEIFFER ET AL. state that nearly 90% of the required measurement tasks can be identified as *dimension and position measurement* [13]. The measurement device the assembly of which is described within this paper will enable better measurements within this significant area, responding to the increasing demand for high accuracy measurement systems.

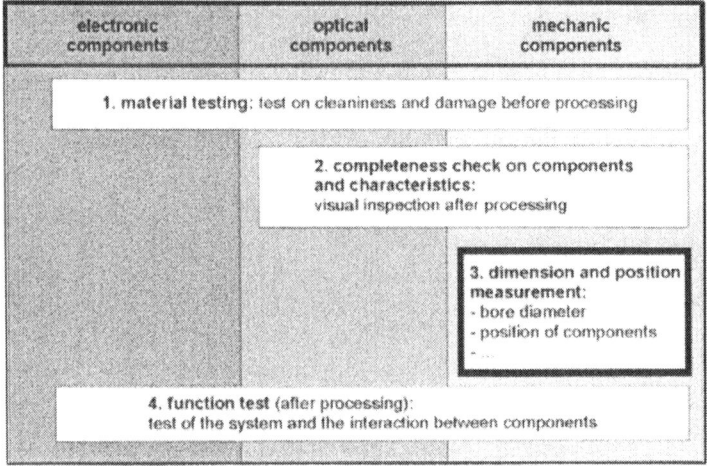

Fig.2. Measurement tasks in the field of microtechnology [13]

3 Assembly of a micro/nano measurement device

3.1 DFµA methodology – conceptual overview

A novel DFµA methodology [7] is used to influence the stylus design and to choose the appropriate assembly processes. The main objective of the methodology is to facilitate improvements in the design process by
- Applying design rules and guidelines which are focused on the microworld to cope with its specific challenges and
- Considering key assembly process features in early design stages.

Another main objective is *to determine the appropriate assembly processes* by considering process related requirements. Offering *qualitative cost analysis* supports the decision making for the assembly system design.

Figure 3 shows the conceptual structure of the DFµA methodology and the underlying models. The first design specifications are based on the product requirements which influence the design the most (mainly functional requirements). Only conceptual drawings are needed. But, like in other DFA methods, the more comprehensive the initial information, the more effective the result. This initial product design are analysed and evaluated by applying the DFµA guidelines.

After feeding the input from the *DFµA guideline model* into the design the *process-product analysis* (which is the key development within the methodology) is carried out. The aim is to consider process related design aspects in the product design and to determine the most appropriate assembly processes. The processes are selected based on the improved design. To decide between several suitable processes a qualitative cost analysis can be conducted to support the decision making.

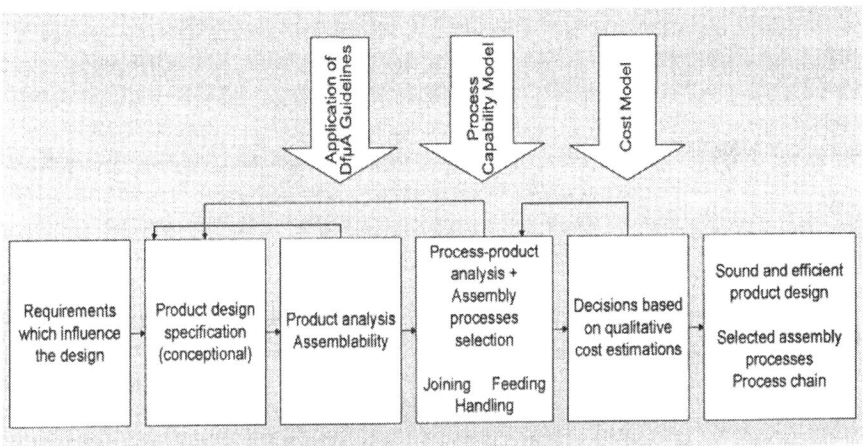

Fig.3. DFµA methodology and underlying models [7]

3.2 Application of the DFµA methodology

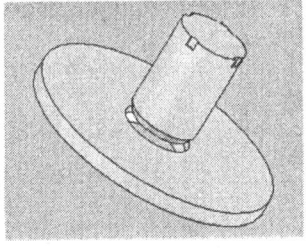

Fig.4. Pin in plate assembly

Figure 4 presents the assembly of a CMM stylus in a schematic way. To assure functioning of the part it is critical to maintain a 90° angle between plate and pin. Further essential and challenging tasks such as *alignment accuracy, sizes/dimensions, surface roughness, functional elements, environmental constraints (e.g. vibration), reversibility of bonds, volume, part feeding restrictions and fragility*, make the assembly of the measurement device unique.

The DFμA methodology is used to influence the stylus design. The figure indicates chamfers on the plate and the pin that aim at easing the assembly and assuring perpendicularity. The four notches in the pin are designed in compliance with the used gripper, guaranteeing the right orientation of the part within the gripper so that the actual insertion process takes place very close to the desired 90° angle. The chosen appropriate assembly processes are considered in this design. The combination of tight tolerances in the parts design with the high characteristics of the selected assembly processes allows the desired accuracy. Other assembly designs and accordingly assembly process chains were evaluated. The following key points are addressed:

- Match between design and assembly processes (high accuracy positioning, gluing, etc.)
- Relevant design rules (according to requirements)
- Selected processes
- Influenced design features (gripper handle, joint design, capillaries for glue deposition, chamfer etc.)

3.3 Implementation of the assembly system

Possible implementations, containing not only the hardware setup and its elements but also relevant validation routes, are described in this section. Figure 5 shows the micro assembly system that is used to assemble the stylus to the plate. The system is characterised by three degrees of freedom realised by three linear piezo-driven stages (X,Y,Z). Furthermore a camera is used to observe the process. A piezo-driven gripper, attached in Z-direction to a force sensor is used to manipulate and compose the pin into the plate. A glue dispenser is used to bond the parts together. Figure 5 gives an overview of the whole system, including network controller, light sources as well as an detailed view of the point of work, displaying the linear stages in X- and Y-direction as well as the gripper (together with force sensor attached to linear stage in Z direction), the camera and the glue dispenser.

 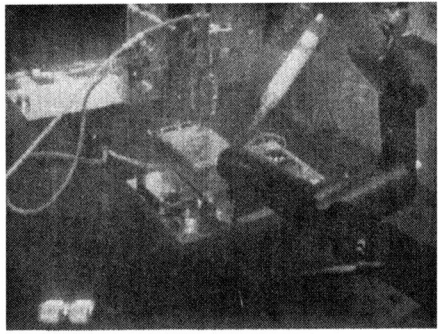

Fig.5. Microassembly station

The chosen system setup allows inspection and supervision of the assembly process. Accurate knowledge on geometry, length, force, pressure and surface roughness of the surveyed parts as well as flow characteristics or the composition of the joining medium (glue) guarantees a high-quality assembly process [13]. The post-assembly evaluation will be extremely challenging and is important to prove the predicted results. For that reason it is envisaged to use the ZEISS F25, for accurate micro/nano metrology. The F25 provides high resolution of 7.8 nm. The tactile sensor is made from the piezoelectric membrane with small probing force at 0.5 mN, and therefore it is non destructive and is ideal for the presented microparts.

4 Conclusion and outlook

The assembly process for a novel CMM stylus is described. The design and assembly process selection procedure is supported by a specific DFµA methodology. The presented assembly challenge was used to evaluate the methodology. It can be stated that the preliminary results and the outlined validation are promising but not finally concluded yet. The methodology still relies on human reasoning and interaction but supports the design and process selection process sufficiently. The assembly system layout is described and illustrated and the validation procedure outlined.

Future work can be divided into three strands. The actual validation of the probe assembly needs to be carried out and the assembly needs to be integrated into a metrology system. Finally the DFµA methodology needs to be further developed, in particular the decision making method.

Acknowledgments

The presented work is part of the ongoing United Kingdom EPSRC "Grand Challenge" research project "3D-Mintegration" which aims to provide radically new ways of thinking for end-to-end design, processing, assembly, packaging, integration and testing of complete 3D miniaturised/integrated '3D Mintegrated' products. The presented assembly challenge is provided by NPL.

References

1. Rampersad, H.K., *Integrated Simultaneous Design for Robotic Assembly*. 1994, Chisester et al.: John Wiley & Sons.
2. Hesselbach, J. and A. Raatz, eds. *mikroPRO, Untersuchung zum internationalen Stand der Mikroproduktionstechnik*. 2002, Vulkan: Essen.
3. Koelemeijer, S. and J. Jacot, *Cost Efficient Assembly of Microsystems*. mst-News, 1999. **January**: p. 30-33.
4. Unknown, *Market Analysis for MEMS and Microsystems III, 2005-2009*. 2005: WTC - Wicht Technologie Consulting (on behalf of NEXUS Association).
5. Dimov, S.S., et al. *A roadmapping study in 4M Multi-Material Micro Manufacture*. in *4M 2006 - SECOND INTERNATIONAL CONFERENCE ON MULTI-MATERIAL MICRO MANUFACTURE*. 2006 A roadmapping study in 4M Multi-Material Micro Manufacture. Grenoble: Elsevier (Oxford).
6. Leach, R., et al., *Advances in traceable nanometrology at the National Physical Laboratory*. Nanotechnology, 2000. **12**: p. R1-R6.
7. Tietje, C. and S. Ratchev. *Design for Micro Assembly - A methodology for product design and process selection*. in *IEEE International Symposium on Assembly and Manufacturing (ISAM)*. 2007. Ann Arbor, USA: Omnipress.
8. Whitney, D.E., *Mechanical Assemblies - Their Design, Manufacture, and Role in Product Development*. 2004, New York, Oxford: Oxford University Press.
9. Van Brussel, H., et al., *Assembly of Microsystems*. Annals of the CIRP, 2000. **49**(2): p. 451-472.
10. Fearing, R. S. *Survey of Sticking Effects for Micro-Parts*. in *IEEE International Conference for Robotics and Intelligent Systems IROS '95*. 1995. Pittsburgh.
11. Ando, Y., H. Ogawa, and Y. Ishikawa. *Estimation of attractive force between approached surfaces*. in *Second Int. Symp. on Micro Machine and Human Science*. 1991. Nagoya, Japan.
12. Tichem, M., D. Lang, and B. Karpuschewski, *A classification scheme for quantitative analysis of micro-grip principles*. Assembly Automation, 2004. **24**(1): p. 88-93.
13. Pfeifer, T., et al., *Quality control and process observation for the micro assembly process*. Measurement. **30**(1): p. 1-18.

A DFA FRAMEWORK FOR HYBRID MICROSYSTEMS

Marcel Tichem[1], Dafina Tanase[2]

[1]Department of Precision and Microsystems Engineering
Faculty of Mechanical, Maritime and Materials Engineering (3mE)
Delft University of Technology
Mekelweg 2, 2628CD, Delft, The Netherlands
m.tichem@tudelft.nl – T: +31 15 2781603

[2]Electronic Instrumentation Laboratory
Faculty of Electrical Engineering, Mathematics and Computer Science (EEMCS)
Delft University of Technology
Mekelweg 4, 2628 CD Delft, The Netherlands
d.tanase@tudelft.nl – T: +31 15 2786432

Abstract This paper presents a framework for Design For Assembly of hybrid microsystems. Hybrid microsystems are microsystems, mainly semi-conductor based, which offer diverse functionality, and which are realised by integrating elements (die, parts) from different material and technology domains. From macro-domain mechanical products, it is known that the design of these products is of major importance for the performance of the assembly process. The paper addresses the question which aspects of assembly-oriented design are relevant for hybrid microsystems. A Design For Assembly (DFA) framework is presented, which distinguishes two main levels of design decisions: the *integration approach level* and the *process and technology level*. The framework is illustrated on the basis of a product case: an optode for medical applications.

1 Introduction

Hybrid microsystems originate from two converging trends in the precision and micro mechanical engineering domain and the semiconductor domain, respectively. In the mechanical engineering domain, there is an ongoing trend for miniaturisation of specific functions (e.g. mechanical, optical and fluidic functions). Applications can be found in consumer electronics, medical instrumentation and (opto-electronic) communication systems. To meet the miniaturisation demands, microsystem technology (MST) is an important enabler. In the semiconductor domain, microelectronics evolves in two directions: *More Moore* (ongoing miniaturisation, leading to nanoelectronics) and *More than Moore*. The latter leads to broadening of functionality on the small scale (e.g. micro-optic, micro-fluidics). Essentially this means that the two domains of miniaturised mechanical engineering systems and microsystems are converging, leading to new applications and product performance possibilities.

This paper focuses on *hybrid microsystems*. As opposed to monolithic microsystems, hybrid microsystems are composed of multiple elements for the primary functions. In addition, they often combine multiple functional domains; next to electronic functionality, optical or fluidic functionality may be included. In hybrid microsystem production, two main steps can be identified: the wafer-based processing of die or chips and the integration of the different elements into composed systems. Next to chips, elements manufactured using other sets of techniques may also have to be integrated, e.g. optical elements made out of glass. A number of levels can be identified in the process of packaging and assembly of hybrid microsystems and the products these systems are to be integrated in, a number of levels can be identified. The level this paper focuses on is the integration of single chips (and possibly other elements) into components, i.e. the lowest levels of packaging.

In hybrid microsystem integration, both technical and economic demands must be met. The technical challenge is in the handling, aligning and joining, sometimes with high (post bonding) accuracy, of small elements. The economic challenge is that this must often be done in large quantities against low cost.

In the domain of microsystems, some considerable research activities are focused on exploring physical working principles, and the realisation of prototype systems for evaluation is often done in an ad-hoc manner, only considering the microsystem's functional requirements. However, large volume manufacturing would require different sets of techniques, as well as redesign of the prototype microsystems, both targeted at high volume manufacturing.

This paper presents a framework for a Design For Assembly approach for hybrid microsystems, which aims at supporting the (re)design of microsystems for volume manufacturing. More specifically, the goals of the paper are to explore the requirements for a DFA framework (Section 2), to identify the relevant design decisions to be made (Sections 3-4), and to illustrate the proposed approach on basis of a case study (Section 5). Section 6 provides conclusions and an outlook on future work.

2 DFA focus for hybrid microsystems

In the design and assembly of macro-domain mechanical products and systems, it has been demonstrated that the key for optimisation of the assembly process lies in the product design activity. Through all the decisions that product designers take, the assembly process as well as the required assembly techniques are largely predetermined. This also implies that this is the phase where alternatives for product design need to be considered and evaluated from the perspective of assembly. Potential assembly difficulties need to be faced and solved during product design. This is what is referred to as Design For Assembly (DFA): design of the product from the viewpoint of optimising assembly.

For macro-domain products, structured sets of design guidelines and structured product analysis tools are available for DFA. In terms of the product structure, they focus on three different levels:

- Product level: structuring of a product and a product family in order to allow easy realisation of product variants and product generations. Modular product design, taking into account the type of function from a user's perspective (basic function, variant function, option) is key to this optimisation step.
- Sub-assembly: structuring of a sub-assembly for efficient assembly. The main goal is to reduce the assembly content, i.e. the amount of work needed to assemble the sub-assembly or module. It is tried to reduce the part count, and to minimise the number of secondary operations like sub-assembly reorientations.
- Part/connection level: optimisation of the handling and mating of parts. Part shapes and part interfaces are defined in such a way that easy handling and part alignment and joining is realised.

Probably the best-known method for DFA for macro-domain mechanical products is the Boothroyd and Dewhurst DFA method [1]. The focus in their method is in particular on the sub-assembly and part/connection level as defined above. More recently, with the emergence of micro-domain products, it has been tried to adopt such methods for this new domain of products. A conclusion is that several of the rules for macro-domain products are still valid for the micro-domain. Rules that are specific for the micro-domain for instance focus on reducing (the effect of) adhesive forces. More recent research on Design For Assembly for meso and micro-products can be found in [2, 3].

Considering the type of products this research focuses on, hybrid microsystems, it is necessary to redefine what the emphasis in a DFA analysis must be, and which aspects of existing approaches can and cannot be reused. The following can be observed:

- A DFA method for hybrid microsystems must consider the lower levels of system integration (chip → component) and the assembly issues related to this level.
- The design freedom for this type of systems is limited compared to macro-domain products. In macro-domain products, the product functionality is a result of material and geometry choices made by the designer, and quite some design freedom exists to adapt part geometry for ease of assembly. The intimate relationship between product functionality and geometry is less dominant for hybrid microsystems. For instance, the many macro-domain rules which aim at preventing part nestling and ease of insertion are less important for the lowest packaging levels.
- Integration technology is less well developed compared to macro-domain product assembly. Many techniques are still under development. DFA for traditional products was developed under the existence of known ways of assembly. Micro-assembly and microsystem integration is a developing field. Although quite a

few techniques exist for micro-electronic packaging, new techniques are needed and under development for the packaging and assembly of hybrid microsystems.

3 Structure of the DFA framework

The proposed DFA framework defines the main decisions to be taken by developers of microsystems and their integration (assembly, packaging) process. Figure 1 shows the framework.

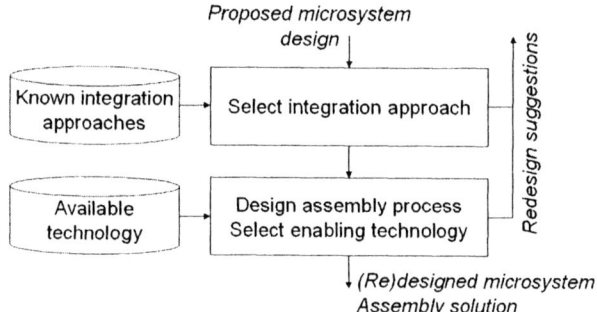

Fig. 1. Proposed DFA framework

For hybrid microsystems, it is useful to distinguish two levels of design considerations: the *integration approach level* and the *process and technology level*. On integration approach level, options for the system architecture and its resulting main sequence of assembly and packaging activities is determined. Decisions on this level have a major influence on the efficiency of the assembly process. This level is detailed in Section 4.

Once the main integration approach is chosen, the assembly process and the enabling technologies need to be detailed. The assembly process is the series of transformations by which (a) die/part(s) is (are) brought from an initial state, defined by the manufacturing process, to a final state, i.e. the final position and fixation in the product. This needs to be done for each die/part to be assembled. The first step to be done is the design of this assembly process. Secondly, suitable techniques for each of the operations (i.e. the transformations) have to be chosen. This includes transfer, alignment and bonding of the elements. The process and technology level is not further detailed in this paper.

4 Architecture level

Figure 2 shows the main approaches to integration. Monolithic integration means that functionality is realised in one chip, using a series of processing steps. As the diversity in functionality increases, this approach usually becomes more difficult. Also, the advantages of using the optimal technology for each function may lead to the choice of a hybrid approach. Finally, reducing the time-to-market may be the decisive criterion; developing a monolithic chip may require a long development lead time.

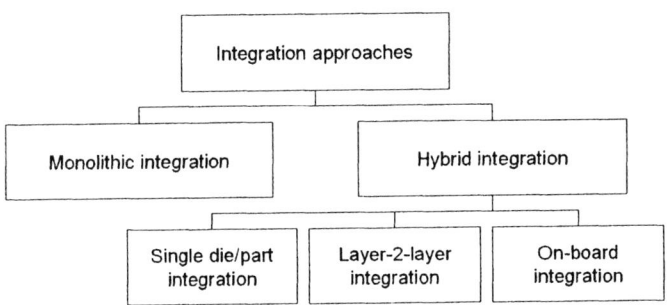

Fig. 2. Overview integration approaches

From a part-logistic view, only a limited number of principally different hybrid integration approaches exists.

- *Single chip/part integration.* Microsystems are realised by the integration of individual chips and parts. The main advantage of this integration approach is its high flexibility in terms of product designs that can be realised. The disadvantage is that each operation is single part oriented, thus increasing the integration effort. This is in contrast to the wafer-oriented approach in front-end processing of the wafers.
- *Layer oriented integration.* Microsystems are realised in batch approaches, by integrating layers containing different types of functionality. The best known approaches are wafer-to-wafer bonding and Wafer Level Packaging (WLP). Individual microsystems are obtained by dicing the bonded stack of functional layers. This approach improves the assembly efficiency compared to the first option mentioned, but it limits the design freedom of the microsystem designer. Also, the technology for this approach is less developed.
- *On-board integration.* The essence of this integration approach is that a toolbox of components is realised, each one having a certain function, which can be combined in arbitrary combinations on a substrate by standardised interfaces. The substrate may also support interconnection of the components: a circuit board. A well-known example is Surface Mount Technology in the context of electronic printed circuit board (PCB) assembly, which has been developed for easy inte-

gration of electronics. In the context of micro-optical systems, Silicon Optical Bench (SiOB) technology is an enabler for this integration approach. The standardisation helps in realising robust integration processes, and reuse of processes and equipment.

For a given microsystem design, a suitable integration approach needs to be selected. In some cases, a *series of integration approaches* needs to be defined, showing the step-wise development of the microsystem and/or its integration approach. This is often specified in terms of a roadmap for the development of the microsystem. The following aspects are identified in defining a microsystem roadmap and a progressing integration process:

- Ongoing microsystem miniaturisation. The need for smaller system dimensions (e.g. area or thickness) may require making new choices. For example, instead of electrically interconnecting microsystems using wire bonds, flip chip technology may have to be chosen to educe system area.
- Increased integration. To bring a microsystem design to the market, one may first choose to integrate the system on the basis of adding individual components. At the same time, a development may be started to realise a monolithic integration of the different chips (from System in Package, SiP, to System on Chip, SoC).
- Changing production volumes. For example, the first prototypes of a new microsystem, used for evaluation and demonstration purposes, will usually be produced in low volumes, for which technology investments are economically not affordable. Later, the focus may shift towards higher volume production, necessitating a redesign of the microsystem and its integration process. Investments in technology (development) may be affordable for this new situation.

5 Case-based illustration

To illustrate the presented DFA framework, the case shown in Figure 3 is considered. The so-called optode is part of an implantable system to be used for photodynamic therapy. Another medical application which also makes use of optodes is tissue viability monitoring after colon surgery. For both medical applications research is currently being performed at the TU Delft, DIMES and the Erasmus Medical Centre, Rotterdam. It is beyond the scope of this paper to describe these medical applications in more detail or to explain the entire functionality of such measurement systems, but for now let us have a look at the optode. For more information on the application and the optode see [4].

The optode is a combination of light sources and detectors with suitable signal conditioning and read-out electronics. For these medical applications, the light sources are chip-form LEDs (Light Emitting Diodes) and the light detectors are photodiodes. While the photodiode and the read-out electronics can be integrated on a silicon wafer using IC (Integrated Circuits) technology, the LEDs are supplied as stand-alone chips, which need to be assembled on the silicon wafer. Therefore, the focus in this case study is on the assembly of the LED chips.

Figure 3a shows only two of the LED chips, but different configuration of optodes can be used depending on the requirements of the medical application. For example, for photodynamic therapy with some advanced monitoring techniques, 4-6 LED chips need to be assembled, each one emitting at a certain wavelength. The size of one LED chip is approximately 200 x 200 µm, resulting in an optode surface area of approximately 1 x 1 mm.

As an initial step, a number of prototype optodes were fabricated and the LED chips are being manually placed on the silicon wafer, so that first technical and clinical tests can be performed. Eventually, the system may have to be produced in millions per year, necessitating an efficient assembly process. It is this future production scenario this case study focuses on.

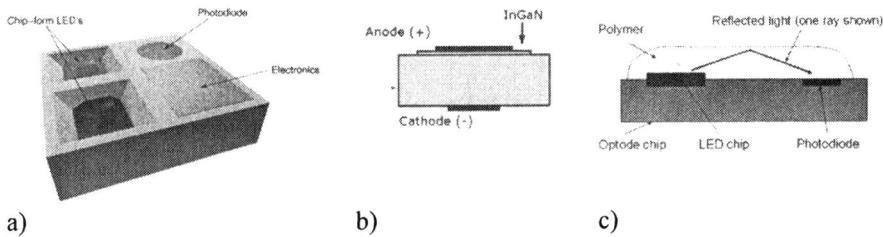

Fig. 3. The optode system: *a)* overview, *b)* LED chip considered in the paper, *c)* polymer on top of the optode chip for reflecting the light to the photodiode.

In order to obtain the functionality for which it was designed, the optode system requires that the LED chips are assembled and electrically interconnected. An LED chip is considered which electrical connections (anode and cathode) are on opposite sides, see Figure 3b [5]. Other configurations for the electrical connections of the LED chip also exist, but for the sake of simplicity those are not considered in this discussion. Furthermore, it is also important to know that the optode system as shown in Figure 3a is to be covered with a polymer, see Figure 3c. The polymer has two functions: it is either used for final packaging of the sensor system or as a gas sensitive membrane.

5.1 Integration approach level

As already indicated above, monolithic integration of the LED chips with the optode chip is not an option. LED chips are standard catalogue elements, produced in very high quantities. Hence, clearly a hybrid integration approach is chosen.

Integration on basis of assembling individual LED chips one by one can certainly be expected to be feasible. The LED chips can be kept in their manufactured order, so that locations of individual LED chips are known, thus preventing an organisation (chip feeding) step. Two options exist: the LED chips are either assembled *before* or *after* dicing and singulating the optode chips. This either results in assembly of single LED chips to the optode wafer or to assembly of single LED chips to single optode chips. The total effort in placement will approximately be identical. The advantage of the chip-to-wafer option is that subsequent processes may still be carried out on wafer level. In this case this would mean that it is worthwhile to investigate whether the electrical interconnections and the adding of polymer can be performed on wafer level.

The layer-to-layer integration approach is worth to investigate for improving the efficiency of the LED chip assembly process. This would require reorganisation of the LED chips. In their manufactured state, the LED chips are contained in a wafer structure, but the pitch between the LED chips does not match to the required position on the optode wafer.

Finally, an integration approach on basis of standardisation is technically possible. For many applications, the manufactured LED chips are packaged in to 1st level components (then referred to as LEDs). This is a standard process. Such LEDs may than be added by standard Surface Mount Technology on the optode chip using pick and place processes and solder reflow. The optode wafer needs to be prepared for this. However, an LED has considerably larger dimensions than an LED chip, so that integrating larger numbers of LEDs (4-6) within the optode chip area of approximately 1x1 mm is not possible anymore. This option is therefore not considered from now on. Figure 4 summarises the three remaining main hybrid integration approaches for this case.

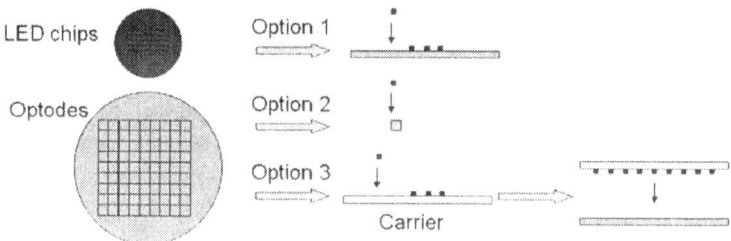

Fig. 4. Summary of options for the main integration approach: single LED chip to optode wafer (1), single LED chip to single optode chip (2), reorganisation of LED chips on some kind of carrier, followed by batch transfer to optode wafer (3)

5.2 Process and technology level

The integration approach is the basis for defining the assembly process, i.e. the sequence of operations to bring the LED chips to the optode chips. For Options 1 and 2 (see Figure 4), pick and place solutions for the handling of LED chips need to be selected. It can be expected that this is technically feasible. Option 3 requires some form of LED chip reorganisation. One option the LED chip supplier indicated is that changing the pitch between the LED chips is possible by using stretchable foils. After manufacturing and dicing of the LED chips, they are transferred to a stretchable foil which is stretched in two perpendicular directions to change the pitch between the LED chips. If larger pitch changes are needed, this process may be repeated several times, transferring the LED chips to a new unstretched foil between each step. The supplier indicated that the required pitch between LED chips of approximately 1 mm with sufficient accuracy is feasible with this approach. A challenge which then needs to be faced is the massive transfer of all LED chips to the electrical contacts on the optode wafer. This is not investigated further in the case study.

An issue that remains is realising the two electrical interconnections between LED chip and optode chip. One connection can be realised by solder reflow. The other connection, however, needs more attention. Three techniques were identified, see Figure 5a. The first option is to apply a wire bond. The disadvantage is that after this step the optode needs to be covered by a polymer, and the wire bond may break due to the adding of the polymer. This solution is not considered to be sufficiently robust. A second option is to build an electrically connecting bridge by means of a series of deposition steps, see Figure 5b. Each step would require a mask, and the mask alignment is critical for creating a good connection. This deposition process is expensive; on the other hand, in this way connections can be made on wafer level, rather then for each LED chip individually. The bridges are much ore robust than the thin wire bonds. The third option uses a glass wafer, in which electronic leads are processed using lithographic techniques. This would require one mask. After processing the electronic leads, the glass wafer is bonded with the wafer containing optode chips, thus batch-wise creating electrical interconnects. The consequence is that in the eventual system, there will be a glass chip between the optode chip and the polymer. It is assumed that the intermediate glass layer will not affect the optical performance of the system, but this needs careful investigation.

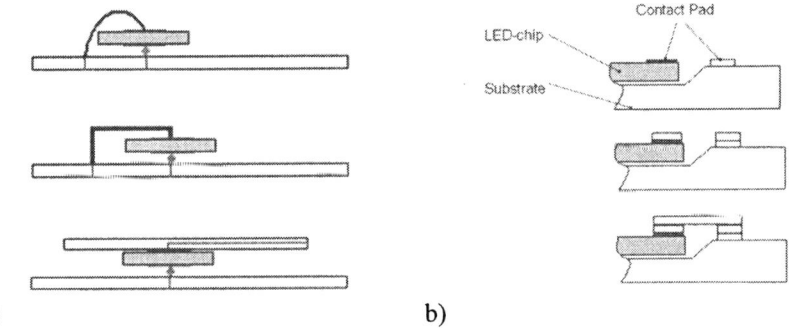

Fig. 5. a) Options for the electrical interconnects (top to bottom: wire-bond, bridge, glass wafer with electric leads); *b)* Process sequence for creating a bridge by a series of deposition steps.

After this step the case study was concluded, although a final decision on the integration approach and the technologies to be chosen could not be made. Further detailed analyses and experiments on various technical options are needed to allow such a choice to be made. A major advantage of applying the DFA approach to the described case was that it raised many issues which the microsystem designers were not considering at the moment, but which helped them in making well balanced decisions, both for the current prototype microsystems and for the future mass-produced microsystems.

6 Conclusions

A DFA framework for hybrid microsystems is proposed, which distinguishes two levels of decision making for designers of such systems: the integration approach level and the assembly process and technology level. The case study shows that it is indeed possible to use these levels of thinking in defining an integration solution for hybrid microsystems. The second conclusion is that such an approach supports the consideration of future production scenarios of the microsystems under development in an early phase of their design.In future work the proposed framework will be tested and further detailed on basis of other case studies. A more fundamental approach to defining the DFA framework is needed. It is expected that with the increasing importance of hybrid microsystems, such DFA methods will become essential for their successful market introduction.

Acknowledgements

This research is part of the Delft Centre for Mechatronics and Microsystems (www.dcmm.tudelft.nl) and is financed in part through the MicroNed programme and the MEMPHIS programme. MSc student Ing. R. Ooms is acknowledged for his contribution to this work.

References

[1] G. Boothroyd, P. Dewhurst, Product design for assembly, 1989
[2] T. Salmi and J. Lempiäinen, First Steps in Integrating Micro-Assembly Features into Industrially used DFA Software, In: Precision Assembly Technologies for Mini and Micro Products, S. Ratchev (ed.), Proceedings of IPAS 2006, February 19-21, 2006, Bad-Hofgastein, Austria, pp. 149-154
[3] C. Tietje, S. Ratchev, Design for Microassembly – A Methodology for Product Design and Process Selection, Proceedings of the 2007 IEEE International Symposium on Assembly and Manufacturing (ISAM), Ann Arbor, Michigan, USA, July 22-25, 2007
[4] E. Margallo-Balbás, J. Kaptein, H.J.C.M. Sterenborg, P.J. French, D.J. Robinson, Telemetric Light Delivery and Monitoring System for Photodynamic Therapy based on Solid-State Optodes, Proc. of Photonics West – BiOS 2008 (accepted)
[5] www.cree.com, CPR3DE Rev. -, Cree® UltraThin™ LED - Data Sheet - CxxxUT200-Sxxxx

STATISTICAL ASSEMBLIES WITH FORM ERRORS – A 2D EXAMPLE

Pierre-Antoine Adragna, Hugues Favrelière, Serge Samper, Maurice Pillet

Symme Lab. Université de Savoie, 5 chemin de Bellevue 74940 Annecy le Vieux
{pierre-antoine.adragna, hugues.favreliere, serge.samper, maurice.pillet}@univ-savoie.fr

Abstract When dealing with precision in tolerancing of assembly systems, the modelling complexity of the mechanism increases. At first, one can distinguish the 1D tolerancing approach that only concerns variations of dimension. Then, several models are defined to set 3D tolerances, considering that the form error is negligible compared to the orientational and translational variations. Finally, some approaches are proposed to take into account the form variations in the tolerancing of mechanisms. However, some modelling approaches considers the form error as a tolerance zone to add to the 3D tolerances as defined by Rule#1 of the ASME standard, or ISO 8015. This paper proposes another point of view, considering the positioning of parts through contact points of their rigid deviation shapes under a defined assembly force and set-up. Rather than considering the positioning of a single part, here is proposed an approach of batch parts assembly by a statistical description of shapes. The result of the method is a statistical positioning error of one part on the other considering the form deviations of parts.

Keywords 3D assembly, form errors, positioning force, statistical assembly

1 Introduction

The tolerancing of an assembly system is based on a model of the mechanism. Depending on the complexity of this model large errors can be obtained on the parts positioning. The most simple modelling that can be identified only considers the dimensional variations as presented by Graves [1] for example. More complex models of the mechanism are developed to exploit the 3D tolerance zone. Many methods are proposed such as the vector chain of Chase [2], the T-Maps of Davidson and his team [3], the small displacement torsor (SDT) of Bourdet [4], and the clearance and deviation domains of Giordano and his team [5]. But as the precision of the mechanism increases, the form deviation of shapes has to be considered. Ameta [6] proposes a model in the T-Maps space to study the influence of the form deviation based on the tolerance zone, but only one form deviation is as yet considered. Radouani [7] presents an experimental study of the positioning of one part on another regarding the size of their form deviations. Neville [8] gives an

---

Please use the following format when citing this chapter:

Adragna, P.-A., Favrelière, H., Samper, S., Pillet, M., 2008, in IFIP International Federation for Information Processing, Volume 260, *Micro-Assembly Technologies and Applications*, eds. Ratchev, S., Koelemeijer, S., (Boston: Springer), pp. 23-33.

algorithm to determine the positioning error of a butting assembly of two components considering their form roughness and dimensional error with or without an auxiliary surface. Our solution [9] is based on the identification of the contact points of the two faces given by a mechanism defined by external face data and a positioning torsor expressed by a F force located on an axis. Our method allows the prediction of the positioning error into the SDT space.

Figure 1 below shows the chosen example for this paper. We reduce the study to a two dimensional profile assembly. The two parts are linked by the "A_i" joint and the functional requirement "Cf" is given by the clearance between the "B_i" faces. B_i faces are supposed to be perfect in this example. The point of interest is their relative displacements according to the assembly variations due to the A_i faces.

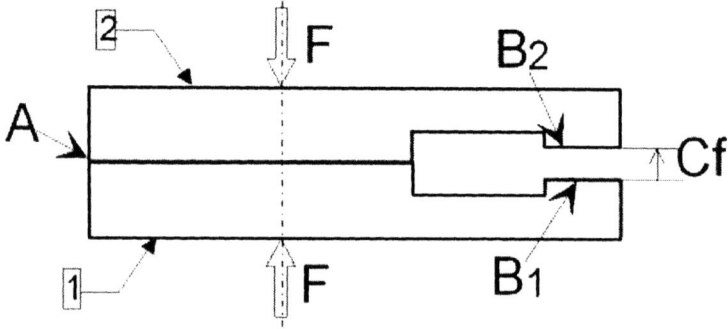

Fig. 1. 2D assembly and its functional requirement

2 The proposed method and background

2.1 A single assembly

Our approach consists of identifying all the possible contact points of the two faces regarding their form deviations and a given pre-positioning mechanism. The final contact points that define the stable positioning are identified by a positioning force.

The first step of the method is the identification of the possible contact points. To do so, the difference surface is introduced. This surface corresponds to the difference of the form deviations, and represents point to point distances. This difference surface can be found by a point to point distance computation, but the modal analysis of form deviation is recommended. Figure 2 presents the difference surface computed point to point. One can observe the roughness of the measured shapes that is filtered by the use of the modal characterisation.

The second step of our approach is the identification of all the possible contact points between the two parts surfaces. The proposed solution is the use of the convex hull that identifies contact faces, hence the contact points. Figure 2 shows the convex surface of the difference surface and the identified potential contact points.

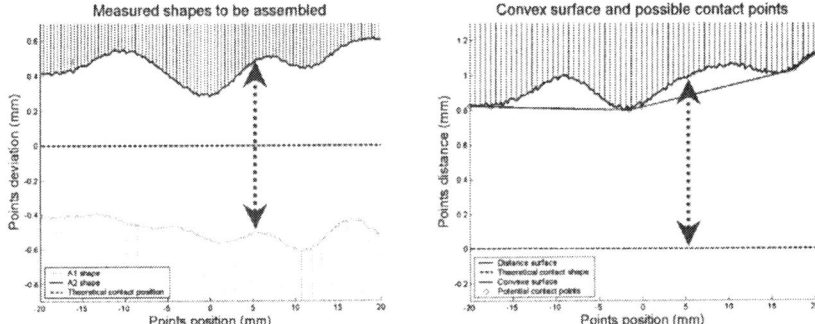

Fig. 2 a) parts form deviations A_1 and A_2, *b)* relative distance surface

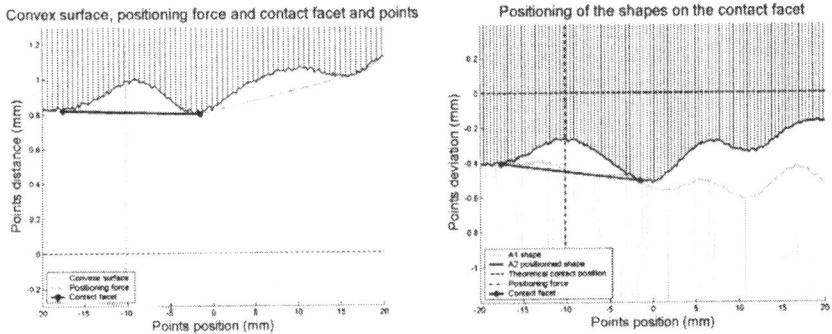

Fig. 3 a) Convex surface, *b)* positioning deviation according to the positioning force location

The third and last step is the identification of the contact facet, giving the contact points. A solution is the use of a positioning force, whose location of the direction identifies the contact facet. As a rigid model is considered, the force intensity is not used. The identified contact facet corresponds to the opposite of the positioning error. Figure 3 shows the identified contact facet for a given position force location and the parts assembly corresponding. The positioning deviation corresponds to the deviation of the identified facet to the theoretical contact position.

2.2 Modal analysis

Introduced by the team of Samper [10-11], the modal analysis of form deviations is a generic approach to building a basis of form errors for any shapes. Hence any form deviation can be analysed in this modal basis and described as a set of coefficient as for the Fourier transform. The modal analysis method is based on the modal shapes of vibrations for the ideal geometry. These shapes have interesting properties such as:
- it is a vectorial basis of form errors,
- modals shapes are naturally sorted by growing complexity,
- modal shapes are easily calculated for any kind of geometry (use of the Finite Element Model to solve complex shapes)

The method is based on the space solution of the natural vibrations of forms. The following equation 1 is the d'Alembert equation of vibrations.

$$\Box(U) = 0 \Leftrightarrow \nabla^2 U = \frac{1}{c^2} \frac{\partial^2 U}{\partial t^2}$$

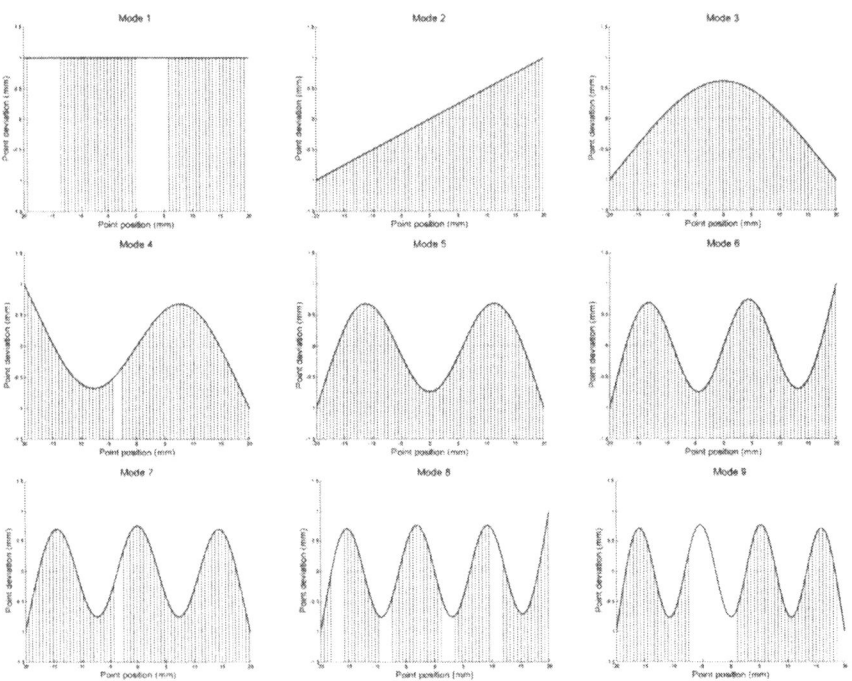

Fig. 4. First nine modal basis of our A_i contact 2D profiles

This equation can be solved by an analytical way for simple cases or by a numerical way for most of cases. The obtained vibration modes are considered as form deviations Φ_i. These Φ_i shapes are modified to obtain a orthogonal basis, and in order to have a metric meaning of the modal coefficients, the modal shapes amplitude are set to the unit [11]. Figure 4 shows the first nine form errors of the modal basis of our case of application.

2.2.1 Analysis of a single shape

The first interest of the use of the form deviation modal analysis is the filtering of the form roughness, and the reduction of the number of characteristics that describing the form deviation. The following figure 5 shows the modal characterisation of an error shape, called the modal signature of the form deviation in the modal basis, and the recomposed form corresponding to the filtered shape. The result of the modal characterisation of a measured form deviation V on the B modal basis is the modal signature Λ. This Λ modal signature is composed of the λ_i modal coefficients calculated by the following relation.

$$\lambda_i = B \cdot \Phi_i / \|\Phi_i\|^2$$

The recomposed shape with the rigid modes (translation and rotation) corresponds to the rigid shape and is equivalent to the Least Square associated shape. The recomposed shape R is obtained by the following relation:

$$R = \Lambda.B$$

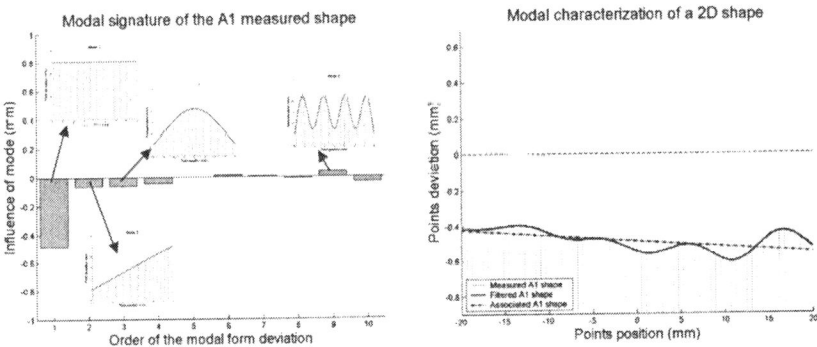

Fig. 5 a) modal signature, *b)* recomposed shape

2.2.2 Simulation of shapes

The modal characterisation of form errors can be used to create simulated shapes. A random draw of the modal coefficients creates a random shape which form complexity depends on the number of considered modes. Based on the observation of shapes analyses, we consider that the amplitude of the modal coefficients is given by the following law:

$A(i) = A_0 / i$

Where A_0 is initial amplitude, i is the order of the modal coefficient and $A(i)$ is the maximum amplitude of the ith modal coefficient. The following Figure 6 shows the amplitude law of the coefficients and a random draw of a modal signature.

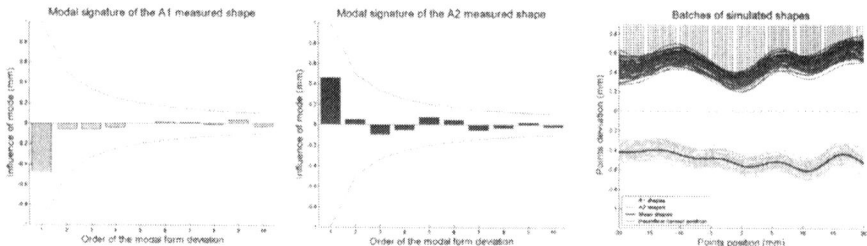

Fig. 6 a and b) random modal signatures , *c)* batches of simulated form deviations

The simulation of a batch of form deviations can be obtained from an initial modal signature, on which are added random modal variations. The amplitude of these modal form variations has the same type of law as the initial form deviation. Another random variation can be added on the on the recomposed shapes that can represent roughness or measurement error. A set of 50 simulated shapes are presented in figure 6.

2.2.3 Statistical analysis of a set of shapes

The second interest of the modal analysis method is the statistical description of a set of deviation shapes. Presented in CIRP-CAT 2007 [12-13], the modal analysis is extended to the qualification of a batch of form deviations. From the batch of modal signatures Λ_i (characterizations of the shapes), one is able to compute the mean modal signature $\mu_{\Lambda i}$ and the covariance matrix $\Sigma_{\Lambda i}$, whose root of the diagonal represents the standard deviation modal signature $\sigma_{\Lambda i}$. It is also possible to have a geometrical representation of this statistical qualification. The following relation links the mean form deviation of the batch to the mean modal signature:

$\mu_V = B.\mu_\Lambda$

The following relation links the covariance matrix of the form deviations to the covariance matrix of the modal signatures:

$\Sigma_V = B.\Sigma_\Lambda.B'$

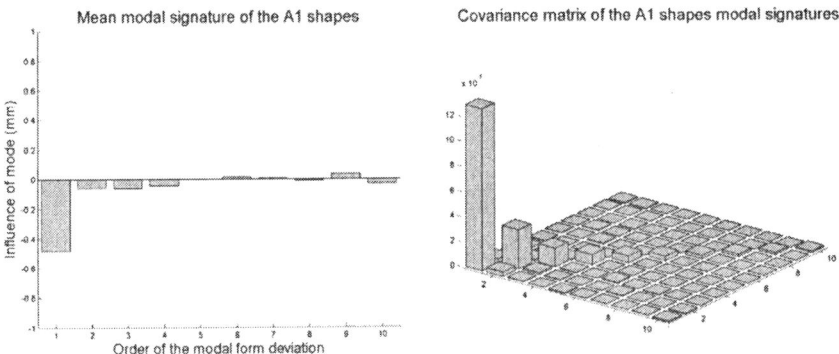

Fig. 7 a) mean modal signature, *b)* covariance matrix of the modal coefficients

3 The statistical assembly of form deviations

This section details our proposed method to deal with the assembly of a batch of components. A first solution is to use the approach of the first part to compute single assembly. This method will be used to confirm the following proposition. The idea of the statistical assembly is to find the mean positioning of the batch, and then the standard deviation around this mean positioning.

3.1 Mean positioning and contact points

The first step of our approach is to determine the mean positioning and the contact points associated. These mean contact points are used to determine the mean positioning and its standard deviation linked to the means and standard deviations of the parts form deviations.
This first step consists of solving the positioning of the two mean shapes of the two parts batches. This is achieved by the previous approach of assembling single parts, except that parts are mean forms. One hence obtains the mean positioning error of the parts batches, and more important is the identification of the contact points.

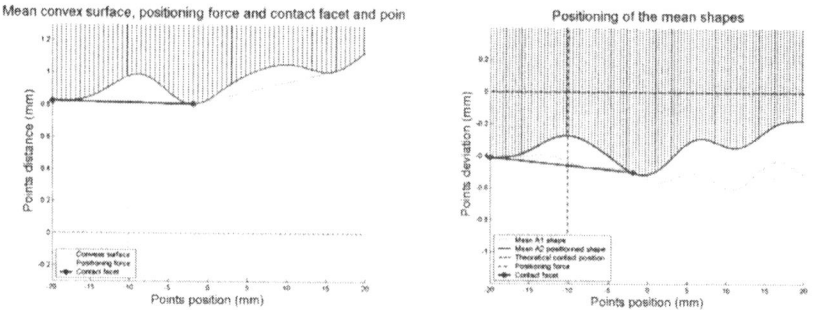

Fig. 8 a) identified mean contact facet, *b)* mean positioning

3.2 Standard deviation of the positioning error

Thanks to the identification of the contact points, standard deviation surfaces of the parts form errors can be linked to the standard deviation of the positioning error. The solution we choose is to consider the covariance matrix of the contact points, given by the covariance of the measured form deviations or by the modal characterisation. As parts are independent, the covariance matrix of the positioning error is the sum of the covariance matrices of the parts form deviations. In our case, the two contact points are identified as the 1st and the 91st measured points. Hence, it is possible to predict the positioning covariance matrix from the covariance matrix of the contact points.

The following Figure 9 shows a particular positioning of shapes from the two batches. It can be observed that the contact points of this particular assembly are not identical to the contact points of the mean shapes positioning.

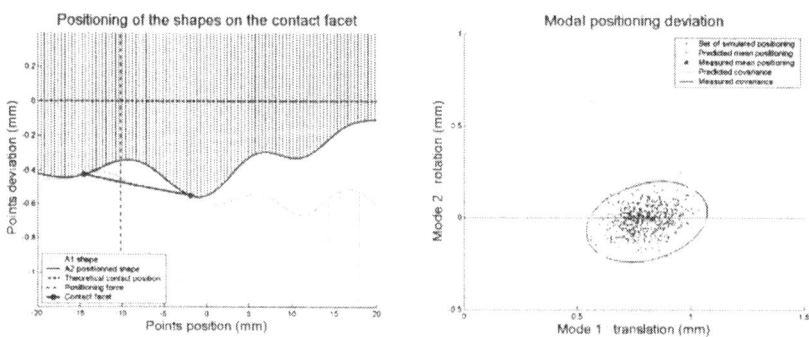

Fig. 9 a) a particular assembly from the batches, *b)* the positioning results and prediction

However, the positioning result of the batches is showed by the blue dots in Figure 9. The red dot corresponds to the mean positioning error of the results and the red ellipse corresponds to 3 times the covariance of the results. The prediction of our approach is represented by the green parts. The green dot corresponds to the predicted mean positioning error, and the green ellipse is 3 times the predicted positioning covariance. It can be observed on this simulation that the predicted results are similar to the simulated ones.

Another simulation with a larger shape variations and another location of the positioning force is showed in the next figure.

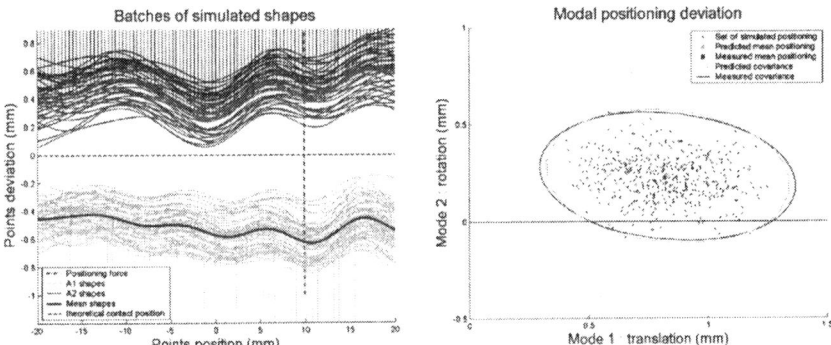

Fig. 10 a) batches of shapes, *b)* associated positioning results and prediction

With the new simulation, as the shape variation is more important, the identification of the contact points is less accurate. However the predicted result is close to the statistical characterisation of the simulations.

The results are presented in the modal coordinates that can be transcribed into a small displacement torsor as follows. Thus, the positioning deviation of the A2 shape can be transferred to the B2 shape to check the respect of the functional requirement "Cf".

4 Conclusion

Surface assembly considering the form error is treated by the use of the modal parameterisation. This method allows us to describe form errors of all geometries in an exhaustive way. It also provides a filtering of the form deviation to keep influent modes and reject roughness.

The problem of non determination of the assembly is solved using the force torsor. Its location allows us to find the contact points of both faces by the use of the distance surface. The characterisation of the contact facet gives positioning deviation of surface 2 on surface 1.

The proposed approach to deal with statistical assemblies of form errors is interesting in the way that form deviation of shapes are transcribed into positioning error. This positioning deviation is expressed in the modal space in our case, but can also be described in small displacement torsor of T-Map.

The next step of our approach is the simulations on more complex cases (more variation, and 3D components), and the confrontation of our theory to measured assemblies.

References

1. S. Graves, Tolerance Analysis Formula Tailored to Your Organization, Journal of Quality Technology, Vol 33, No 3, July 2001
2. K.W. Chase, Tolerance Analysis of a 2D and 3D assemblies, ADCATS Report, No 94-4, 1999
3. J.K. Davidson, J.J. Shah, Geometric Tolerances : A New Application for Line Geometry and Screws, ImechE Journal of Mechanical Engineering Science, Vol 216, Part C, 2002 pp. 95-104
4. P. Bourdet, L. Mathieu, C. Lartigue, A. Ballu, The Concept of Small Displacement Torsor in Metrology, Proceedings of the International Euroconference, Advanced Mathematical Tools in Metrology, Oxford, September 27-30 1995
5. M. Giordano, B. Kataya, E. Pairel, Tolerance Analysis and Synthesis by Means of Clearance and Deviation Spaces, Proceedings of the 7th CIRP International Seminar on Computer Aided Tolerancing, 2001, pp. 345-354
6. G. Ameta, J.K. Davidson, J. Shah, Influence of Form Frequency Distribution for a 1D Clearance Wich is Generated from Tolerance-Maps, Proceedings of the 10th CIRP Conference on Computer Aided Tolerancing, Specification and Verification for Assemblies, Erlangen Germany, March 21-23 2007
7. M. Radouani, B. Anselmetti, Contribution à la Validation du Modèle des Chaînes de Cotes – Etudes Expérimentale du Comportement de la Liaison Plan sur Plan, Congrès International Conception et Production Intégrées, CPI 2003, Meknès Moroco, October 22-24 2003
8. K.S.L. Neville, Y. Grace, The modeling and analysis of butting assembly in the presence of workpiece surface roughness part dimensional error, International Journal of Advanced Manufacturing Technologies, Vol. 31, 2006, pp. 528-538
9. P.-A. Adragna, S. Samper, H. Favrelière, M. Pillet, Analyse d'un assemblage avec prise en compte des défauts de forme, Congrès International Conception et Production Intégrées, CPI 2007, Rabat Morocco, October 22-24 2007
10. F. Formosa, S. Samper, I. Perpoli, Modal expression of form defects, in Models for Computer Aided Tolerancing in Design and Manufacturing, Springer series 2007, pp. 13-22
11. P.-A. Adragna, S. Samper, M. Pillet, Analysis of Shape Deviations of Measured Geometries with a Modal Basis, Journal of Machine Engineering: Manufacturing Accuracy Increasing Problems – Optimization, Vol. 6, No 1, 2006, pp. 95-102

12. P.-A. Adragna, S. Samper, M. Pillet, Inertial tolerancing applied to 3D and form tolerancing with the modal analysis, Proceedings of the 10th CIRP Conference on Computer Aided Tolerancing, Specification and Verification for Assemblies, Erlangen Germany, March 21-23 2007
13. H. Favrelière, S. Samper, P.-A. Adragna, M. Giordano, 3D statistical analysis and representation of form error by a modal approach, Proceedings of the 10th CIRP Conference on Computer Aided Tolerancing, Specification and Verification for Assemblies, Erlangen Germany, March 21-23 2007

Chapter 2

Methods and Solutions for Micro-Product Design

A CLASSIFICATION AND CODING SYSTEM FOR MICRO-ASSEMBLY

Amar Kumar Behera, Shiv G. Kapoor and Richard E. DeVor

Department of Mechanical Science and Engineering
University of Illinois at Urbana-Champaign
Urbana, IL 61801-2906, USA
sgkapoor@uiuc.edu
Tel: +1-217-333-3432, Fax: +1-217-244-9956

Abstract This paper presents the development of a standard classification and coding system for micro-scale assembled devices that can link micro-assembly concepts and technologies to the attributes of the assembly components and their interrelationships that define the assembled device/product. The proposed system is based on an n-digit coding of both individual parts as well as the assembly. The coding scheme identifies a form code, a material code and a process code for both parts and the assembly and then partitions each into the relevant attributes. A specific example of coding of the parts of a miniature spin bearing has been considered to illustrate the applicability of the coding system.

Keywords coding, classification, micro-parts, micro-assembly, spin bearing

1 Introduction

Assembly has taken on considerable importance in the field of micro-systems manufacturing with a drive towards increasing miniaturization and function integration. IC fabrication and packaging and MEMS precipitated the need for assembly of very small parts and this movement rapidly spread into various other applications such as microfluidic systems, optical MEMS, and many others. As researchers strive to transform nanoscience and technology into useful engineering systems through integration that requires the bridging scaling laws across the nano-micro-meso domains, new design tools must be forthcoming to carry out micro-assembly of critical components in a wide range of telecommunication, medical, defense, etc., applications.

Developing a systematic architecture for assembly has been a driving force of research efforts as seen in the work of Hollis et al. leading to development of tabletop precision assembly systems [1]. Likewise, Kuo et al. [2] have carried out assembly of a

150 μm diameter pin with a 200 μm thickness plate under the guidance of three laser fibers creating a Molten Separation Joint (MSJ). Another interesting example of micro-assembly is a commercial unit by Klocke Nanotechnik with four nanorobotics axes (X, Y, Z-stages plus microgripper) and a joystick, which is capable of carrying out assembly of glass fibers, laser diodes, RF-mixer, micro systems or the handling of thin wires, gears, SMD-chips and micro sensors [3]. Furthermore, considerable research has been done on the development of specific micro-assembly tools and methods including work on visual servoing, sensor-based assembly and a vacuum handling tool [4-13]. At the micro-assembly systems design and planning level, Kurniawan et al. have reported an attempt at developing a morphological classification of hybrid microsystems assembly [4] and several research works dealing with the planning for assembly have also been reported [5-6].

Although there has clearly been considerable work on micro-assembly in the last several years, much of the work in this field is still ad-hoc with solutions presented to very specific problems. Furthermore, the proliferation of microsystems technologies suggests that it will become increasingly important to approach the problem of micro-assembly more systematically to insure that all the relevant available technologies can be brought to bear on a given problem. To this end, this research deals with developing a comprehensive design tool for micro-assembly. This tool will be comprised of three essential elements: a) a hierarchical classification and coding system for a micro-assembly and its parts that captures their basic attributes; b) a database of technologies based on the current state-of-the-science that can be used to address micro-assembly requirements; and, c) a set of design rules and algorithms that link assembly and part attributes to the database for the selection of appropriate technologies. Figure 1 illustrates this concept in detail. This paper addresses the first of these essential elements: an assembly/part classification and coding system.

In this paper, we consider a multi-tiered framework involving coding of assembly and coding of parts. Individual parts of the code deal with form, material and process parameters. The digits in the coding scheme are justified with mathematical and/or logical analysis as required, such as analysis of forces, geometry, current and past experimental results in micro-assembly, etc. A case study on assembly of a miniature spin bearing is used to demonstrate the applicability of the coding scheme.

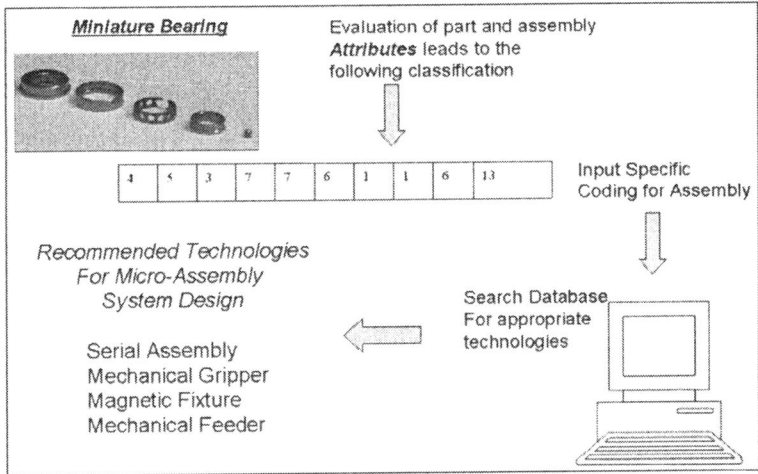

Fig. 1. Classification and coding system for micro-assembly applications

2 Overview of Coding of Assembly and Parts

In developing the coding system, it was helpful to identify several micro-scale system/device families of parts that required assembly. These "case studies" were useful in creating meaningful attributes that define the system. A study of each part family led to the development of a set of parameters that uniquely define the part characteristics/requirements and are relevant to the micro-assembly process. Both parts and the assembly were found to have separate unique characteristics. These characteristics are geometric, materials, and process-based. Therefore, the coding for both parts and assembly was classified as having a form code, a material code and a process code, each with the generic structure **1234 56789 AB...**, which uses up to nine digits and some number of letters. Each digit represents a part or assembly attribute and each attribute may ultimately be defined by a range of numbers, e.g., 0-9, 0-99 etc. The form code is used for storing geometry-related information, the material code is used for storing information related to material properties of critical parts of the assembly and the process code is used for storing information related to assembly environment and surface properties of the individual components. Table 1 outlines the scheme for coding of assembly, while Table 2 outlines the scheme for coding of parts.

Table 1. Coding of Assembly

Assembly Form (Code digit)	Assembly Material (Code digit)	Assembly Process (Code letter)
Distinctly different parts (1)	Material of most critical part (3)	Assembly environment (A)
Type of assembly (2)	Number of dissimilar materials (4)	Surface properties of assembly exterior (B)
	Tolerance of most critical part (5)	
	Tolerance of most critical joint/interface (6)	
	Weight of assembly (7)	

Table 2. Coding of Parts

Part Form (Code digit)	Part Material (Code digit)	Part Process (Code digit)
Part Symmetry (1)	Material (4)	Positioning and Alignment (7)
Geometrical Shape (2)	Mass (5)	Surface Roughness (8)
Characteristic Dimension (3)	Specific Stiffness (6)	

3 Identification of Attributes for Coding

An extensive study of part/assembly attributes for micro-assembly yields insights into the parameters that are needed to serve as a basis for identification of the key technologies necessary for assembly as shown in Fig. 2. These parameters can be used as the basis for development of the classification and a coding system. Specifically, different sub-assembly operations for parts of different dimension ranges were related to the gripping, positioning, sensing and forces during assembly. The table is drawn up based on several micro-assembly experiments that have been performed and understood over the years [2-9]. The first column shows sub-assembly operations that are related to the geometry and symmetry of the parts. The second column deals with part size. The remaining columns outline the technologies needed to carry out the assembly for the given parts.

Consider the manipulation of rectangular blocks as shown in the third example. For the given dimension ranges, it is seen that ortho tweezers on a 3-DOF stage are helpful in manipulating the parts for assembly. It is further seen that orthogonal tweezers work well on only polygonal objects such as rectangular blocks and not for spherical parts. This helps us establish geometry as an important part attribute for coding. Now, considering the fourth example, which is a pin-plate assembly, we see that a molten separation joint using μ-EDM and Nd-YAG laser is needed for assembly with CCTV being used for seeing the parts [2]. The dimensions of the parts are of the order of a few hundred microns and, hence, we do need a specialized vision system but it need not be of a very high resolution such as SEM or TEM. Further, the need for a joining process arises due to the specific geometries which are a cylinder and a plate with contrasting symmetries, viz., rotational and reflective, respectively.

Subassembly Operations	Part Size	Gripping	Positioning	Sensing	Procedure/Forces
○ ○ ○ ○	10-100 μm	Needle shaped handling tool	3-DOF stage (to bring the balls in the focus of the microscope)	SEM, Optical	Mechanical – Pick-and-place
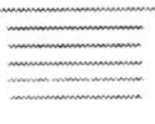	~100-1000 μm x 100-1000 μm	Optically transparent electrostatic	4-DOF stage (to align wafers and bring them in focus)	CCD	Electrostatic
	~100-1000 μm x 100-1000 μm x 100-1000 μm	Ortho tweezer	3-DOF stage	--	Mechanical
	~50-500 μm dia, 50-500 μm thick	Clamping	μEDM+Nd.-YAG Laser	CCTV	Joining – Molten Separation Joint
 ○ ○ ○ ○ ○ ○ ○ ○	100 – 500 μm	Vacuum gripper	3-DOF	CCD	Mechanical
	1-2 mm x 1-2 mm	Vacuum gripper	--	--	Non contact handling
	10 nm – 100 μm	Vacuum gripper, Optical gripper	--	SEM, TEM, Optical microscope (plant cells)	Non contact handling
○ ○ ○ ○	100 μm - 500 μm	Capillary gripper, Two fingered microgripper with thermal glue	--	--	Pneumatic forces
 ⏦	few mm – 1 cm	Friction gripper	--	--	Mechanical
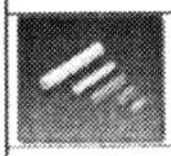	1 – 5 mm	Pneumatic gripper	--	--	Mechanical
	10 μm – few mm	Electrostatic gripper	--	--	Electrostatic
		Magnetic gripper	--	--	Magnetic
	--	Bernoulli gripper	--	--	Hydraulic
	--	Air cushion gripper	--	--	Non contact handling
	100 μm – few mm	Ultrasonic gripper	--	--	Non contact handling

Fig. 2. Micro-Assembly Technologies Organized by Part/Assembly Attributes

Likewise, for the twelfth example concerning assembly of semi-conductor wafers, the use of a magnetic gripper is facilitated by the part being made of magnetic material. This helps us establish part material as an important attribute for coding. This analysis thus leads to the identification of the parameters that are important as both part and assembly characteristics such as symmetry, shape, dimensions, material, alignment and assembly environment, which then leads to broader generalization as form, material, and process codes for both parts and assembly.

4 Coding of Parts

4.1 Part Form Code

The part form code is required to provide key geometry-related information about the part including part symmetry, part shape and part size.

1st digit – Part Symmetry

A part is said to have "exact rotational symmetry" if it can be defined by an envelope that is a solid of revolution or it has n-fold rotational symmetry, "exact reflective symmetry" if it is a prismatic solid or thin-wall part with symmetry about a plane through the center of gravity of the part, and "partial symmetry" if it has portions of the boundary that can be identified as symmetric. It is seen that cylinders, spheres, cones, wafers, rings, gears, etc., can be classified as having rotational symmetry, while rectangular boxes and prisms can be classified to have reflective symmetry. Irregular objects have rotational, reflective and partial symmetry or no symmetry at all. Symmetry can be exploited for gripping, part insertion, fastening, orienting and feeding.

2nd digit – Geometrical shape

The geometrical shape determines the surface(s) for pick-and-place operations, visibility of the part under a microscope and the gripping system. For instance, orthogonal tweezers can easily pick up rectangular parts but not spherical. Moreover, part shape also determines mating surfaces for assembly. Various common geometrical primitives have been identified and linked to their respective symmetries as shown in Fig.3.

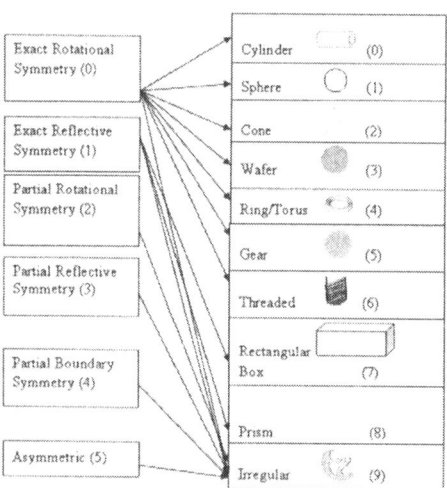

Fig. 3. Part Symmetry and Geometry

3nd digit – Characteristic Dimension

The size of a micro-part determines the magnitude of different forces in play during assembly, visibility, ease of handling and manipulation, particularly part gripping and release, and often, the overall sequence of assembly as well. However, it is useful to find a single measure for the dimension of a part, which may often have different dimensions in different planes. This would not only give an idea of the overall size of a part, but also make the part code more compact. Figure 4 shows an illustration of projected dimensions for a rectangular box. Several possible measures for part dimension (Eq. 1-4) include the surface area (S)-to-volume (V) ratio ($D_{part\_S/V}$), geometric mean ($D_{part\_geom-mean}$), arithmetic mean ($D_{part\_arith-mean}$) and maximum projected dimension (D_{part_max}). A weighted measure, D_{part} (Eq. 5), was found most suitable as representative of overall dimensions of a micro-part. The minimum part dimension was not considered here as the surface area-to-volume (S/V) ratio is already a measure that yields a characteristic dimension that is less than the smallest projected dimension.

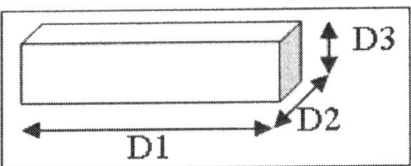

Fig. 4. Projected dimensions of piece part

$$D_{part\_S/V} = \frac{1}{(S/V)_{part}} \quad (1)$$

$$D_{part\_geom-mean} = \begin{cases} \sqrt[3]{D_1 D_2 D_3}, & 3D \text{ for description} \\ \sqrt{D_1 D_2}, & 2D \text{ for description} \\ D_1, & 1D \text{ for description} \end{cases} \quad (2)$$

$$D_{part\_arith-mean} = \begin{cases} \dfrac{D_1 + D_2 + D_3}{3}, & 3D \text{ for description} \\ \dfrac{D_1 + D_2}{2}, & 2D \text{ for description} \\ D_1, & 1D \text{ for description} \end{cases} \quad (3)$$

$$D_{part\_max} = \begin{cases} Max(D_1, D_2, D_3), & 3D \text{ for description} \\ Max(D_1, D_2), & 2D \text{ for description} \\ D_1, & 1D \text{ for description} \end{cases} \quad (4)$$

$$D_{part} = w_1 * D_{part\_S/V} + w_2 * D_{part\_geom-mean} + w_3 * D_{part\_arith-mean} + w_4 * D_{part\_max}, \quad (5)$$

where the weights w_1, w_2, w_3, w_4 are chosen by the user. Note that the sum of all weights is equal to 1.

It is useful to keep measures of the projected dimensions separately at a level further down in the hierarchy. Hence, a 3-digit code is used for the three projected dimensions. For the special primitive geometries, however, these three dimensions are chosen to help evaluate the surface area and volume. This helps maintain the relation between the projected dimensions and the evaluated characteristic part size.

Suitable ranges for part dimensions must be defined to carry out the coding. The ranges were calculated by carrying out a mathematical analysis of forces in micro-assembly combined with a statement-inference method for establishing critical dimensions [4, 7-11]. The force analysis was first carried out to evaluate gravitational, electrostatic, Van-der-Waals and capillary forces for a spherical part-planar surface system as described by Fearing [10] and Enikov [11]. Several simulations were run to predict forces by varying material properties of part and gripper, dielectric, distance between sphere and plane, contact angle and surface tension of liquid. The work was further extended to different geometrical primitives discussed earlier by using a measure for an equivalent spherical radius, r_{eq}, based on the surface area, S, as:

$$r_{eq} = 0.5\sqrt{\dfrac{S}{\pi}} \quad (6)$$

Based on the results of this force analysis and the statement-inference method, different ranges for coding the parts dimensions were established. For example, if D_{part} < 1 µm, the code assigned is 1, which takes care of sub-micron part and feature sizes and the dominance of Van-der-Waals force. For 1 µm <D_{part}<10 µm, it is found that this range is not very suitable for dry manipulation, and to identify this unique characteristic, the code assigned is 2. The next code 3 is assigned to parts in the range 10 μm – 100 μm to identify the range where electrostatic, capillary and Van-der-Waals forces have similar order of magnitude for many cases of simulations. For the range 500 μm – 700 μm, the code assigned is 6 and this helps identify the critical dimension above which metallic vacuum grippers are found suitable and also the range where gravitational force comes within one order of magnitude of the remaining forces. The remaining codes are assigned with a similar logic.

4.2 Part Material Code

The part material code is designed to provide information related to key material properties that influence the assembly process.

4th digit – Material

Part materials were classified broadly as metals, polymers, ceramics, composites and non-metallic minerals. This hierarchy was further broken down with metals split into ferrous and non-ferrous alloys, polymers into thermoplastic, natural and thermosetting, ceramics into electronic, constructional, natural, glasses, engineering and composites into natural, particulate, fiber and dispersion types. Simplifying the classification further finally yielded specific engineering materials such as plain carbon steel, PVC, silicone, graphite, etc. These materials in the last level were selected for developing the coding scheme. This classification along with coding in parentheses is illustrated in Fig. 5.

5th digit – Mass

Peschke et al. [12] mention that the behavior of the objects involved in handling operations is, to a great extent, determined by their weight. This changes as the size of the objects decreases, i.e., as the volume of an object reduces by the third power, the surface area decreases only by the second power. Consequently, the force affected by a part's weight decreases more steeply than adhesion forces caused by the surface area. If the size of an object decreases below a critical value, the weight-dependent force is insufficient for overcoming the existing adhesion forces. In micro-assembly, this can result in the object sticking to the gripper. This sticking would create problems in a system for automated assembly [12]. Hence, it is imperative to know the weight of a micro-part. To identify some critical masses, magnitude of gravitational force with the extreme densities of water and tungsten was tabulated for different geometries and part sizes. This yielded limits for deciding ranges for various codes. For instance, parts with a mass between 10^{-4} mg and 10^{-2} mg are given code 1, while heavy parts with mass greater than 100 mg are given a code 9.

6th digit – Specific Stiffness

Prusi et al. [13] mention that the stiffness of the micro-part is a possible factor affecting the success of a precision assembly work cycle. The Fraunhofer Institute in its 2000 annual report [14] mentions that a high specific stiffness and low specific weight of the micro-parts allows for high dynamics and low power consumption of the drives that actuate the positioning stages. Accordingly, specific stiffness, which is defined as Young's modulus divided by density (E/ρ), can be expected to play an important role in deciding micro-part handling and manipulation. On studying the specific stiffness of various materials, with units MNm/kg, it is found that certain types of materials fall in different stiffness ranges. For instance, rubbers generally give a value of less than 0.1 while ceramics fall in the range of 30-200. Likewise, polymers yield values in the range 0.1-10. These observations help develop a coding scheme where different ranges correspond to low, moderate and high stiffness values.

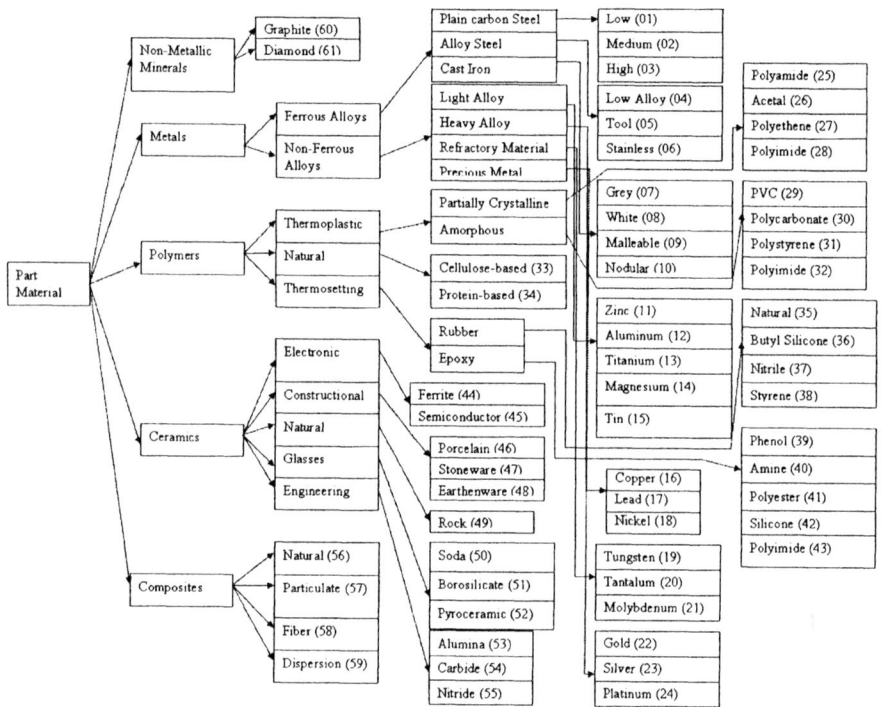

Fig. 5. Classification and coding of part material

4.3 Part Process Code

The part process code is included to provide important information that is related to the assembly process directly.

<u>7th digit – Part Positioning and Alignment</u>

This attribute describes the relative orientation of one part to another. It is based on the number of degrees of freedom required for alignment. It is further classified into coarse alignment techniques, viz., manual and external tool, and fine alignment techniques, viz., self alignment using surface tension, electrostatic force, magnetic force, external tool, separated constraint, and integrated constraint. The alignment directions and desired accuracy are also coded. All of these are stored at a level lower in the hierarchy.

8th digit – Surface Roughness

There are different measures for surface roughness. Of these measures, it was found that the average roughness, R_a is the most suitable measure for the coding scheme [15]. A non-dimensional ratio for coding the surface roughness can be defined as:

$$\Re^* = \frac{R_a}{D_{part}} \quad (7)$$

Here, $\Re^*$ is the relative surface roughness obtained by normalizing the average roughness with respect to the characteristic part dimension, D_{part}, which was defined by Eq. 5. For example, for a part with 1 Å average roughness and part dimension of 10 mm, $\Re^* = (10^{-10}/10^{-2}) = 10^{-8}$. On the other hand, although a part would never be expected to have roughness that exceeds its characteristic dimension, giving a limit of $\Re^* = 1$, some parts may actually have such a possibility since the characteristic dimension is evaluated taking all the dimensions of the object into account. Hence, a coding scheme for surface roughness can be defined as:

$$\Box = ceil(-\log_{10}(\Re^*)) \quad (8)$$

Here, the function ceil(x) is defined as the smallest integer greater than x. With this coding, a $\Re^*$ value between 10^{-5} and 10^{-4} would yield a code 5. The codes vary from 0 to 9 covering $\Re^* \leq 10^{-8}$ on one end with a code 9 and $\Re^* \geq 1$ with a code 0 on the other.

5 Coding of Assembly

Several attributes of an assembly have been identified that are key to deciding the steps in the assembly process. Form, material and process codes relating to an assembly were formulated as discussed below.

5.1 Assembly Form Code

The assembly form code is required to provide key geometry-related information about the assembly, specifically, the number of different types of parts and the type of assembly.

1st digit – Distinctly different parts

For an assembly consisting of 'k' parts, part 'i' is considered distinct if either of these conditions is satisfied:
i) The geometrical shape of part 'i' is different from all parts 'j', where j=1,2,...,k, j≠i
ii) The *characteristic* dimension of part 'i' is different from all parts 'k' having same geometrical shape as 'i'.

2nd digit – Type of assembly

The type of assembly is often a decision that has to be made by the design engineer based on his experience of what is exactly needed for the application. This choice in turn influences the selections of assembly system elements. Several different types of assembly are identified such as pick-and-place, peg-in-hole, joining, sacrificial layer, etc.

5.2 Assembly Material Code

The assembly material code is required to provide information related to materials and their properties that directly affect the assembly process. It is seen that the most critical part imposes the maximum constraints on a successful assembly. Every new type of material necessitates specific decision taking with regards to its properties such as brittleness, ductility, yield strength, etc., which influence handling and other procedures related to the assembly of the piece part. The tolerance requirement on the most critical part determines the most stringent constraints in the manufacture and assembly of the piece parts. Similarly, the tolerance requirement on the most critical joint determines the most stringent constraints during the assembly process. The mass of the assembly determines the overall manipulation and packaging of the final assembly and is not necessarily the same as the mass of all the parts put together since there could be additional weight due to joining processes such as weld deposit, glue, etc. Further analysis is being carried out on these attributes.

5.3 Assembly Process Code

The assembly process code is required to provide process related information about the assembly. The assembly environment impacts the assembly process significantly in terms of selection of grippers such as vacuum grippers, electrostatic grippers, etc., which can work well only in specific environments. Important environment variables are assembly medium (vacuum, clean room, etc.), temperature and humidity. The surface properties of the assembly such as roughness and presence of external features such as chamfers, grooves, etc., determine its manipulation, visibility and packaging. A quantitative measure for this can be established based on an average taken on the surface roughness of the surfaces on the exterior. The codes can then be established similar to the code for surface roughness.

6 Example Application

One example of a micro-scale system/device family is the class of miniature bearings used in missile and space-craft guidance systems, e.g., spin bearings and gimbal bearings. Figure 6 shows the components of a miniature spin bearing assembly consisting of outer and inner races, a retainer and twelve balls. Each of the piece parts was analyzed to yield the part coding as shown in Table 3. All the parts have rotational symmetry giving a code 0 for the first digit. The balls are spherical with a code 1, while the remaining parts are close to cylindrical but not quite due to curvature and other features giving a code for irregular shapes, which is 9. Also, the inner race has dimensions: outer radius of 3486 µm, height of 2032 µm and inner radius 2977 microns yielding a characteristic dimension code of 898 corresponding to these three dimensions, explained in detail in Table 4. Corresponding dimensions for the outer race are 5164 µm, 2771 µm and 4615 µm, while for retainer are 4254 µm, 1981 µm and 3486 µm. It is seen that the races and retainer are fairly large and hence, gravity is expected to dominate. The balls are medium-sized with diameter 1587.5 µm giving a value for $D_{part\_S/V}$ of 264.58 µm and the remaining three measures for characteristic dimension as 793.75 µm. With equal weights of 0.25, the characteristic dimension is obtained as 661.45 µm. Hence, it might be important to take care of sticking effects. The races and balls are made of low alloy steel which corresponds to the code '04' as seen in Fig. 5, opening up the possibility of magnetic manipulation. The retainer is made of delrin with a low specific stiffness of 2.19 giving a code 3, and therefore, this part requires care in handling. The geometry of the retainer requires alignment for the balls to enter the holes. The directions of the alignment for the assembly process forms the first digit of the position-align code and is established from the assembly process as 0 for x-axis and 1 for y-axis, respectively, for the two degrees of freedom for the races and retainer. The coarse and fine alignment procedures give codes of 1 and 4, respectively for the retainer and the races. Similarly, the required accuracy of ~60 nm gives a code for the last digit of the position-align code as 2. Also, the inner race is manufactured with an average surface roughness specification of 60 nm, which gives a value of $\mathfrak{R}^*$ equal to 2.51×10^{-5}.

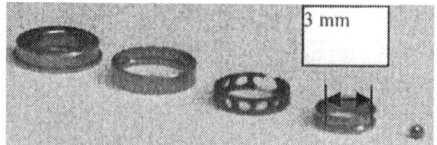

Fig. 6. Spin Bearing Assembly Components

The assembly code for this part was also developed as shown in Table 5. It was seen that this assembly is moderately complex with four types of parts, giving a code 4 for the first digit of the assembly code. An ad-hoc assembly type is required calling for a new innovative design (code 9). The balls are identified as the most critical part corresponding to a code '04' and can be considered for parallel manipulation due to their large number. It is a fairly heavy assembly weighing 724.27 mg (code 9) and gravity is going to dominate overall removing the need for any special manipulation for sticking effects while removing the assembly from the workspace.

Figure 7 shows the actual assembly design solution for the spin bearing. The assembly procedure uses a pusher which is used to push the inner race, which is fed from the magazine, into the central chamber. A plunger then comes down to force it onto the post. This centers the inner race. As the plunger pushes down, the inner race goes through a set of spring actuated doors that also help to center it. Similar steps are then repeated for the retainer. The only difference is the geometry of the pusher that helps align the retainer with respect to the ball feeder. Next, the ball feeder is raised to a pre-determined height. Then, balls load on their own by gravity feeding. With the help of external vibration or some similar external effort, the balls enter the grooves of the retainer. A cylindrical magnet at the bottom of the structure keeps the balls in place. Simultaneous with the lowering of the ball feeder the outer race is fed using the placer assembly. The outer race comes down to snap on to the remaining parts of the assembly. Next, the placer assembly is lifted up with the fully assembled spin bearing held in the doors. The plunger comes down to force out the assembled bearing.

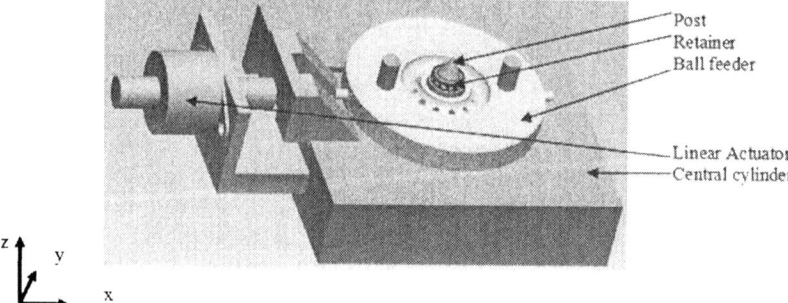

Fig. 7. Assembly of Spin Bearing

Table 3. Example coding of parts for miniature spin bearing assembly

Part	Description	Code	Characteristic Dimension Code	Position-Align Code	Inferences for assembly
Inner Race	Rotational symmetry (0); Irregular (9); D_{part} = 2391.4 μm (8); Low Alloy Steel 52100 (04); Mass = 135.55 mg (9); Sp. Stiffness = 26.15 (4); 2 DOF alignment, using magazine and pusher, plunger and doors (2); $\Re^*$ = 2.51x10^{-5} (5);	098049425	898	01421142	Fairly large size part; Gravity could be dominant; Possible need of tool control; Smooth part; At least one axis of rotation; Magnetic manipulation possible;
Outer Race	Rotational symmetry (0); Irregular (9); D_{part} = 3488.84 μm (9); Low Alloy Steel 52100 (04), Mass = 365.86 mg (9); Sp. Stiffness = 26.15 (4); 2 DOF alignment, coarse using placer and design constraint (2); $\Re^*$ = 1.72 x10^{-5} (5);	099049425	998	01421142	Large size part; Gravity dominant; Possible need of tool control; Smooth part; At least one axis of rotation; Magnetic manipulation possible;
Retainer	Rotational symmetry (0); Irregular (9); D_{part} = 2797.46 μm (8); Delrin (26); Mass = 26.42 mg (7); Sp. Stiffness = 2.19 (3); 2 DOF alignment using magazine and pusher, plunger and doors (2); $\Re^*$ = 2.14 x10^{-5} (5);	098267325	998	01421142	Fairly large size part; Gravity could be dominant; Possible need of tool control; Smooth part; At least one axis of rotation;
Balls	Rotational symmetry (0); Spherical (1); D_{part} = 661.45 μm (6); Low Alloy Steel 52100 (04); Mass = 16.37 mg (6); Sp. Stiffness = 26.15 (4); 2 DOF alignment using ball feeder and external vibration (2); $\Re^*$ = 9.07 x10^{-5} (5);	016046425	700	21525152	Medium size part; Electrostatic, adhesive, surface tension forces could be important; Possible need of tool control; Smooth part; Magnetic manipulation possible;

Table 4. Coding of part dimensions for miniature spin bearing assembly

Part	D1	D2	D3	Dpart_S/V	Dpart_geom	Dpart_arith	Dpart_max	Dpart	Code	Characteristic Dimension Code
Inner Race	2976.88	3486.15	2032	165.88	2912.09	3001.43	3486	2391.39	8	898
Outer Race	4615.18	5163.82	2771.14	228.98	4196.30	4366.26	5164	3488.84	9	998
Retainer	3486.15	4254.5	1981.2	140.96	3297.67	3496.73	4255	2797.46	8	998
Balls	793.75	--	--	264.58	793.75	793.75	793.75	661.45	6	700

All dimensions are in microns

Table 5. Example coding of assembly for miniature spin bearing assembly

Sub-assembly	Description	Code	Inferences for assembly
Spin bearing	4 parts (4); New assembly technique (9); Most critical part – balls (low alloy steel) (04); 2 dissimilar materials (2); Tolerances – 60 nm (1,1); Mass = 724.27 mg (9); Assembly environment – clean room (0); <$\Re^*$> = 7.68x10^{-5} (5);	4904211905	Moderately complex assembly; ad-hoc technique; magnetic manipulation possible; minimal influence of material properties; low tolerances; fairly heavy assembly; normal micro-scale assembly environment; smooth exterior

7 Conclusions

This paper has described a comprehensive framework for coding and classification for micro-assembly. Some of the key conclusions from this paper are:
  i. Piece-parts for micro-assembly can be classified and coded using the proposed coding scheme for form, material and process parameters of the individual parts.
  ii. The final assembly can also be classified and coded using the proposed coding for assembly attributes.
  iii. Key steps in the assembly of a spin bearing are outlined by analyzing the codes for the individual parts and final assembly and linking them to the design process.

With continual pressures to reduce product life-cycle development times, this classification and coding system procedure can serve as a useful tool in the product design process as well as the process of designing/reconfiguring the micro-assembly system. Also, such a coding system can help quickly assemble and disassemble components in a reconfigurable assembly station where different types of grippers, manipulators, vision systems and force sensing equipment is available.

Acknowledgements

The authors gratefully acknowledge the funding provided for this research by Mircolution, Inc., and Office of Naval Research – Small Business Technology Transfer Program. The authors further acknowledge the contributions of UIUC graduate assistant Nicholas Stephen Fezie in the development of the spin bearing assembly system.

References

1. Hollis, R. L., Gowdy, J., Rizzi, A. A. (2004), "Design and Development of a Tabletop Precision Assembly System" Mechatronics and Robotics, (MechRob '04) Aachen, Germany, September 13-15, 2004, pp. 1619-1623
2. Huang, J. D., Kuo, C. L. (2002), "Pin-plate micro assembly by integrating micro-EDM and Nd-YAG laser", International Journal of Machine Tools & Manufacture 42 pp. 1455-1464
3. http://www.nanomotor.de/aa_production_system.htm (2006), Last accessed Nov. 29, 2006
4. Kurniawan, I., Tichem, M., Bartek, M. (2006), "Morphological Classification of Hybrid Microsystems Assembly", Proceedings of the Third International Precision Assembly Seminar, pp. 133-148
5. Fatikow, S., Faizullin, A., Seyfried, J. (2000), "Planning of a Microassembly Task in a Flexible Microrobot Cell", Proc. of IEEE Intl. Conference on Robotics and Automation, pp. 1121-1126

6. Mukundakrishnan, B., Nelson, B. H. (2000), "Micropart Feature Design for Visually Servoed Microassembly", Proc. of IEEE Intl. Conference on Robotics and Automation, pp. 965-970
7. Nelson, B., Zhou, Y., Vikramaditya, B. (1998), "Sensor-Based Microassembly of hybrid MEMS devices", Control Systems Magazine, IEEE, pp. 35-45
8. Feddema, J.T., Simon, R.W. (1998), " Visual Servoing and CAD-driven microasssembly" Robotics & Automation Magazine, IEEE Vol. 5, Issue 4, pp. 18-24
9. Zesch, W. Brunner, M. Weber, A. (1997), " Vacuum Tool for Handling Microobjects with a Nanorobot", IEEE International Conference on Robotics and Automation, pp. 1761-1766
10. Fearing, R. S. (1995), "Survey of Sticking Effects for Micro Parts Handling", Proceedings of the International Conference on Intelligent Robots and Systems-Volume 2, pp. 2212
11. Enikov, E. T. (2006), "Micro- and Nano-assembly and Manipulation Techniques for MEMS", *Microsystems Mechanical Design*, Springer
12. Weck, M., Peschke, C. (2004), "Equipment technology for flexible and automated micro assembly", Microsystem Technologies, Vol. 10, pp. 241–246, Springer-Verlag
13. Prusi, T., Heikkila, R., Uusitalo, J., Tuokko, R. (2006), "Test environment for high performance precision assembly – Devclopment and Preliminary Tests", Proceedings of the Third International Precision Assembly Seminar, pp. 93-100
14. "Fraunhofer Annual Report" (2000), Institut Angewandte Optik und Feinmechanik
15. "Characterizing Surface Quality: Why Average Roughness is Not Enough", Available online at http://www.veeco.com/pdfs.php/246, Last Accessed June 08, 2007

A METHOD FOR THREE DIMENSIONAL TOLERANCE ANALYSIS AND SYNTHESIS APPLIED TO COMPLEX AND PRECISE ASSEMBLIES

Frédéric Germain, Dimitri Denimal, Max Giordano

Symme Lab. Université de Savoie, 5 chemin de Bellevue 74940 Annecy le Vieux
{frederic.germain, dimitri.denimal,max.giordano}@univ-savoie.fr

Abstract The tolerancing process for precise mechanical systems in a context of short or long run in industrial production requires a rational method from the specification of the functional requirements until the final products are checked. Nowadays the optimization of the tolerances is generally carried out empirically. Compromises must be made between the functional requirements in the design process and the manufacturing step. The limits of accuracy imposed by the process must be taken into account. The need for a rational method is particularly necessary for new products in the field of traditional mechanisms as well as of micromechanics or even of micro-systems. In the design process, in the case of an assembly, the functional requirements must be defined in geometrical terms, in order to satisfy the customer requirements. Then, these geometrical requirements must be translated into specifications on the various parts so that on the one hand the assembly can be carried out under well defined conditions and on the other hand, after assembly, the functional requirements are strictly respected. This transfer of specifications with creations of new specifications of the assembly requires a three-dimensional geometrical analysis taking into account the geometrical deviations- form defects, position and orientation- and size deviations. Clearances in the assemblies also will intervene. They are necessary to ensure the mechanical motions but also to compensate for variations of geometry in the case of hyper constrained assemblies. In the phase of industrialisation, the geometrical and dimensional tolerances will be necessary for the choice of machines, for manufacturing planning process and for the measurement processes during the production but also for the final quality control of the parts and of the system. The methods of assemblies are also strongly conditioned by the functional requirements.

1 References to related works and literature

Many works about dimensional and geometrical tolerances are presented in literature, but few of them offer a general method for three dimensional tolerance analysis and synthesis. The synthesis can be broken down into two parts: the qualitative and quantitative aspects. Initially, the question is to determine which surfaces are to be toleranced, which kinds of tolerances must be used, to define the references or datum reference frames. The quantitative synthesis consists in calculating the values to be assigned to the various tolerances in order to guarantee the functional requirements. Two approaches are possible: worst case tolerancing and statistical tolerancing. O. W. Salomons presents special software to help the designer to analyse and specify the tolerances [1], [2]. This tool is based on the concept of TTRS (Technologically and Topologically Related Surfaces) suggested by A. Clément [3] and taken again by other researchers [4]. In this method starting from the functional technological analysis the functional surfaces of the parts are linked between each other creating a binary tree. Several solutions are often possible and no rule is given to associate various surfaces. It seems that it is only the ability of the designer that allows one to choose the associate parts. B. Anselmetti also proposes an approach based on the expertise of the designer and he defines a named method "method CLIC". The rules, which lead to a choice of specifications, are not always justified but they give a solution in conformity with the ISO standards or with the ASME standard [5]. He shows the importance of the assembly method for the judicious choice of the tolerances and the datum systems under the qualitative aspect. On the other hand the calculation of the tolerances is based on the traditional chains of dimensions. The 3D problem is translated to a one-dimensional problem thanks to the concept of direction of analysis but the variations of orientation remain difficult to take into account. This last concept is also presented by L. Laperrière [6]. The concept of "Jacobean torsor" in fact is the same concept as the small displacements torsor used to model the variations of position and orientation.

In the statistical approach to analysing tolerances, generally one considers a functional condition "y", function of parameters characteristic of the geometry of parts xi: y=f(xi). It is assumed that each of these parameters is affected by independent dimensional tolerances. One distinguishes the analytical methods, which consists in calculating statistical characteristics of the variable y while those of variables xi are known and the simulation of Monte-Carlo which consists in generating a great number of random events. The first often encounters complex mathematical difficulties except in the particular case of linear functions. The second is greedy in computing times to obtain accurate results.

The analytical method uses the first order development in the vicinity of the configuration targets y=f(xi0)+Σ $\partial F/\partial xi$(xi-xi0). If the variables xi are random independent variables of average m(xi) and variance v(xi), then the mean value and the variance of the y variable are obtained by m(y)= f(m(xi)) and v(y)= Σ ($\partial F/\partial xi$)² v(xi). S. Nigam and J. Turner present a review of the statistical approaches for the analysis of tolerances starting from this parameterised form [7]. C. Glancy and K.

Chase propose a development with the second order for the analysis of tolerances of the assemblies in the nonlinear cases and use the statistical moments until order 4 [8]. The simulation of Monte-Carlo is more often used. It applies even if the relation is not linear and for any type of laws of distribution of the variables. Let us quote for example R.J. Gerth, W. Mr. Hancock who use the method of Monte-Carlo on a parameterized one and three dimensional model and use the data of manufacturing process to adapt the choice of the tolerances [9].

2 The research method

The method suggested here is based on some tools to implement in a systematic process and generally applicable to the majority of the mechanical systems, more or less complex assemblies and mechanisms. The stages are as follows:
- First the functional requirements of the assembly must be defined in geometrical terms.
- Secondly the qualitative tolerances of the parts are built in accordance with these requirements.
- Then the relations between the tolerances of the parts and the assembly requirement are established.
- Finally the values of the tolerances can be analysed or optimised.

Before detailing each of these steps, an example is presented that will be used as a discussion thread with the course of the method. Each step will be applied to the example and will enable to follow the logic of the method without losing its generic aspect. It is about a double hydraulic cylinder permitting to operate a shutter in an airplane. The cylinder acts in a classical mechanism with four principal solids according to the scheme fig.1a. The cylinder is double for safety conditions. Two independent hydraulic systems feed the two bodies so that in the event of escape in a circuit or one body, the other cylinder can still control the movement.

2.1 Functional requirements of the assembly in geometrical terms

In a general way a functional requirement is expressed by the acceptable limits of a geometric feature compared to another element belonging to the same part or another part of the mechanism or compared to a reference datum system built from several geometric features of the parts. One understands by geometric feature a particular point, a set of part surface points or an element built from a surface, the axis of a cylinder for example. Thus, the tolerances can be the allowed limits of a parameter (distance or angle) or can be a tolerance zone. The ISO standards on geometrical specification of products (GPS) can be extended to the functional conditions of an assembly.

58 A Method for Three Dimensional Tolerance Analysis and Synthesis applied to Complex and Precise Assemblies

For the suggested example the system made up of the two cylinders is considered. It is linked on the other parts of the mechanism by spherical joints. The system is connected to the other parts by four spherical surfaces. The functional conditions on this mechanism concern the relative positions of these four spherical surfaces. There are external requirements for the system represented on the Fig. 1.

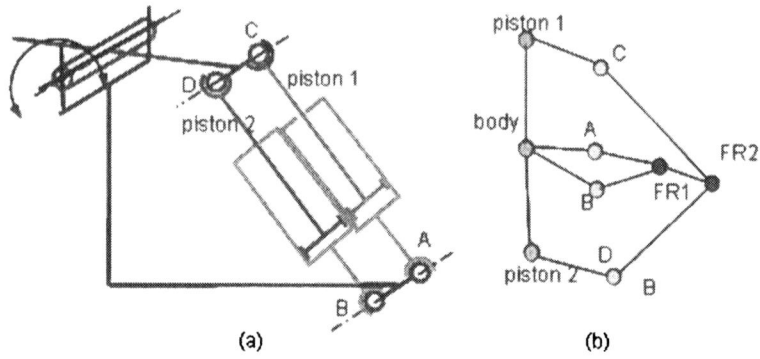

Fig. 1. Double cylinder mechanism (a) and minimal graph with the functional requirements (b)

The two spheres A and B rigidly linked to the cylinders must be in precise relative position. One can express this condition by a tolerance of localisation Fig. 2. The functional requirement relating to the two other spheres can be expressed in the following way: the two centres of sphere C and D must each be in a cylindrical diameter t tolerance zone of which axes are parallel and passing respectively by A and B and perpendicular to AB. Although the standards do not envisage expressing tolerances on assemblies, one can express this functional condition by using conventions of the ISO standard as well as possible supplemented by a comment (see Fig. 2).

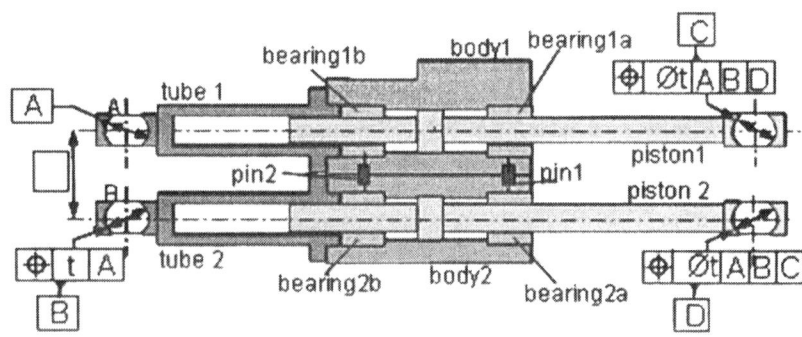

Fig. 2. Double cylinder and functional requirements

2.2 Qualitative aspect for tolerances

The problem is to define the dimensional and geometrical tolerances parts under qualitative aspect. The method to carry out the transfer of specifications under the qualitative aspect is based on graph representations presented in [10]. Two kinds of specifications then will appear: those which allow for carrying out the joints with given conditions of assembly and those which make it possible to guarantee the functional requirements after assembly. The kinematical graphs, the contact graphs and the tolerance graphs are imbricated graphs, which allow the qualitative synthesis even for complex cases comprising tens of surfaces of contact. In a kinematical graph, the vertices of the graph represent the parts and the edges are the joints. They are used for example for a kinematical or dynamic study. In graphs of contact, each vertex represents a Cartesian frame associated either with a contact surface, or built from several surfaces, or it is an arbitrary reference frame attached to each part. The graph edges characterise the geometrical deviation between reference frames in comparison with a perfect theoretical configuration of reference. There are two types of edges: between the reference frame of the same part, they characterise the geometrical deviations due to the manufacturing defects and surface of the joints, which characterise the possible contacts and gaps. The graphs of tolerance rise from the graphs of contact. The edges are oriented and enable to represent the toleranced surfaces and the datum system used form geometrical tolerances. These various graphs can be simplified or developed as a whole or locally. The development of the graph on a local level permits to analyze in detail the realization of a joint and the influence of a functional requirement over the feature of this joint. Specifications GPS ISO constitute an efficient tool for the representation of the tolerances because it is closed to the functional needs. One will define in this stage the toleranced surfaces, the surfaces or group of surfaces to be taken as a datum, the type of necessary tolerances: form orientation, localization or size.

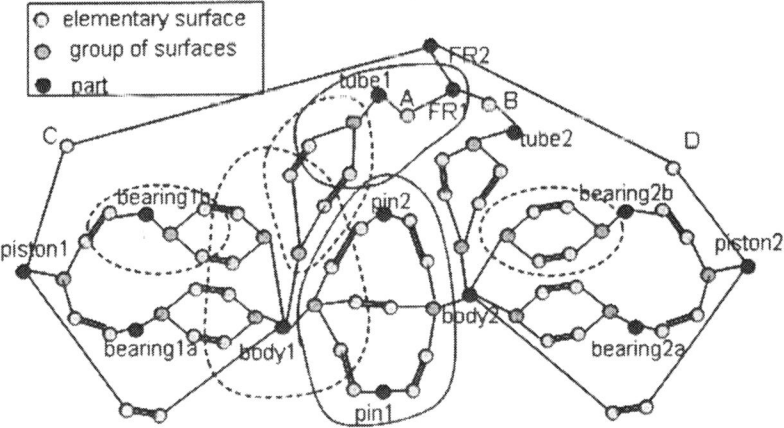

Fig. 3. Contact graph of the double cylinder

Let us take again the example of the double cylinder. On the one hand, the minimal kinematical graph, the more possible simplified is represented Fig.1b. The external functional conditions were represented. On the other hand in the complete graph above Fig. 3, all the contacts and all the parts appear. From this graph, one can extract sub-graphs representing the parts or the joints (closed curves in Fig. 3).

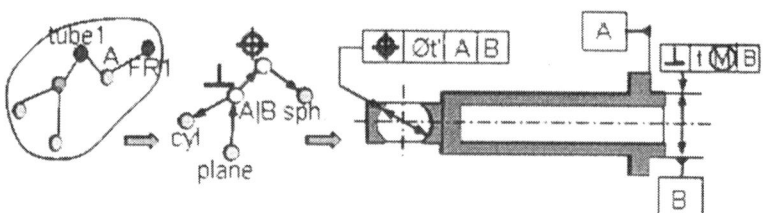

Fig. 4. Sub-graph for the tube 1, graph of the tolerances and standardized specifications

From this graph one carries out two types of simplifications:
- the joints in parallels are replaced by only one equivalent joint, the assembly requirement generate tolerances between the contact surfaces of the joint. The surfaces can be ordered, a priority surface will be a datum for the tolerance of another element. (Fig. 4).
- the joints in series are replaced by only one joint, the functional requirement implies tolerances on the intermediate part (**Error! Reference source not found.**).
- two parts with no relative motion are replaced by one equivalent part.
Thus gradually, one passes from the graph Fig. 1b to the graph Fig. 3 by generating the geometrical specifications on each part.

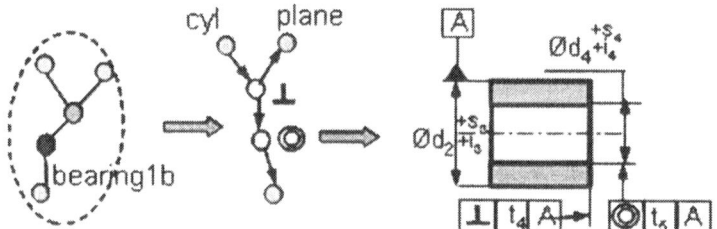

Fig. 5. Tolerance scheme for intermediate part

2.3 The quantitative step

This consists in establishing the relations between the data that characterize the specifications of the parts, and the variable that characterize the functional requirements. These relations can be established either starting from the worst case approach or according to a more realistic statistical approach in a context of interchangeability and industrial production in series. The geometrical variations of orientation and position are characterised by small displacements torsors. A linear approximation of the angular deviations supposed small, leads to representing the rigid displacement by a torsor. The deviation of a surface will thus comprise three angular components and three linear components of displacement for a particular point and following the three orthogonal directions of the frame. According to the statistical approach, these relations require us to know a priori or to impose data relating to dispersions on the geometrical characteristics in production. The geometrical deviations are represented by the components of the small displacements torsors. They are characterised by an average vector and a covariance matrix.

In the case of the functional requirement results from a simple stack up tolerances it is easy to compute the mean vector and the covariance matrix of the resulting deviation torsor which is the sum of the torsors component by component. It is the traditional chains of dimensions method extended to the three dimensions.

When clearances in the joints intervene, one distinguishes a resulting deviation torsor and a resulting clearance torsor. The resulting deviation torsor is determined as in the preceding case without taking into account the clearances. Its components are random variables. The resulting clearance torsor is unspecified. The limits of his components are fixed by real dimensions which are random variables. For example, for a shaft in a boring, the limits of the displacements permitted by the clearance depend on the two diameters. One calls clearance domain the sets of displacements limited by these random variables. For a given set of parts, the sum of the clearance domain and the deviation torsor is called shifted clearance domain. For a whole of parts, one will be able to then determine by the Monte Carlo method the whole of the possible variations due to the deviations and the clearances.

For joints connected in parallel, and a given assembly, the intersection of all the shifted clearance domains gives the resulting deviation after assembly from each branch in parallel. The assembly is possible only if this intersection exists. One will there obtain the dispersion of the variations of positions after assembly by taking into account the geometrical defects and the clearance by the Monte Carlo method. One studies first the assembly of the two bodies. The tolerances of the diameters of the two pins-holes are given in table 1. It is supposed that the values of dimensions and geometrical defects in the manufactured production of the parts are independent random variables centred on their target value. It is also supposed that the distribution is Gaussian with a standard deviation equals to the sixth of the tolerance.

Pin-hole	Ø10H6/h6	mean	tol	σ
hole	Ø10H6	10.0045	9 μm	1.5 μm
pin	Ø10h6	9.9955	9 μm	1.5 μm

Table 1. Fit of pin-hole

The tolerances of localisation are t1=0.01 mm (see Fig. 6). Fig. 7 gives the distribution of the variations of the relative position of the two bodies after assembly. It is noticed that in the worst case the assembly is not possible without constraint, whereas nearly 97.5% of the assemblies are possible according to the assumptions of statistical distribution. In 95% of the cases, the defect of setting in position of the two bodies does not exceed 9 microns. One hundred thousand samples were carried out for this simulation.

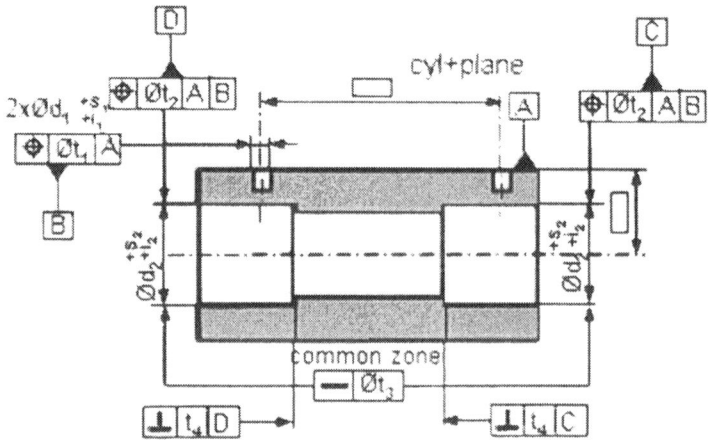

Fig. 6. Qualitative tolerance specification for body 2.

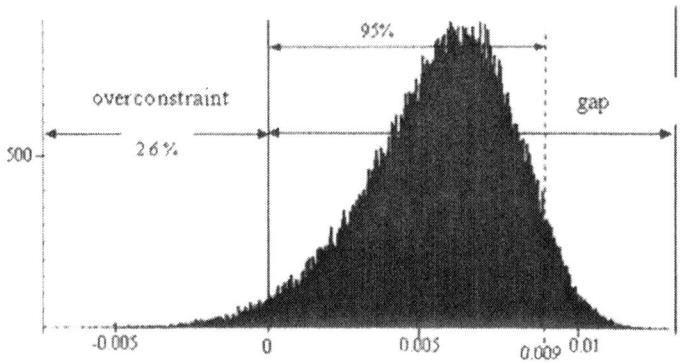

Fig. 7. Histogram for the gap or subconstraint in the pin-hole assembly.

For all the assembled bodies, the variation concerning each of the two functional requirements can be simulated.

Shaft/bearing	Ø36H6/h6	mean	tol	σ
Bearing	Ø36H6	36.008	16 µm	2.7 µm
shaft	Ø36h6	35.992	16 µm	2.7 µm

Table 2. The fit of the tube and the body

For requirement 1, the variation of distance between the centres of the spherical surfaces A and B results from a set of variations in series from surface A to B while passing by tube 1 then body 1, body 2 and the tube 2. The resulting variations due to the joints linked in parallel are initially computed. Then one deals with the stacking of the equivalent joint in series. The localisation tolerances are 0.01 millimetres and the dimensional tolerances are defined in Table 2. Fig. 8 gives the histogram of the obtained deviations.

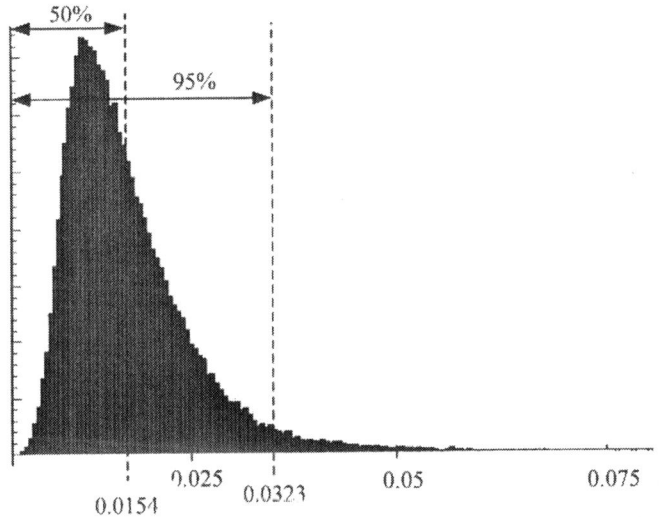

Fig. 8. Statistical repartition of functional requirement 1 by taking into account the statistical clearances and statistical deviation

3 Major results

The relations between the values which characterise the tolerances of parts and those which characterise the functional requirements are not easy to determine in the case of a very complex graph. But it is always possible to analyse the functional requirement from given part tolerances by the Monte Carlo method. Then the method permits us to guide a synthesis of tolerance in an objective of optimisation, i.e. choice of the tolerances which guarantee the functional requirement but without over-quality. The tools of assistance to simulation make it possible to define the structure of the mechanism by the graphs. After the qualitative determination of the tolerances defined automatically starting from the decomposition of the contact graph, one will be able to introduce the values of the adjustments and the tolerances defined a priori. The computing of geometrical stack up tolerances and the combination of the motions due to the clearances allow us to check the fitting and to see the influence of the various parameters when regenerating the computing after modifying the data. An example of a hydraulic cylinder with two bodies assembled by two axes is investigated. The dispersions obtained after assembly enable designers to check the choices of the fits and the geometrical tolerances and to question the technological choices in order to optimise the system before prototyping.

The method suggested provides an important benefit in comparison with the conventional methods of analysis. On the one hand one proposes here a method of synthesis of the tolerances under the qualitative aspect which allows a coherent quantitative analysis, whereas no general method of synthesis does exist. In addition, the geometrical defects such as coaxiality, symmetry, etc, and the tolerance size are taken into account in our model. Finally the majority of the models of simulation are applied on mechanisms with particular and simple structures: open or closed loops while the method suggested here is more general and can be applied to complex mechanisms.

References

1. O.W. Salomons, and al, A computer aided tolerancing tool I : tolerance specification, Computers in Industry Vol 31 1996 pp.161-174
2. O.W. Salomons, and al., A computer aided tolerancing tool II : tolerance analysis, Computers in Industry Vol 31 1996 pp.175-186
3. A.Clement, A. Riviere and M. Temmerman Cotation tridimensionnelle des systèmes mécaniques, théorie et pratique PYC Edition, 1994
4. A.Desrochers and A. Clément A dimensioning and tolerancing assistance model for CAD/CAM systems, International Journal of Advenced Manufacturing Technology, Vol.9, 1994, pp.352, 361.
5. B. Anselmetti, K. Mawussi, Tolérancement fonctionnel d'un mécanisme : identification de la boucle de contacts, IDMME, May 14-16 2002 Clermont-Ferrand, France.
6. L. Laperrière, W. Ghie, A. Desrochers, Projection of Torsors : a Necessary Step for Tolerance Analysis Using the Unified Jacobian Torsor Model. 8th CIRP International Seminar on Computer Aided Tolerancing, April 28-29, Charlotte, North Carolina, USA.
7. S. D. Nigam and J. U. Turner, Review of statistical approaches to tolerance analysis, Computer-Aided Design, Vol 27, N° 1, pp.6-15, 1995.
8. C.G ; Glancy, K. W. Chase, A second-order method for assembly tolerance analysis, Proceedings of the ESME Design Engineering Technical Conferences, sept. 12-15, 1999, Las Vegas, Nevada.

9. R. J. Gerth, W. M. Hancock, Computer aided tolerance analysis for improved process control, in Computer & Industrial Engineering, Vol. 38, 2000, pp. 1-19.
10. P Hernandez, M Giordano, G Legrais, A new method of design integrated tolerancing, 14th CIRP 2004 Cairo, Egypt.
11. M. Giordano, E. Pairel, P. Hernandez, Complex Mechanical Structure Tolerancing by Means of Hyper-graphs, in Models for Computer Aided Tolerancing in Design and Manufacturing, J.K. Davidson Editor 2007, Springer, pp.105-114.

NEW DESIGNS FOR SUBMILLIMETRIC PRESS-FITTING

Ludovic Charvier, Fabien Bourgeois, Jacques Jacot

Laboratoire de Production Microtechnique, EPFL
CH-1015 Lausanne, Switzerland
ludovic.charvier@epfl.ch

Grégoire Genolet, Hubert Lorenz

Mimotec SA, Blancherie 61
CH-1950 Sion, Switzerland
info@mimotec.ch

Abstract Press-fitting is a very common assembly process used in many fields, especially in the watchmaking industry. It consists of introducing a shaft in a hole (also called receptor) with a certain value of tightening (or interference). However, this process is not straightforward and it is often observed that some assemblies do not manage to hold out the time and function they are meant to. In order to solve this problem, influent parameters had been searched and are now studied. The shape of the receptor has appeared as a critical one. Indeed, new machining processes have been developed this last decade and we are now able to make new designs which could cope perfectly with this application. The article will deal with the influence of the two main parameters involved in a press-fitting assembly: the maximal press-fitting force, measured axially; and the resistive torque exercised between the shaft and the receptor. The main purpose will be to find which kind of design allows the contact to become less stiff, which means less internal constraints. Thus, the tolerances, especially those of the receptor (they are less well controlled than those of the shaft) could be enlarged and this would be a great improvement for the industrial world.

1 Introduction

Press-fitting is a joining process which is very simple to set up. Nevertheless, when the dimensions are smaller than one millimetre, everything becomes much more complicated to control. The main difficulty is that the tolerances demanded on the shaft and on the hole are extremely thin (about 1 µm). By consequence, the manufacturers are no longer able to cope with the requirements. Bourgeois showed that the assembly force is proportional to the interference and the stiffness of the receptor. The main idea of this study is so to reduce the stiffness of the design that the tolerances acceptable for the assembly could be enlarged. This idea is illustrated in Figure 1.

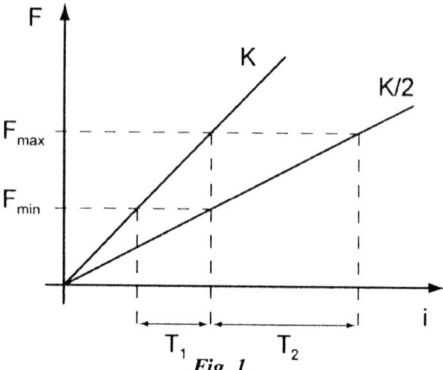

Fig. 1.
Evolution of the interference range when the stiffness of the contact is divided by two

Indeed, if we divide the stiffness of the contact (K on the picture) by a factor two, we obtain a range for the tolerance T which is about two times larger, for the same F_{min} and F_{max}. These two forces represent the maximum and minimum axial forces the contact has to support.

We also observe that the interference range is moved, which may raise problems for some applications. However, we better understand now the interest in trying to reduce as much as possible the stiffness of the assembly.

Moreover, another point has to be taken into consideration. Press-fitting can be used so that the shaft resists to an axial force but in some other cases, it also has to resist to a minimum axial torque. The consequence is that the contact surface has to be increased while the resistive torque directly depends on this factor. Here is the formula of the resistive torque given by the Lamé-Clapeyron's model [1] which explains that dependence:

$$M = F \cdot \frac{D}{2} = \frac{\pi}{4} \mu \cdot i \cdot D \cdot E \cdot L$$

Fig. 2.
Resistive torque formula given by Lamé-Clapeyron's model

where M is the resistive torque, μ the friction coefficient, i the interference, D the shaft diameter, E the Young modulus and L the length of contact.

Here $\pi.D.L$ represents the contact surface for a proportion of contact of 1 but which can be modified as we will present later on.

2 Ideas of the experiment

The main idea of this series of experiments is to reduce the stiffness of the contact in keeping a good contact surface for the resistive torque. For that, we decided to design new receptor shapes while shaft fabrication is better controlled. We can know the diameter of a shaft with a tolerance of ±1μm, which is completely impossible for a cylindrical hole. This is due to manufacturing processes and measuring tools which have about the same precision as the tolerance we want to obtain.

First of all, we have chosen to make our new receptors with a LIGA process. Thus, we have almost no constraints for the shape we want to design. Moreover, we can obtain a thickness up to 400 μm which is typically the interesting range for our application.

Concerning the shape, we have tried to focus on the fact that the matter is less constrained when it works in traction/compression instead of flexion. With this approach, the receptor material would remain in its elastic field and the stiffness of the contact would be reduced. In order to show this influence, we decided to make designs with and without flexion oblongs. The following figures show this aspect:

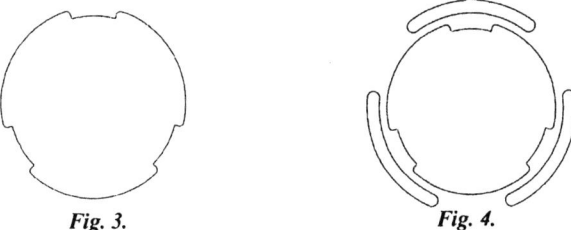

Fig. 3. *Fig. 4.*
Designs with and without flexion oblongs

Another point which had been considered as primordial was to study the influence of the contact surface. For this, we decided to experiment with designs with different proportions of surface in contact with the pin. For instance, in Figures 5 and 6 we can observe a proportion of a half and of three thirds.

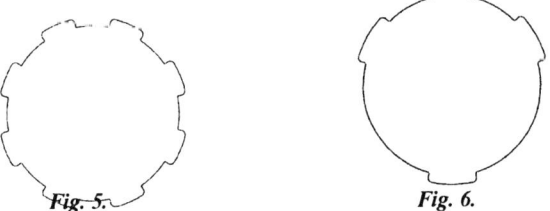

Fig. 5. *Fig. 6.*
Designs with different proportions in contact with the pin

And at least another parameter we would have studied was the micro oxydations. Indeed, in a previous study lead by Fabien Bourgeois [2], we noticed that some of these phenomena could appear during the press-fitting. We decided to test some shapes, with cannels, the goal of which would be to evacuate these microweldings. The following figures show the designs we developed to investigate this field:

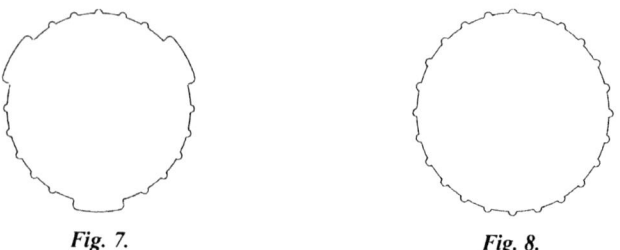

Fig. 7.   *Fig. 8.*

Designs with channels to evacuate the eventual microweldings

3 Experimentation description

The experiments are made on a Promess press whose maximal delivered force is 1000N. In this part, we will study the maximal press-fitting force obtained by each design and also the resistive torque associated with each particular shape.

Figure 9 shows the "posage" used to align the shaft with the design to test. This device is also used to determine the maximal press-fitting force.

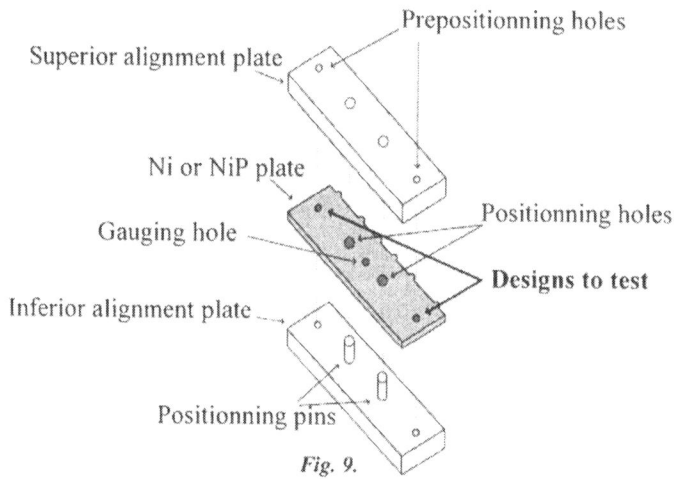

Fig. 9.
Device for the determination of the maximal press-fitting force

The shaft enters in the prepositioning hole. We align the axis of the press with the axis of the shaft and we process to the press-fitting (strictly speaking).

In the test for the determination of the resistive torque associated with each design, we process it differently. Figure 10 presents the system:

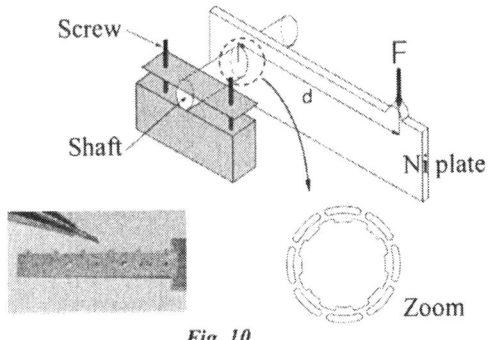

Fig. 10.
Device for the determination of the resistive torque

With the press, we come in contact with the plate at a distance d from the shaft. Thus by knowing the force necessary to make the plate turn and the distance d, we manage to obtain the resistive torque associated with the design tested.

4 Results

Due to the different parameters we would like to analyse, we made the experiments on seven different designs:

- 8 fluting, with flexion, r = ½
- 8 fluting, no flexion, r = ½
- 24 fluting, no flexion, r = ¾
- 3 fluting, with flexion, r = ¾
- 3 fluting, no flexion, 15 channels, r = ¾
- 3 fluting, with flexion, r = ¼
- 3 fluting, no flexion, r = ¼

NB: r = ratio between design surface and shaft surface

As the production of samples by LIGA process takes quite a long time, we do not have many points of measurement. Nevertheless, we managed to obtain the global trend for each design. We specify that each point on the following graphs represents one measure.

4.1 Press fitting force

<u>Experimental conditions:</u> temperature: 21°C±2°C, press-fitting speed: 1mm/s, L = 0.4mm, E = 155GPa, µ = 0.2, D = 994µm±0.5 et i = 2-27µm

Figure 11 presents the results of the press-fitting test:

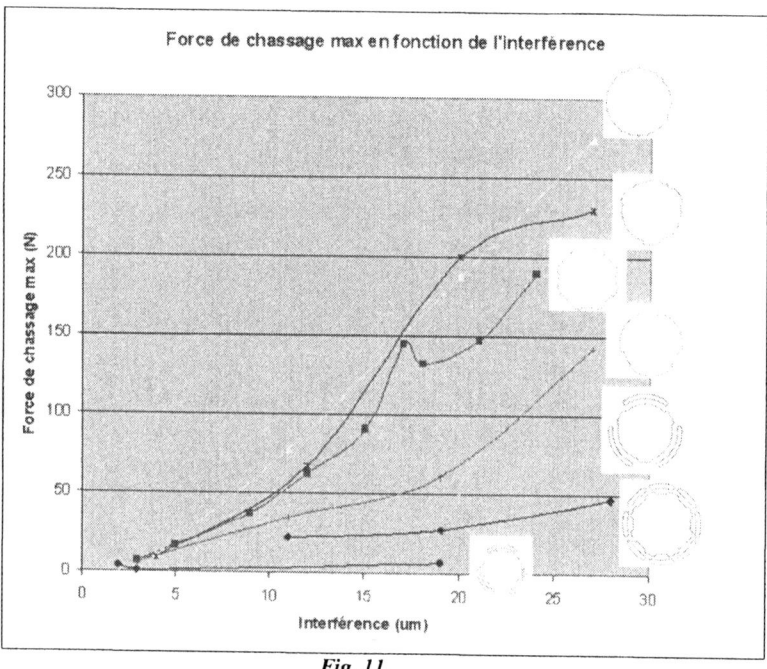

Fig. 11.
Evolution of the maximal press-fitting force with the interference

4.1.1 Results analysis

The goal here was to observe the maximal press-fitting force value for each design and for a range of interference from 2 to 27 µm.

As we expected, designs with flexion oblongs reach a lower value than the others, which means that the matter is less constrained and probably does not reach the plastic field. This can be interesting for reversible applications or for devices which require a very high precision in axial positioning.

Conversely, designs from the top of the graph require a bigger force to obtain the same results, which means that the contact will be stronger but the precision in the axial positioning will probably be more uncertain.

4.2 Resistive torque

The experimental conditions remain the same and we chose a distance d equal to 5 mm. Fig. 12 presents the results of the resistive torque test:

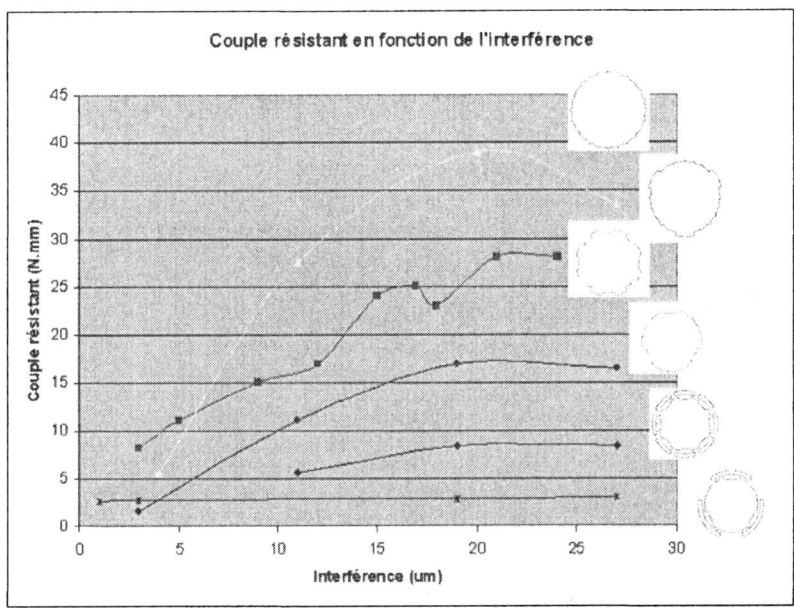

Fig. 12.
Evolution of the resistive torque with the interference

4.2.1 Results analysis

We notice here the designs with the higher r ratio have a better resistive torque that the others, which seems logical as the torque directly depends on the contact surface between the shaft and the receptor. Moreover, we observe a kind of saturation for each design over an interference of about 20μm. This could be explained by the fact that the matter has reached a certain deformation (plastic or not) and the surface in contact remains fairly constant despite the increase in the tightening. Nevertheless, this phenomenon has to be confirmed by other tests.

NB: We do not have results for the design with 3 fluting, with flexion, r = ¾ because it did not even manage to support the manipulation to be tested. We can simply ensure that its comportment is very bad and its performance may be inferior to the worst case tested.

5 Perspectives

Despite the very few designs we have already tested, some guidelines appeared to us. First of all, the flexion oblongs play a great part in the stiffness of the contact. For the same design, the maximal of the press-fitting force is divided by four due to these empty spaces. Secondly, the proportion of the receptor surface also plays a great role. When we pass from a r ratio of ¼ to ½, we increase the resistive torque from about 33%, and it is almost the same from ½ to ¾.

Another approach which could be a lead for such a study would be the influence of the material. It could be interesting to build a table which would gather these kinds of values for the main materials used in industry. In particular we think of silicon which has a lot of particularities that could be very promising in this field of study. However, to begin to press-fit in silicon, we first have to determine a design which will reduce drastically the value of the maximum press-fitting force in order not to start a crack in the monocrystal of this material. The shape with 8 fluting, with flexion, r = ½ would seem to be interesting for this application but it still has to be proved.

To determine these designs, we based our reflection on common sense, the fact that flexion is stiffer than traction/compression and that contact surface ratio would not be necessary equal to one. Yet we could also ask a finite elements software to give us its own optimised shape, given the right boundary conditions application. Thus, we would be able to compare what is theoretically calculated by a soft, with the shape we have chosen intuitively. The next step would be to adapt the form given by the computer to obtain something feasible, and to test this design to measure its performance. With such an approach, we would have a solid basis to work on, especially if the values obtained in concrete terms match with the results given by the simulation.

References

1. E. Fortini, Dimensioning for Interchangeable Manufacture, *Industrial Press* (1967)

2. F. Bourgeois, Vers la Maîtrise de la Qualité des Assemblages de Précision (EPFL, Lausanne, 2007)

DESIGN AND TESTING OF AN ORTHO-PLANAR MICRO-VALVE

O. Smal[1], B. Raucent[1], F. Ceyssens[2], R. Puers[2], M. De Volder[3], D. Reynaerts[3]

[1]Université Catholique de Louvain, Dept. of Mechanical Engineering, Div. CEREM, Place du Levant 2, 1348 Louvain-la-Neuve, Belgium

[2]Katholieke Universiteit Leuven, Dept. of Electrical Engineering, Div. MICAS, Kasteelpark Arenberg 10, 3001 Leuven, Belgium

[3]Katholieke Universiteit Leuven, Dept. of Mechanical Engineering, Div. PMA, Celestijnenlaan 300 B, 3001 Leuven

Abstract This paper presents the production and testing of an ortho-planar one-way micro-valve. The main advantages of such valves are that they are very compact and can be made from a single flat piece of material. A previous paper presents and discusses a micro-valve assembly based on a spider spring. The present paper focuses on the valve assembly process and the valve performance.. Several prototypes with a bore of 0.2 mm have been built using two manufacturing techniques (µEDM and stereo-lithography) and tested for pressures up to 7 bars.

1 Introduction

Micro-valves are one of the most important components of micro-pumps which exist in hundreds of configurations [1]. They exist in various working principles and designs as presented in [2-4]. One of the smallest check valves available commercially is from the Lee Company which is a poppet-type and has a diameter of 3.2 mm and a length of 13.2 mm. A list (non exhaustive) of commercial miniature valves is given in Table 1.1.

Table 1.1. Industrial check valves

manufacturer	principle	diameter	length	-
Check-all Valve	Poppet-type	22 mm	60 mm	[5]
Clippard minimatic	Poppet-type	9.5 mm	19 mm	[6]
Command controls corp.	Ball	14.2 mm	25 mm	[7]
Deltrol Fluid Products	Ball	11.2 mm	24.5 mm	[8]
Halkey-Roberts	Ball	7.3 mm	21.6 mm	[9]
Lee Company, The	Poppet-type	3.2 mm	13.2 mm	[10]
O'Keefe Controls Co	Ball	8 mm	11.6 mm	[11]
Sterling Hydraulics	Ball	19 mm	27.2 mm	[12]

Most of these commercial miniature valves are either poppet-type or ball-type valves. These two principles are illustrated in Figure 1.1. The forward flow is left to right. Note that in these versions there is a spring to maintain the ball or the cone on its seat at zero or very low negative pressure. One can see in Table 1.1 that industrial check valves are big regarding micro-pumps. This paper discusses on a small valve that can be built from a single flat piece of material.

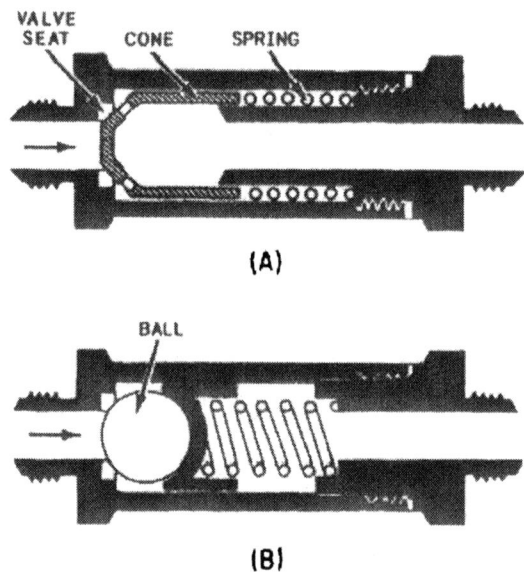

Fig. 1.1. Valve principles: (A) poppet-type valve, (B) ball valve (by courtesy of Integrated Publishing [13])

A previous paper [14] presents and discusses a micro-valve assembly based on a spider spring. It discusses the advantages of ortho-planar springs, and describes a new analytical model of the particular spider spring. The model is validated by a FEM and by measurements. The present paper focuses on the valve assembly process and the valve performance.

The first section (section two) of the paper sums up the situation with respect to the previous paper and presents the second ortho-planar spring realization. In section three, the valve assembly is presented and analysed. Section four presents the sealing tests and the flow rate measurements for low and high pressures.

2 Spider spring manufacturing

This section recalls and summarizes what had been presented in [14] concerning the spring realisation. The spider spring is produced using a micro electro discharge machining (µEDM) milling machine; the clamping system is presented in figure 2.1. The stainless steel foil is 10 µm thick and the cutting electrode has a diameter of 150 µm.

The second technique of manufacturing that was used is SU8 lithography. This process is presented and characterized in various papers [15, 16]. The main limitation of this method concerns the mask needed to manufacture the parts.

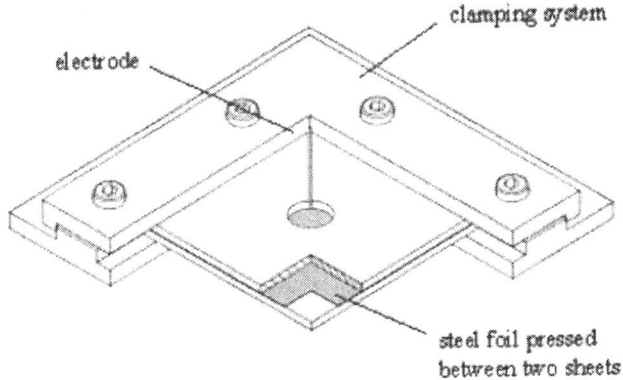

Fig. 2.1. Clamping system for EDM manufacturing

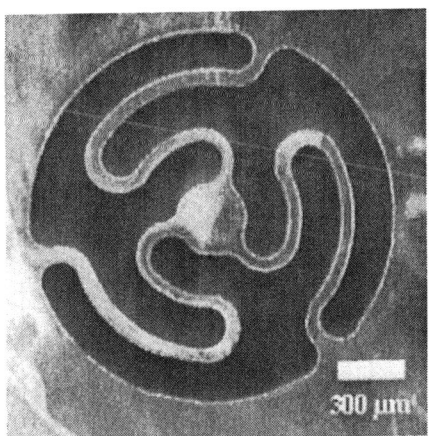

Fig. 2.2. Valve spring manufactured by EDM (150 µm electrode)

The resolution of this mask, which depends on the method used (laser, ink-jet printing), will directly influence the accuracy of the final spider spring. An example of a mask is shown in Figure 2.3. This shows an array of spider springs, which consist of a central disc connected to an outer ring (1) via spider spring arms. The outer ring is connected to plate (3) via break-away beams (2). Figure 2.4 presents some realizations of spider springs made from a resin epoxy-type material called SU8.

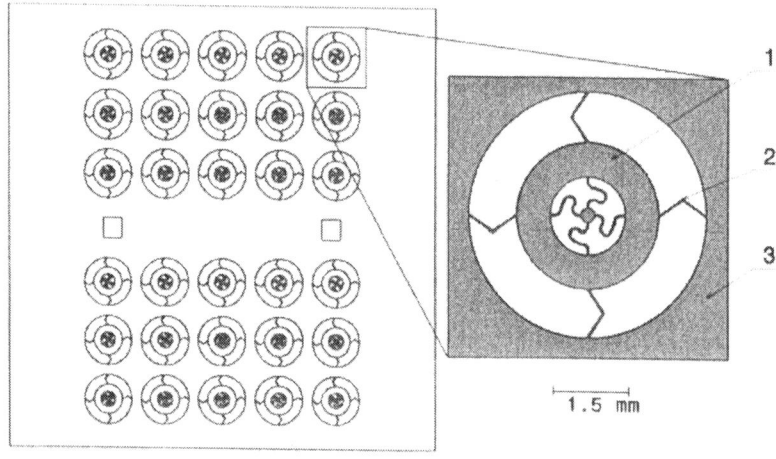

Fig. 2.3. Mask for lithography process

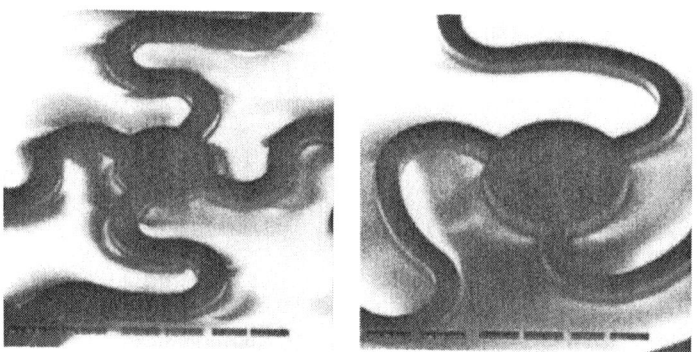

Fig. 2.4. SU8 end products (mask accuracy 2 μm)

This method has several advantages:

- it uses materials with **lower** Young's moduli (compared to materials used with the EDM technique), in particular epoxy type - i.e. SU8: E = 4 GPa,
- possibility to manufacture an **array** of spider springs in one step,
- possibility to manufacture a **multi-layered** spider spring. By varying the thickness of the spider spring along the diameter, it is possible to pre-stress the valve (see figure 2.5).

The advantage of pre-stressing the valve is to prevent backflow (so to enhance sealing) at very low pressure. As shown in Figure 2.5, there are two main methods to pre-stress the spider spring: either to modify the seat, or to modify the spring. So, the possibility to manufacture a multi-layered spider spring leads directly to the pre-stress of the valve.

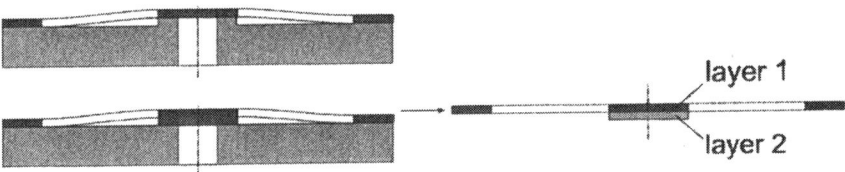

Fig. 2.5. Valve pre-stress principle (lateral view)

3 Assembly of the micro-valve

This section presents all the parts included in the valve and the assembly of all these elements. Figure 3.1 illustrates the principle of the valve (not to scale). The inlet of the valve is at the bottom and outlet at the top. The spider spring (3) is placed on the aluminium seat (5): since the central disc is over the input hole, an o-ring (2) is placed on top of the spring and an aluminium ring (4) is placed around the spider and the o-ring. The role of the o-ring is to uniformly distribute the force over the external part of the spider spring. Finally, a lid (1) is placed on top of everything. The reason for separating parts 4 and 5 is that in that way, the top surface of part 5 (i.e. the seat of the valve) can be easily polished to obtain a better quality surface. The crucial point of this assembly is to correctly align the spider spring (3) and the seat (5). In the first version of the valve, the spring is aligned visually (using a microscope) with the seat, and then the o-ring is placed on the spring. This operation is quite difficult because of the small dimensions of the spring.

80 *Design and Testing of an Ortho-Planar Micro-Valve*

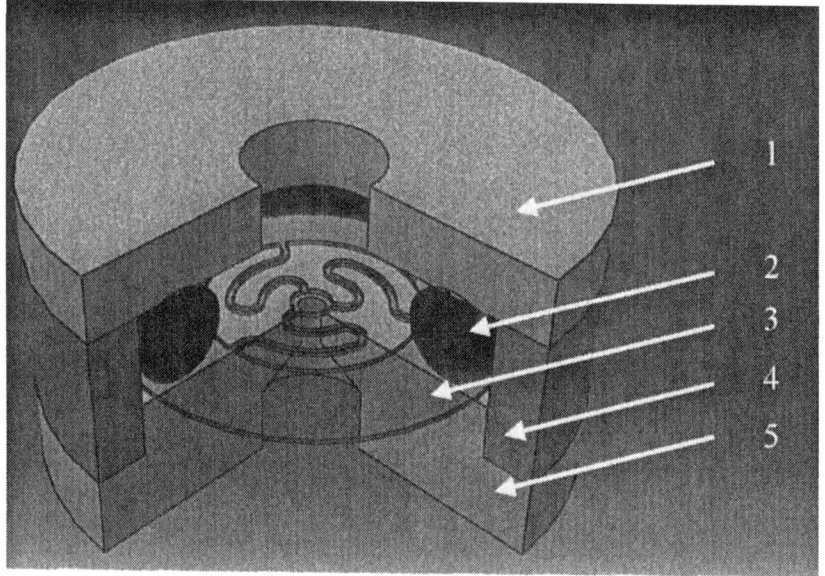

Fig. 3.1. Principle of assembly of the valve

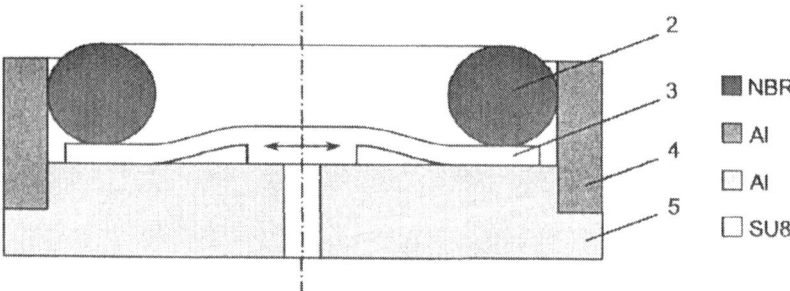

Fig. 3.2. Spider spring placement

All the parts are held together using screws (not represented on the diagrams) passing through parts 1, 4 and 5. Another solution would have been to glue all the elements together but in that case, we would have to realise as many complete valves as springs to be tested.

Starting from theoretical considerations, we can find rules for design for manual assembly at classical dimension in [17]. These are listed in decreasing order of importance:

1. reduce part count and part types,
2. strive to eliminate adjustments,
3. design parts to be self-aligning and self-locating,
4. ensure adequate access and unrestricted vision,
5. ensure the ease of handling of parts from bulk,
6. minimise the need for reorientation during assembly,
7. design parts that cannot be installed incorrectly,
8. maximise part symmetry if possible or make parts obviously asymmetrical.

For our application, we will discuss rules 3 and 4. In the design presented in Figure 3.2, part 3 is not self-locating on part 5 (rule 3 is not respected). Furthermore, due to the clearance between part 3 and 4, adjustment of part 3 has to be made through the o-ring (2). This decreases the access and restricts the vision. A simple way to improve the situation is to design part 3 to be self-aligned. In That way, we will try to provide parts with leads, lips, tapers, chamfer, etc. so that alignment is built into the design. However several limitations should be considered:

- tapers or chamfers are not possible in our stereo-lithography process
- lips on the seat makes the polishing of the seat difficult
- pilot on spider should be deep, but manufacturing time increases with the deepness of a layer and a layer is limited in thickness...

Several solutions were considered. These impose the need to either modify the spider spring, or to modify the seat or to manufacture an extra intermediate part.

name	Part modified	Schemes	Advantages/drawbacks
pilot	spider	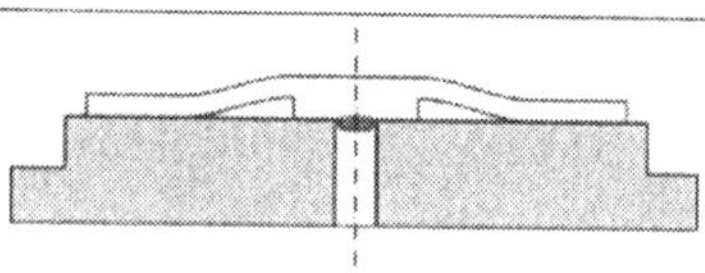	- chamfers not possible to manufacture - 3 layers means very long manufacturing time - risk of wedging
lips	seat	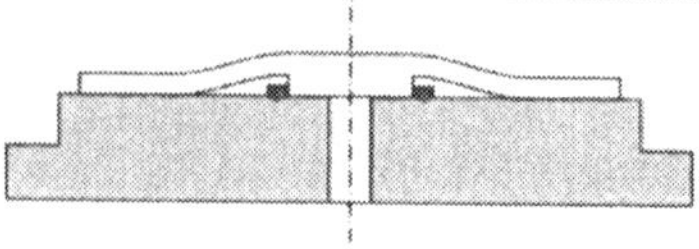	- requires precision micro-milling machine - polishing of the upper surface of the seat not possible anymore - risk of wedging
		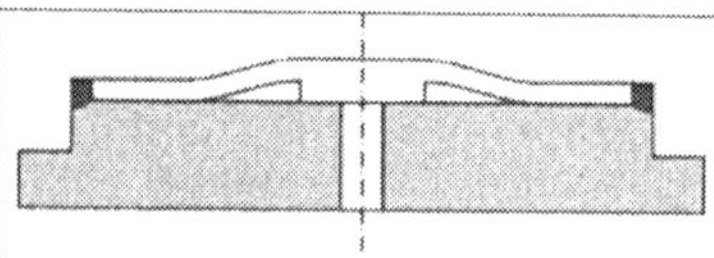	- requires precision micro-milling machine - polishing of the upper surface of the seat not possible anymore
	Extra part (PDMS)	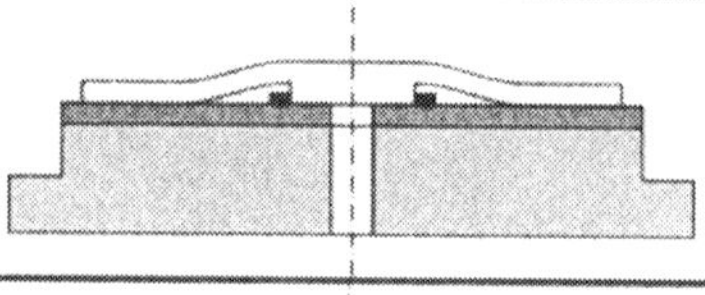	+ Better surface finish of the upper surface of the seat - risk of wedging
		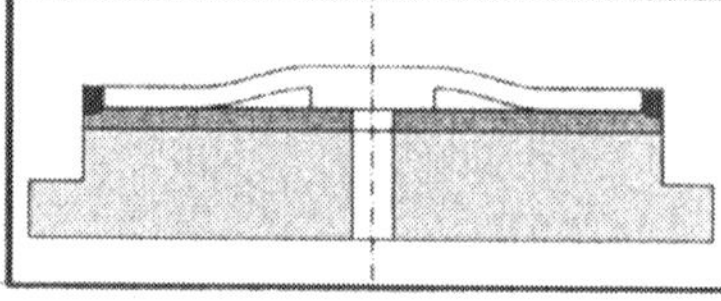	+ Better surface finish of the upper surface of the seat ✓

Fig. 3.3. Solutions for alignment problem

The adopted solution presented on figure 3.4 has two advantages: the first is to allow an accurate centering of the spider spring, and the second one is to improve the surface quality of the seat.

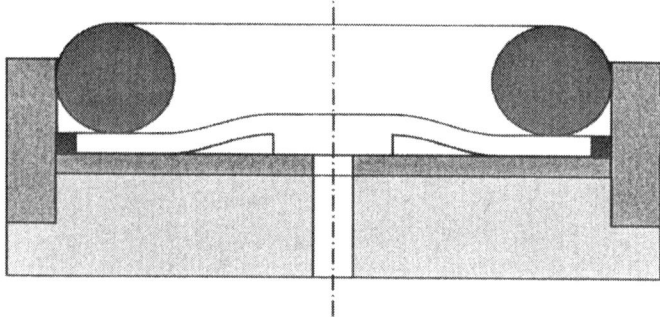

Fig. 3.4. Drawing of principle of the solution adopted for the micro-valve

4 Valve tests

This section presents the flow rate measurements as a function of the pressure difference. Two test beds were realised, the first one for low pressure measurements, where the pressure is generated by a water column, and the second one, where a pressurised tank is added. In both cases, the flow rate is measured using a graduated tube and a chronograph.

4.1 Low pressure tests

The tests bed is presented in Figure 4.1. Several spider springs were tested using this test bed. The maximum pressure was 0.1 bar (a water column of 1 meter). Measurements for spider spring with no pre-stress are very bad (high leakage in backflow) and are not presented here. Figure 4.2 presents results for pre-stressed spider springs. The main characteristics of these springs are presented in table 4.1. There is practically no backflow and we see that, in general, the flow in the forward direction is higher when the stiffness of the spring is smaller.

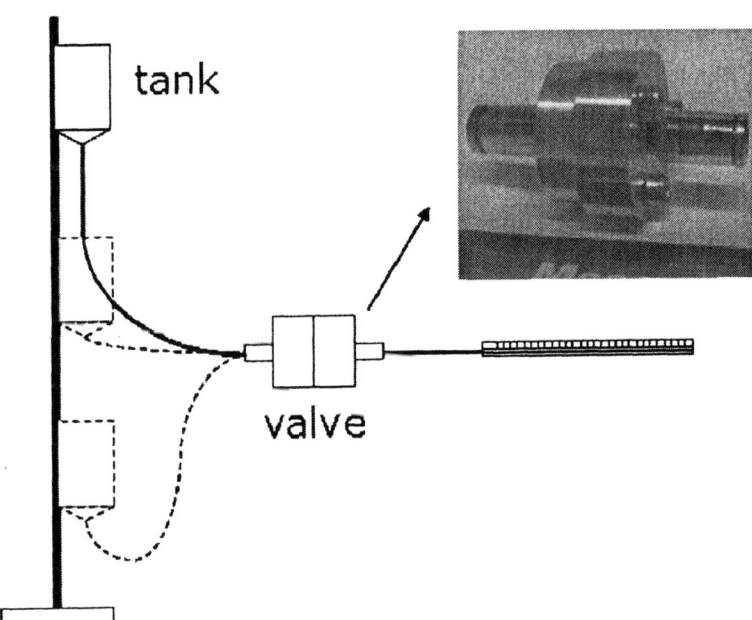

Fig. 4.1. Test bed for flow rate measurements

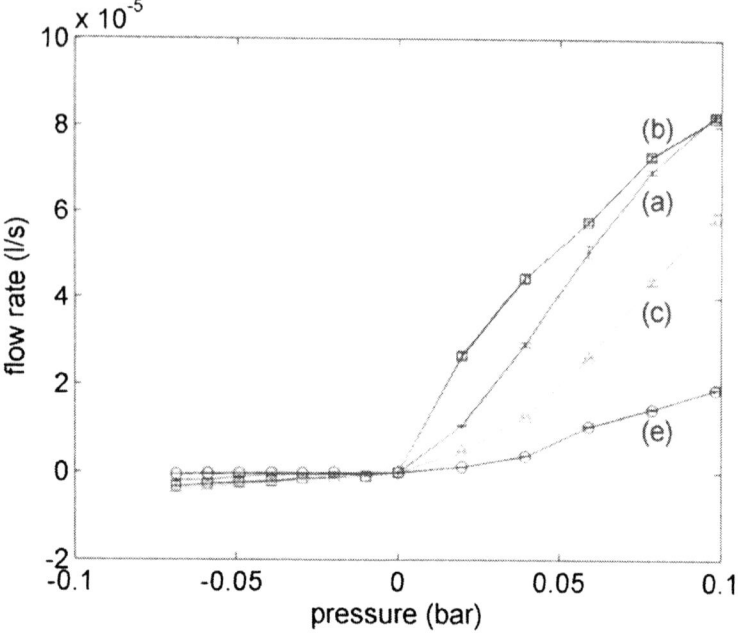

Fig. 4.2. Flow rate measurements for the parameters presented in table 4.1

4.2 High pressure tests

To test the valve under higher pressures, we used a tank with two compartments. One side is filled with air and the other with water. A pump was used to increase the air pressure (and hence the water pressure up to 5 bars). Flow rate measurements are presented in Figure 4.3. The dotted line represents the theoretical behaviour of the same test bed without any spider spring. The conclusion of that graph is that, for higher pressures, the behaviour is quite similar for the various spider springs. At these pressures, all the valves are fully open and the presence of the spider spring introduces an additional pressure drop with respect to the free circulation dotted curve.

Table 4.1. Parameters for spider springs

-	a	b	c	d	e
material	SU8	SU8	SU8	SU8	SU8
Nbr of arms	3	3	4	4	4
Sheet thickness	30 μm	30 μm	30 μm	30 μm	30 μm
b_{width}	50 μm	60 μm	40 μm	50 μm	80 μm
$Angle_1$	65°	65°	35°	35°	25°
$Angle_2$	46°	46°	16°	16°	6°
Stiffness	9 N/m	12 N/m	19 N/m	24 N/m	44 N/m

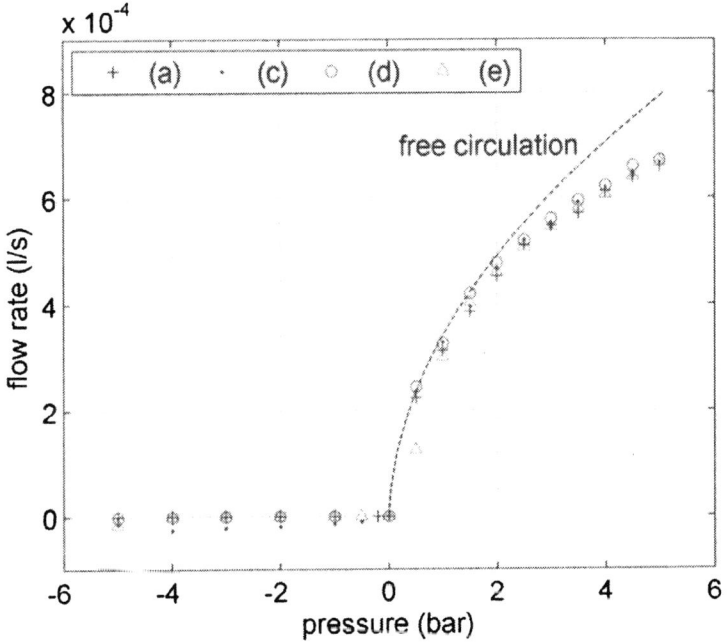

Fig. 4.3. High pressure tests for the parameters presented in table 4.1

5 Conclusion

This paper presents two manufacturing techniques for ortho-planar springs. It emphasises the importance of a pre-stress on the valve in order to improve sealing and prevent backflow. It discusses the micro-valve assembly regarding classic DFA rules. The chosen solution to align and correctly locate the spider spring is an extra part made of PDMS. Finally it gives a full characterization of ortho-planar valves for low and high pressure tests.

References

1. D.J. Laser and J.G. Santiago, (2004). A review of micropumps. JMM, 14, R35–R64.
2. B. Li, Q. Chen, D.G. Lee, J. Woolman and G.P. Carman, (2005). Development of large flow rate, robust, passive micro check valves for compact piezoelectrically actuated pumps. SAA, 117, 325–330.
3. N.T. Nguyen, T.Q. Truong, K.K. Wong, S.S. Ho and C.L.N. Low, (2004). Micro check valves for integration into polymeric microfluidic devices. JMM, 14, 69–75.
4. X. Yang, C. Grosjean, Y.C. Tai and C.M. Ho, (1998). A mems thermopneumatic silicone rubber membrane valve. SAA, 64, 101–108.
5. Check-All Valve Manufacturing Company, West Des Moines, Iowa, USA (October 2, 2007); http://www.checkall.com
6. Clippard Minimatic, Cincinnati, Ohio, USA (October 2, 2007); http://www.clippard.com
7. Command Controls Corp., Elgin, Illinois, USA (October 2, 2007); http://www.commandcontrols.com
8. Deltrol fluid products, Bellwood, Illinois, USA (October 2, 2007); http://www.deltrolfluid.com
9. Halkey-Roberts, St. Petersburg, Florida, USA (October 2, 2007); http://www.halkey-roberts.com
10. The Lee Company, Westbrook, Connecticut, USA (October 2, 2007); http://www.theleeco.com
11. O'Keefe Controls Co., Trumbull, Connecticut, USA (October 2, 2007); http://www.okcc.com
12. Sterling Hydraulics Ltd., Crewkerne, Somerset, UK (October 2, 2007); http://www.sterling-hydraulics.co.uk
13. Integrated Publishing, (October 2, 2007); http://www.tpub.com
14. O. Smal, B. Dehez, B. Raucent, M. De Volder, J. Peirs, D. Reynaerts, et al. (2006). *Modelling and characterisation of an ortho-planar micro-valve*. Precision Assembly Technologies for Mini and Micro Products, Springer, 2006, ed S. Ratchev, pp 375-326.
15. R. Feng and R.J. Farris, (2003). Influence of processing conditions on the thermal and mechanical properties of su8 negative photoresist coatings. *JMM, 13*, 80–88.
16. H. Lorenz, M. Despont, N. Fahrni, N. LaBianca, P. Renaud and P. Vettiger, (1997). Su-8: A low-cost negative resist for mems. *JMM, 7*, 121–124.
17. G. Boothroyd and P. Dewhurst, (1989) Product Design for Assembly, Boothroyd Dewhurst, Inc., pp 2-12 ,2-16

ROBUST DESIGN OF A LENS SYSTEM OF VARIABLE REFRACTION POWER WITH RESPECT TO THE ASSEMBLY PROCESS

Ingo Sieber, Ulrich Gengenbach, Rudolf Scharnowell

Institute for Applied Computer Science
Forschungszentrum Karlsruhe GmbH, Germany
Ingo.Sieber@iai.fzk.de

Abstract The aim of this paper is to show that a compensation of manufacturing tolerances by means of the functional design is reasonable. The approach is discussed exemplarily in the application of a lens system of variable refraction power considering the assembly tolerances. This approach is based on the steps sensitivity analysis, tolerance analysis, and design optimisation and will result in a robust design with respect to the assembly process.

1 Introduction

A monolithic setup of optical systems is not possible, as this would give rise to topological and geometric problems. In addition, a uniform material system is lacking. The modular setup of these hybrid systems results in an isolated manufacture of the individual components and their later assembly in a single system. An important aspect of construction is to ensure a certain functionality of the combined system, which is closely linked with the geometry of the structure and the application conditions. To maintain the overall function of a system under the given manufacturing conditions, the system design has to be robust with respect to the expected tolerances. Hence, simulation and optimisation of hybrid systems as a function of the tolerances induced by the assembly process play a crucial role. The concept presented in this paper is the compensation of tolerance effects induced by the assembly process by means of the functional design (robust design).

1.1 Application example

By way of example the approach described above will be applied to the design of a functional model of an intelligent implant to recover the accommodation ability of the human eye. The ability to focus on a close-by object (accommodation) gets lost while ageing due to a stiffening of the human lens. A possible approach to recovering this ability is an intelligent implant on the basis of a mechatronic system, the so-called Artificial Accommodation system [1]. This system is built up of the functional units of signal acquisition, signal processing, active optics, control, and energy supply. The presented work addresses the compensation of tolerances induced in the optical system by the assembly process. The optical system is very sensitive with respect to tolerances, since small position errors will already result in noticeable degradation of the performance of the total system.

The active optical system consists of a lens system which varies its refraction power by means of a lateral movement of one of its components and an actor which performs the movement in a defined manner.
As an optical element, a so-called Alvarez-Humphrey lens [2] is used (Fig. 1).

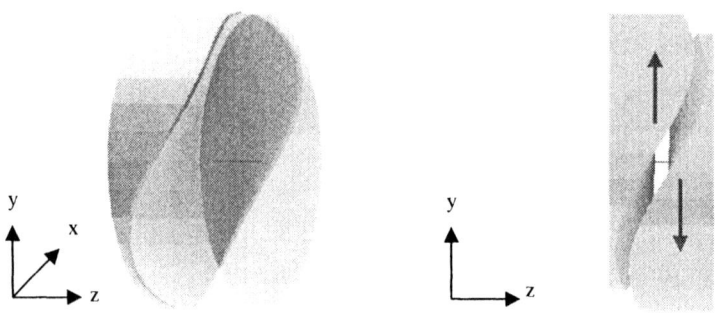

Fig. 1. Functioning of an Alvarez-Humphrey optics with a variable refraction power.

The Alvarez-Humphrey lens consists of two parts, the opposing surfaces of which are conjugated. The surface sag of the opposing surfaces is given by equation (1) with the lateral movement v and the parameter A:

$$z(x,y) = A((y-v) x^2 + 1/3(y-v)^3) \tag{1}$$

The relation between the lateral movement v and the change in refraction power ΔD is given by equation (2):

$$\Delta D \sim A\, v \tag{2}$$

The functional model of the artificial accommodation system among others serves to demonstrate the functioning of the total system and the interaction between the individual components. To reach this goal, it is not necessary to keep the dimensions, but it is allowed to scale the system. The functional model of the implant is designed to fit in an eye model the size of a grapefruit. The simulation model of the optical subsystem is shown in Fig. 2.

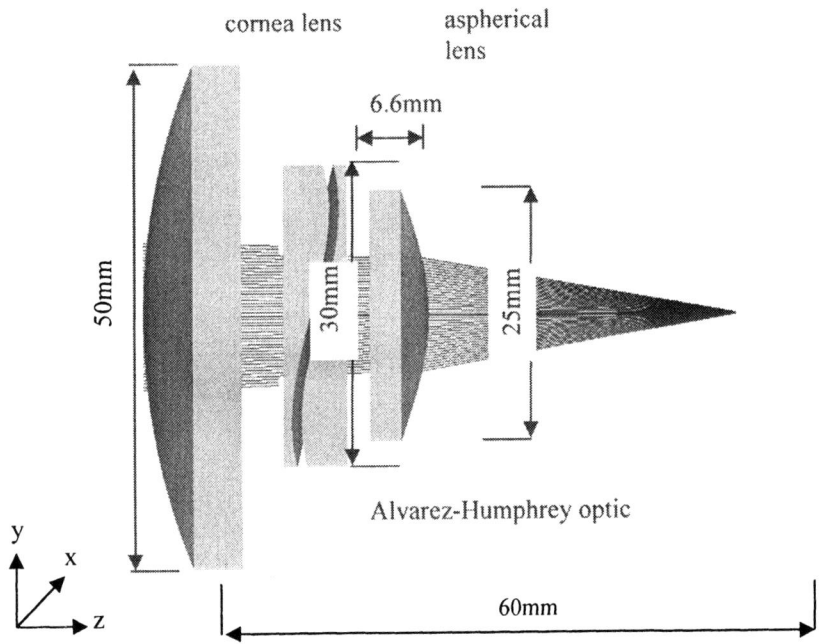

Fig. 2. Simulation model of the optical subsystem of the eye model.

2 Assembly tolerances

Optical refraction power is varied by means of a lateral movement of the first Alvarez-Humphrey surface in y-direction. To yield a good optical performance, both Alvarez-Humphrey surfaces must be well adjusted. An error in positioning these two parts may have a severe impact on the performance of the total system. Consequently, the most interesting part for a tolerance analysis of the optical subsystem is the Alvarez-Humphrey optics.

Three main types of assembly tolerances may be identified:
- lateral position error in x-direction (Fig. 3, left)
- wedge error (Fig. 3, middle)
- orientation error (Fig. 3, right)

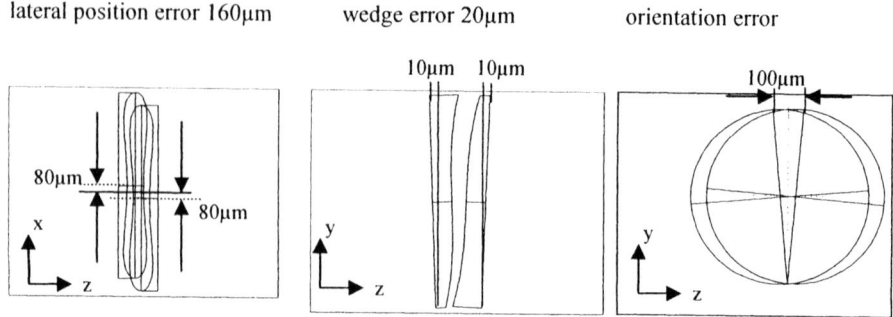

Fig. 3. Different types of assembly tolerances identified for the Alvarez-Humphrey optics.

3 Tolerance analysis and optimisation

One main feature of a robust design is that the difference in the performance criterion is minimised over the total tolerance range. This approach results in a design which ensures a defined functionality with a minimum deviation of performance over the total tolerance range.

As a criterion for the evaluation of the system's performance, the modulation transfer function (MTF) is chosen. The MTF, also known as spatial frequency response, is a direct measure of how well the various details in the object are reproduced in the image. The MTF is well-established to specify and judge the optical imaging quality [3].

The first step in tolerance analysis is a so-called sensitivity analysis. The sensitivity analysis is used to investigate the influence of the individual tolerances on the evaluation criterion. Hence, the worst-case value of the individual tolerance is chosen and the MTF is calculated. The simulations were carried out using the optical simulation tool ZEMAX-EE [4]. Fig. 4 depicts the MTF versus the spatial frequency for the three different types of assembly tolerances identified and the reference (from top left to bottom right: reference, lateral position error, wedge error, orientation error).

Comparison of the curves of the assembly tolerances with the reference curve leads to the following statements:

- the lateral position error has the most severe influence on the MTF. A breakdown of the curve below 25 cycles/mm shows the decrease in the system's resolution due to this tolerance

- within the estimated tolerance range, the wedge error has no influence on the imaging quality of the system

- the orientation error has a minor effect on the system's performance. Apart from the angular fraction, this tolerance also contains a fraction of lateral displacement in x-direction. This lateral fraction causes the degradation in the trend of the MTF

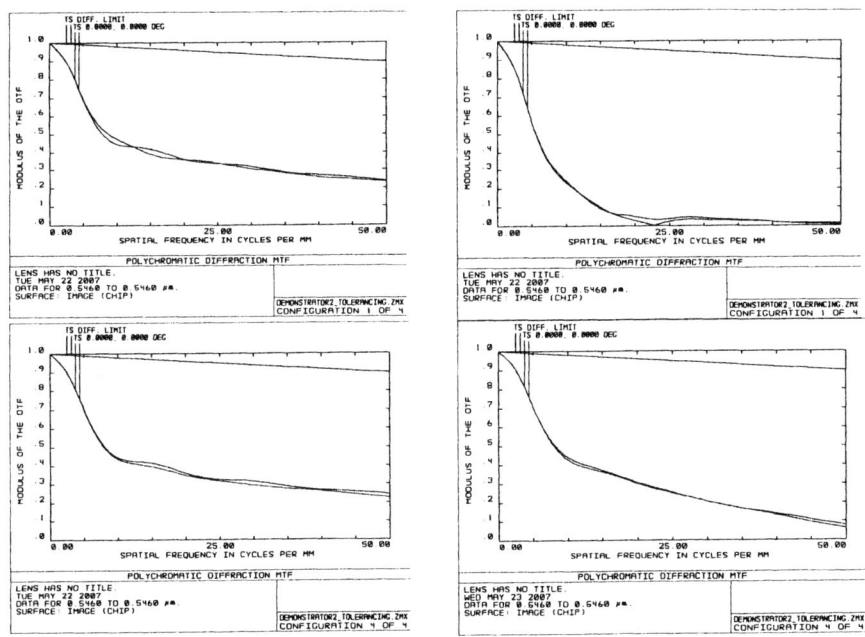

Fig. 4. MTF versus spatial frequency for the three different types of assembly tolerances identified and the reference (from top left to bottom right: reference, lateral position error, wedge error, orientation error).

The lateral position error is identified to be the tolerance with the most severe influence on the system's performance. On the basis of this position error a tolerance analysis is performed taking the estimated tolerance range into account.

To analyse the impact of the lateral position tolerance on the system's performance, a Monte Carlo analysis was performed. In a Monte Carlo analysis for a given tolerance range the change of the criterion is determined. For this purpose, the parameters having specified tolerances are set randomly using their defined range and a statistical model of the distribution of the parameters over the specified range. The statistical model used follows a normal distribution with a total width of four standard deviations between the extreme minimum and maximum allowed values. As the criterion, the MTF value at 12.5 cycles/mm is used. The results of 500 Monte Carlo runs are shown in Table 1:

Nominal	0.43242355
Best	0.43242698
Worst	0.22679682
Mean	0.40122752
Std Dev	0.03666711
98% <=	0.43238019
90% <=	0.43199561
50% <=	0.41444874
10% <=	0.35321565
2% <=	0.26861138

Table 1. Result of 500 Monte Carlo runs for the original design. The criterion is the MTF@12.5cyc./mm.

Under the assumption of a normal distribution of the tolerances, the MTF of the system is worse than 0.353 at 12.5 cycles per mm with a likelihood of 10%. The value of the standard deviation also is of interest. This parameter provides information on the oscillation of the system's performance within the tolerance range.

3.1 Optimisation with respect to a robust design

On the basis of the tolerance analysis an optimisation is carried out, the goal being to achieve a robust design. As before in the tolerance analysis, again the MTF value at 12.5 cycles per mm is chosen as the main criterion. The parameters of the optimisation are the coefficient A of the Alvarez-Humphrey surfaces (equation (**1**)) and higher-order coefficients of the polynomial equation of the surface sag (see equation (**3**)):

$$z = a_1 x^2 y + a_2 y^3 + a_3 x^3 y + a_4 y^4 \, a_5 x^4 y \qquad (3)$$

The method used for optimisation is an actively damped least squares algorithm provided by the optical simulation tool ZEMAX-EE.

4 Results and conclusions

The optimised coefficients of the active lens serve as input for another Monte Carlo analysis. The results of this analysis allow for a comparison of the impacts of the tolerances on the system's performance between the original and the optimised design.

Table 2 shows the comparison of the results. The data of the original design are depicted on the left hand side, the data of the optimised design on the right.

Original design		Optimised design	
Nominal	0.43242355	Nominal	0.45563401
Best	0.43242698	Best	0.45582751
Worst	0.22679682	Worst	0.30953146
Mean	0.40122752	Mean	0.43430906
Std Dev	0.03666711	Std Dev	0.02611409
98% <=	0.43238019	98% <=	0.45567880
90% <=	0.43199561	90% <=	0.45527413
50% <=	0.41444874	50% <=	0.44379105
10% <=	0.35321565	10% <=	0.39642523
2% <=	0.26861138	2% <=	0.35649227

Table 2. Comparison of the results of 500 Monte Carlo runs for the original and the optimised design. The criterion is the MTF@12.5cyc./mm.

An analysis of the optimisation with respect to tolerances yields the following results:
- a slight enhancement of the nominal value (about 5.4%), caused by the addition of further coefficients in the surface description.
- enhancement of the worst value by 40.0%
- enhancement of the mean value by 10.7%
- enhancement of the standard deviation by 40.1%
- looking again at the MTF value of 0.353: as seen, using the original design the performance of the system is worse than that value with a likelihood of 10%. Using the optimised system the likelihood of being worse in performance shrinks to 2%.

The results show clearly that a compensation of influences caused by manufacturing tolerances is reasonable by means of the functional design. The approach was derived and applied to compensate for assembly tolerances by means of the functional design of an Alvarez Humphrey optic. This approach is based on the following steps:
- sensitivity analysis to identify the worst offenders
- tolerance analysis on the basis of the Monte Carlo method
- design optimisation of the functional components

The result is a robust design with respect to the assembly process.

References

1. U. Gengenbach, G. Bretthauer, and R. F. Guthof, Künstliches Akkommodationssystem auf der Basis von Mikro- und Nanotechnologie. Mikrosystemtechnik Kongress, Freiburg. VDE Verlag Berlin Offenbach (2005) 411-414
2. L. W. Alvarez, and W. E. Humphrey, Variable power lens and system. US Patent 3507565 (1970)
3. G. H. Smith: Practical Computer-Aided Lens Design, Willmann-Bell, Inc. 1998
4. ZEMAX: Software For Optical System Design, http://www.zemax.com

PART II

Micro-Assembly Processes and Applications

Chapter 3

Process Modelling for Micro-Assembly

PRODUCT-PROCESS ONTOLOGY FOR MANAGING ASSEMBLY SPECIFIC KNOWLEDGE BETWEEN PRODUCT DESIGN AND ASSEMBLY SYSTEM SIMULATION

Minna Lanz, Fernando Garcia, Timo Kallela, Reijo Tuokko

Institute of Production Engineering (IPE)
Tampere University of Technology (TUT)
P.O BOX 589
33101 Tampere, Finland
minna.lanz@tut.fi

Abstract The aim of the present paper is to introduce a feature-based *Product-Process-System model* and ontology in order to retrieve and share knowledge for simulation of manufacturing systems. Product knowledge is the combination of product specific information, such as functionality, colour and product variants, and the corresponding product model. The model provides understanding of the product structure, rules, constraints and assembly-specific information in relation to the product model. In the design and modeling of assembly processes, features form the foundation for analysis and knowledge acquisition of the product. This knowledge includes geometric and non-geometric information. In the present paper, an approach is proposed to share platform independent product-process knowledge between the assembly process and system design and even with the simulation environment.

Keywords product-process ontology, knowledge management, feature-based modeling and analysis

1 Introduction

Within the production framework, assembly plays an exceptional and specific role. In this phase of the value adding chain the different components lead to the function determining unit. The high significance of assembly to a company's success is given by its function and quality determining influence on the product at the end of the direct production chain. Rationalisation of assembly process is still technologically impeded by high product variety and the various influences resulting from the manufacturing tolerances of the parts to be joined. As a result, considerable disturbance rates are leading to a reduced availability of the assembly systems and a delay of the assembly operations [3].

Collaborative design systems exist, yet offer little or no help in the re-use and distribution of knowledge. Without a bilateral knowledge share the design of assembly systems is as much an educated decision as it is a gamble. The advent of social software for collaboration (i.e. wikipedia) and return of XML-formats with semantic web have provided means of transferring expert knowledge into information that can be reused by many, at any given point in time. However, information systems have to cope with the ever-faster innovation cycle, in which new processes, models and systems are being developed.

Ontologies represent a tremendous opportunity for the representation of knowledge to both human and "intelligent" information systems. Ontological classification of assembly systems offer a great approach to standardise the way people and machines communicate, embedding the meaning and the context within the message being sent.

A knowledge management system (KMS) for the design of re-configurable manufacturing systems has to provide tools for storing and retrieving tacit information and strict hierarchical classifications from different domains. The retrieval and share is not the only challenge but KMS must be able to parse and form the retrieved information into meaningful knowledge that aids the generation of manufacturing systems. Ontological classification of manufacturing systems with semantic web offers a great approach to improve the way people and machines communicate, embedding the meaning to the context.

The authors propose a method for integrating product and process specific knowledge via a platform independent *Knowledge Base* (KB), neutral XML-formats and *Generic Product-Process Ontology*. The following chapters go through the theoretical approach (currently under development in the IP-PISA project) to a feature-based product ontology, KB structures and interfaces for enhanced simulation environment.

2 Theoretical Background for the Use of Features in Assembly Process Definition

The geometric definition of the part relation is a result of the design process. The product designer intentionally defines the part relation to meet the functional specifications of the product, and, as a consequence, at the same time the majority of the assembly process is defined. [3, 1]

A feature is any geometric or functional element or property of an object useful in understanding its function, behaviour or performance. Features can be seen as the base elements, detailed description of product properties, analysed and categorised based on the product and/or process ontologies. Including the features in the product model enables the representation to be in a higher abstraction level than just a pure geometrical model. [5, 7]

In the design and modeling of assembly processes, understanding the relations between components is essential, but it is not enough for representation of the assembly process, since it requires also non-geometrical assembly-specific information. Van Holland in his research realised the importance of non-geometrical features for assembly process definition. He divided geometrical and non-geometrical features into their own categories. The assembly specific information is carried over with assembly features. [5, 7, 10].

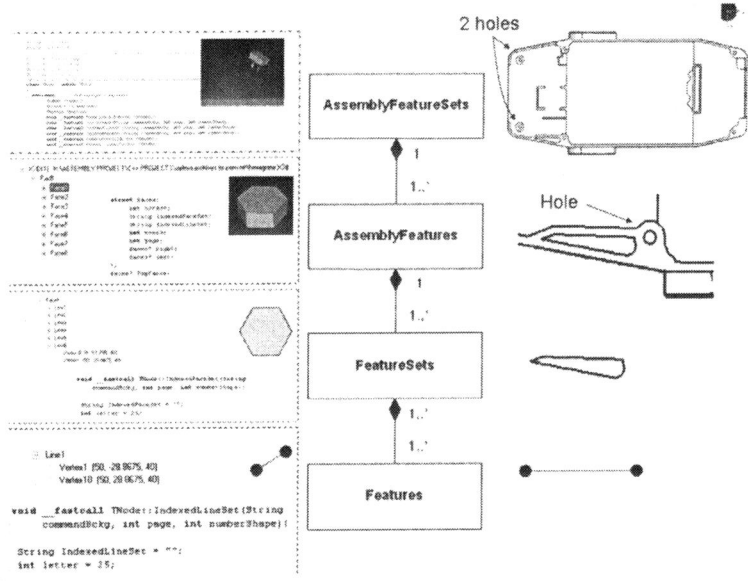

Fig. 1. Feature Layers

In the present model the assembly specific feature information is divided into four layers; *Features, FeatureSets, AssemblyFeatures* and *AssemblyFeatureSets*. Features in each layer can be either geometrical or non-geometrical features or combination of those. This hierarchical representation of feature classes will help in the definition of the structure of the knowledge base in relation to product knowledge.

3 Generic Product-Process-System Model and Ontologies

Figure 2 introduces how the product and process models are integrated. The product specific information is connected into the processes through two classes *Reasoning Machine* and *Reasoning Results*. Linking of the product model and process model is done by a reasoning machine which calculates similarities and processes both models by given mating rules. The assembly rules are used to define all the necessary assembly processes for the product to be assembled. This process information is used to define the assembly system requirements to aid in the decision making before the manufacturing equipment is deployed. The actual reasoning is done inside the Knowledge Base, briefly introduced later in this paper.

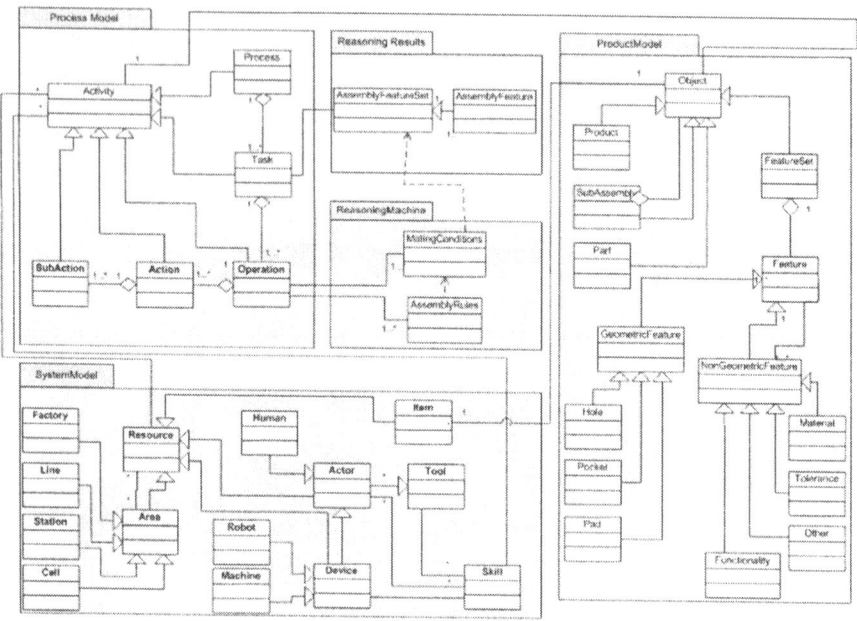

Fig. 2. Product-Process Model

The structure of the product model is a multi-layer structure, represented in the left side of Figure 2. The topmost layer is the product layer and it is for encapsulation of the product information. The layer consists of the information related to manufacturing of the current product. The second layer from the top is the subassembly layer, being a pseudo-class between single parts and the product. Each product can have several subassemblies, and a subassembly can belong to several products. This will reduce redundant information in the knowledge base and will improve reusability of the knowledge.

The third layer, *Parts*, is for individual parts of the product tree. A part cannot consist of any subassemblies or other parts, and it will only contain *PartFeatures* which are geometrical and/or non-geometrical *FeatureSets*. Geometrical *FeatureSets* are pads, pockets or holes. Geometrical *FeatureSets* can be multileveled and they simulate the parts structure by pads, pockets or holes.

The process model is visualised in the left side of Figure 2. Compared to the product model, the process model is similarly multi-layer architecture. The Process model is divided into five primary layers. The first and highest layer is called *Activity*. The second highest class is *Processes*. This class contains of the operations done in the manufacturing unit. Instances of manufacturing processes are such as assembly, part manufacture, test or packaging.

The third layer is for tasks related to manufacturing, such as fix, grasp, join, release and move. The third and forth layers are operations and actions. Operations for example in joining are snap-fit, screw, clue and weld. Each of the operations can contain a script or command sequence, which runs when needed. Actions are the most fundamental actions of movement; translation and rotation.

3.1. Product Ontology

The PISA Product Ontology, figure 3, has been defined to formalize further the product model structure and to enable general rules. The rules are defined as relations between different objects. The rules also define the type of relations and restrictions for the relations.

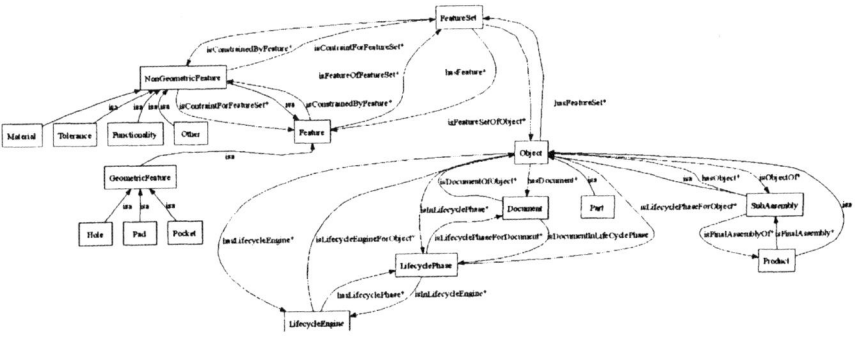

Fig. 3. Product Ontology

In the product ontology the class *Object* is defined to be in the top of the ontology. It serves as a super-class for *Product, SubAssembly* and *Part*. The *Object*-class enables the addition of the feature information into the product structure, when the parts of the product are not known. In the normal case the used features are linked in to the parts and not to the products. The *Part* is the lowest level class in the product structure and it has a defined geometry. A case example of the implemented product ontology can be seen in Figure 4.

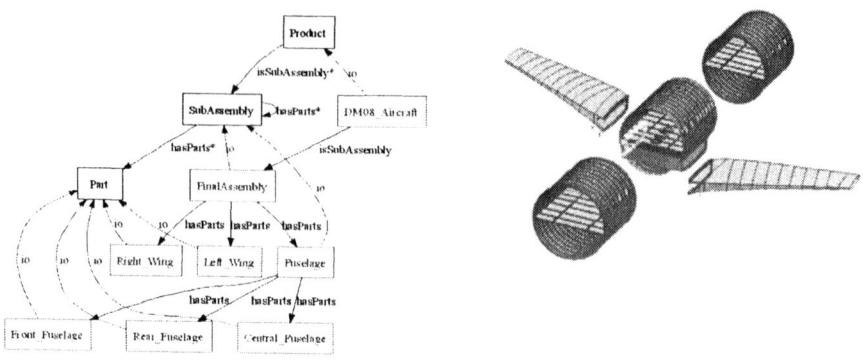

Fig. 4. Example of the product ontology of the DM08-aircraft [6]

3.2. Process Ontology

Manufacturing processes are described with process ontology. The process ontology is the key for combining the product and the system knowledge. The process ontology, in Figure 5, is designed based on the requirements given by *Product - Process model* in Figure 2.

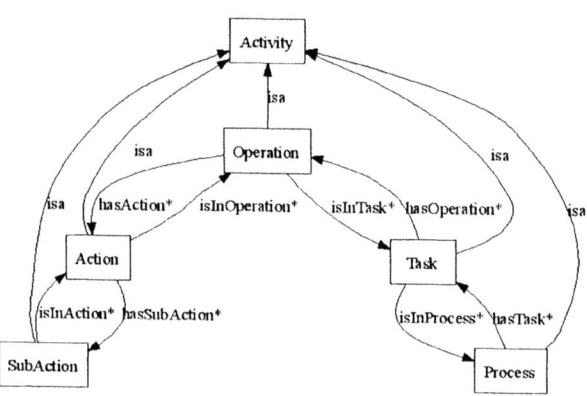

Fig. 5. Process Ontology

In the process ontology the highest class is *Activity*. The second class is named as Processes as the model in Figure 2 required. The process ontology defines classes for tasks and operations. The *Action* and *SubAction* classes are reserved for the most basic functions; rotate and translate and combinations thereof. There can be multiple sub-actions in the *Action* class.

3.3. System ontology

The system ontology, illustrated in Figure 6, is used to describe the manufacturing environment and its characteristics. The system ontology defines a structure for resources, areas and actors. The class *Item* was defined to be a connection between products and system. The class *Actor* is a super class for *Tool*, *Human* and *Device*. Actor has a class named *Skill* for defining further *Actor*'s characteristics. By separating tools, humans, devices and skills we can answer better for the re-tooling requirements. The *Resource* class has a connection to the *Activity* in the process ontology side.

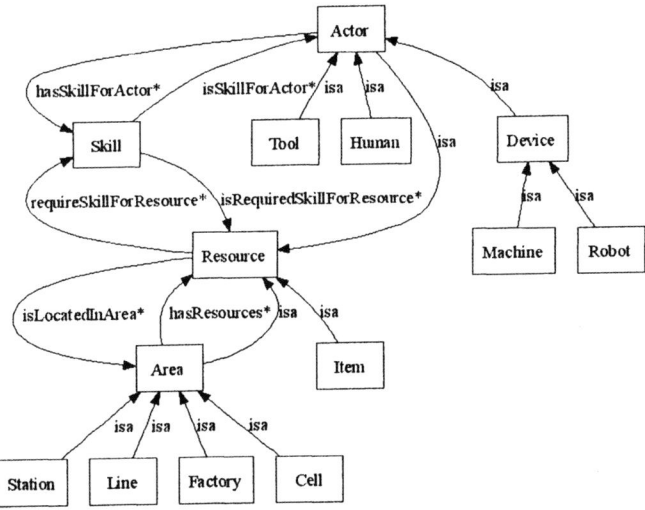

Fig. 6. System Ontology

3.4. X3D representation

The chosen knowledge visualisation language for the project is X3D, which is XML based language and a standard for 3D visualisation in web context. The OWL (Web Ontology Language) is used add the semantic meaning to raw XML and X3D information.

The PISA X3D structure is based on including and linking several files to form a product tree. The active product configuration is divided into several files. On the top is the product file, which has linking to a single sub-assembly (named as product-assembly). The product file consists of additional information about the product, its properties and lifecycle. Also, some information related to packaging or non-assembly processes can be included.

The lowest level file is the part file. Each part file contains a single part, its geometrical and non-geometrical features. At this level, the file content is based on lowest level information in the product model. In the middle of these two levels are the subassembly files. Subassemblies are like pseudo-classes and in X3D syntax they really don't have any other function than to include several lower level parts or subassemblies and to tell manufacturing instruction related. The subassembly files also provide a link to the system-specific process-files. The files are XML-based providing process requirements for the system. These files can also include a sequence or script which will be sent to an agent.

4 Knowledge Base as an Integration Solution between Clients

The knowledge of the product model needs to be stored in specific databases. These knowledge bases, serve multiple roles. It can be said that *KBs* are repositories of shared knowledge. However, overcoming the challenges in interoperability and reuse of components and declarative knowledge are crucial for the further development of the knowledge-based system. Unfortunately, it is hard to get components to interoperate and even harder to reuse other people's work. These difficulties are often a result of incompatibilities in the knowledge models (the precise definition of declarative knowledge structures) assumed by the various components. [4]

The main requirement for the *KB system* is the platform neutrality. This is indispensable in order to fulfill the objective of the product and process knowledge system. The system platform encapsulates all specific parts of a control system in order to provide a neutral interface to the application software. The system software, which provides the functionality needed to support openness for the application software, should be located on the top of the platform. Knowledge acquisition is also accessed by humans, as part of a job that doesn't require previous experience of knowledge based systems [4, 9]

There are at least three stages to the construction and use of a knowledge base: ontology must be defined, reasonably static and long-lasting instances must be acquired, and run-time data must be entered (often when an application is running). For example, in the Assembly process knowledge domain, these steps might correspond to gathering of a design vocabulary, acquiring specific rules for joining parts together, and obtaining the details of assembly sequences. [4]

The knowledge system contains three knowledge domains based on the defined ontologies; product, process and system domain. The product-process ontology, constraints and geometrical information would be stored and accessed from the product domain. The process domain serves as a repository for process related information, such as assembly sequences, activities, tools, assembly paths etc. The system domain serves as information storage for equipment and workcell related information [9].

Figure 7 shows the *KB Architecture* which consist of four access layers, the detailed description of the tools used can be found from Figure 7b. These access layers have the intention to separate the model of the system and create groups according to its responsibilities.

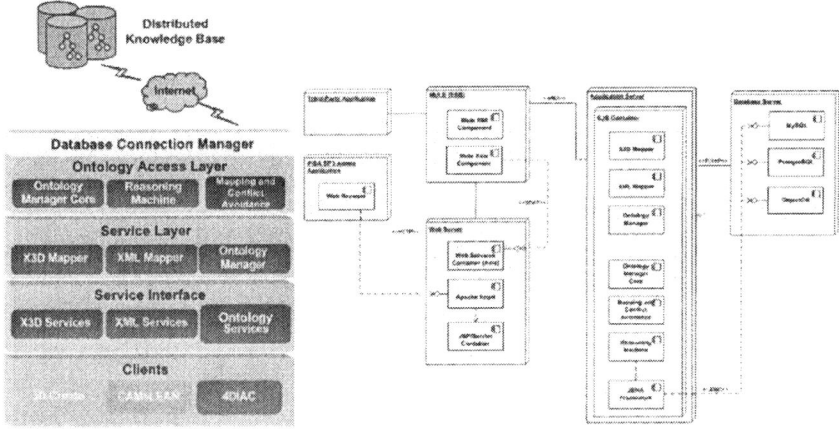

Fig. 7. a) Knowledge Base Architecture and *b)* Deployment Diagram of the KB

The three knowledge domains are implemented using semantic networks, with concept relations to be able to connect the knowledge from one domain to the next. This structure is flexible enough to accept the inclusion of other knowledge domains, such as logistics, supply chain management, order management, and others. The storage of information of each concepts or entity is implemented with X3D and XML-based files. This allows the knowledge base to be distributed in case of a need for collaborative design of the products and systems.

5 Discussion

The ontological approach and the structure of the KB system presented in the paper solve the problem of knowledge representation and sharing between the product-, process-, and system design domains. An effective way to acquire knowledge and share it internally and with outside strategic sources is a needed in today's rapidly changing markets. However, the system providers are not fully agreeable to sharing knowledge created in their proprietary systems with possible future competitors, even though customers are more and more often requiring knowledge capture and exchange between systems. Another major challenge is to provide a robust and acceptable solution for knowledge capture from different sources. Companies rightfully insist on reusing their information structures and familiar ways to use their categorisation system.

Acknowledgement

The development of the methods and actual tools discussed in this paper is done under the European commission funded IP project PISA – Flexible Assembly Systems through Workplace-Sharing and Time-Sharing Human-Machine Cooperation.

References

[1] H. Bley, C. Franke, *Integration of Product Design and Assembly Planning in the Digital Factory*, Institute for Production Engineering/CAM Saarland University, Germany, 2004 p. 6,
[2] D. Deneux, *"Introduction to assembly features: an illustrated synthesis methodology"*, Journal of Intelligent Manufacturing, 1999/10, pp. 29-39
[3] K. Feldman, H. Rottbauer, N. Roth, "Relevance of Assembly in Global Manufacturing", *Annals of the CIRP,* vol. 45/2/1996 pp. 545-552
[4] W.E., Grosso, J.H., Gennari, R.W., Fergerson, M.A., Musen *When Knowledge Models Collide (How it Happens and What to Do)*, Technical Report, Stanford Medical Informatics, Stanford University Stanford, 1998
[5] W. van Holland, *Assembly Features in Modelling and Planning*, Ph.D. thesis, Delft University of Technology, 1997 p.160
[6] E. Järvenpää, Enhancement of Process Planning and Simulation of Final Assembly for Aerospace Industry, M.Sc thesis, Institute of Production Eng., Tampere University of Technology 2007, p. 104
[7] T. Laakko, Incremental Feature Modeling: Methodology for Integrating Features and Solid Models, *Acta Polytechnica Scandinavica*, 1993 p.85
[8] M. Lanz, *An Approach to Feature-based Modelling and Analysis for the Final Assembly*, , M.Sc. Thesis, Institute of Production Eng., Tampere University of Technology, 2005, p.100
[9] E. Lim, V. Cherkassky, "Semantic Networks and Associative Databases – Two Approaches to Knowledge Representation and Reasoning", *IEEE Expert: Intelligent Systems and Their Applications*, volume 7, issue 4, August 1992
[10] T. Tallinen, R. Velez Osuna, J. L. Martinez Lastra, R. Tuokko, *"Product Model Representation Concept for the Purpose of Assembly Process Modeling"*, in Proceeding of the International Symposium on Assembly and Task Planning ISATP 2003, Besançon, France, July 2003, p.6

BRIDGING THE GAP – FROM PROCESS RELATED DOCUMENTATION TO AN INTEGRATED PROCESS AND APPLICATION KNOWLEDGE MANAGEMENT IN MICRO SYSTEMS TECHNOLOGY

Markus Dickerhof, Anna Parusel

Forschungszentrum Karlsruhe
Phone: +49 7247 82-5754
Fax: +49 7247 82-5786
Email: dickerhof@iai.fzk.de

Abstract Knowledge about the relations between process steps or dependencies of materials, technologies, designs, tools and machines is mostly stored in the product developers mind. The same effect takes place when one talks about product related knowledge and the referring technological requirements It is lost if the person leaves the company. The following paper describes a knowledge management approach for a product independent description of Micro System processes and presents a software prototype for decision support in early product development phases.

Keywords process knowledge management, design for X, decision support systems

1 Knowledge Management in MST Industry

Planning development and production today means proceeding along economic and technical parameters, derived from product requirements and customer oriented business processes. During the last few years much effort has been made to push "Knowledge Management" in industry. At the least, the result often represented no more than a new type of document management with additional "meta" information on top to support the requirements of a company's quality management. Documentation of technological capabilities in Microsystems Technology (MST) often leads to a non-reusable, product specific documentation of a production design according to the quality management guidelines. Hence, a lot of information about the subjacent fabrication know-how - or relations between application requirements, physical effects, technological constraints and costs, - is so far not documented, or documented in an insufficient manner. Insufficient in this context means irretrievable or only customer project related so that it cannot be reused for new projects in terms of Nonaka and Takeuchi's Seci model [01]. Knowledge about the relations between process steps or dependencies of materials, technologies, designs, tools and machines is mostly stored in the product developers mind. The same thing occurs with regard to the relation of product requirements and referring technological capabilities. It is lost if the employee leaves the company. Also the "neutral" transfer of information or knowledge about risks from one project to another can even be difficult for a person, because of the complexity that has to be considered. A fortiori know-how transfer is extremely expensive and time intensive for a company, and is therefore worth handling in a well-structured economic manner. The following paper describes an approach allowing us to overcome some of the limitations in present product development through the use of a decision support system based on a MST-Ontology.

1.1 Knowledge and knowledge management in the industrial production context

The terms "knowledge" and "knowledge management" are frequently used in information technology in many respects. Their scope of use ranges from the mapping of simple structural knowledge to the representation of knowledge by means of semantic networks or methods of artificial intelligence.

Application of these approaches is certainly feasible in an enterprise and useful in various fields where enterprise knowledge is generated. The present contribution, however, will focus on product development or fabrication knowledge and in particular on the necessary process know-how especially in the MST context (Fig 1). As MST is an interdisciplinary field of technology, special attention needs to be drawn on influencing factors, coming from market / application requirements and related physical principles.

The percentage of work done in that direction demonstrates the early state of industrial implementation, especially when considering that the numbers do not represent the availability of knowledge within the enterprise.

In industry, work related to the storage of knowledge has advanced to different levels. Large enterprises with complex fabrication processes undertake considerable efforts in acquiring knowledge and integrating it in a "learning" enterprise (which also covers the "supply chain", mostly according to a specific product development and production process).

As far as small and medium-sized enterprises are concerned, this aspect so far has met with less attention, as their in-house process chains are less complex. With increasing product complexity and virtualisation of production chains, the necessity of joining development and production capabilities is essential for small enterprises to be able to produce complex products.

As with large companies, the key for success in such a company's overlapping development is also the process-overlapping knowledge management. This particularly applies to micro systems technology, where description standards are rare.

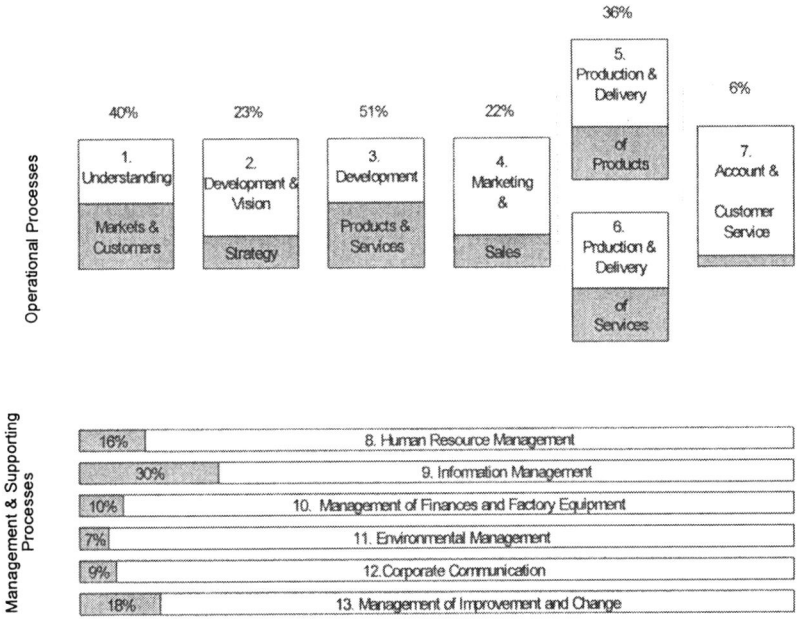

Fig. 1. Knowledge Management in an Enterprise [02]

A number of software systems today claim to map enterprise knowledge. However, no system is known (especially for non silicon MST) which processes the data collected during the product and production development process in such a way that the "evolution" of the process parameters can be assessed or patterns and regulations (design rules) can be derived from the process-overlapping connections of the interdisciplinary partial areas of micro systems technology in analogy to the rules defined in silicon micro systems technology.

1.2 Knowledge management in the development and production phase

The basic idea of the ProWiDa concept is to support product developers with the identification of the best fitting technological approach for a given product requirement (in this specific context, the term *technology* shall be considered in a wider context: technology, AND manufacturing). It is based on the assumption that for reusability of information - collected during a product development process - the acquisition of MST development and fabrication phase should proceed in a order -, application- and product independent manner.

As micro systems are typically integrated as parts, components or subsystems into macro systems, one has to divide between parameters that are of relevance for the product and technological parameters that are dedicated to a technology itself.

The former mostly cover economic constraints requirements or describe capabilities of a technology, e.g. related to the physical effects to fulfill the task given (e.g. surface roughness as a factor of influence for optical properties). The latter allows the user to identify the technology itself (material, technology, design, equipment, tools, etc.)

Anyway, such a methodology for the description of relations between technological parameters and application requirements can be seen as a tool towards "Design for X" approach. To reach the goal of a reusability of information it should be based on an identification of application-independent parameters as a basis for reusability of development and industrialisation results gathered in another context.

Industrial activities in this field are under discussion in white papers and other documents [03]. Scientific activities such as the pretzel model present models for the Silicon-MST field [04]. Sadly the approaches are not easily transferable though the increased complexity coming from the huge flexibility in materials [05]. First activities on a coordinative level into this direction are currently started under the roadmapping activities of the EU Coordination Action Microsapient, attempting to link application requirements with technological capabilities on a very high level. Figure 2 gives a schematic overview on this relation.

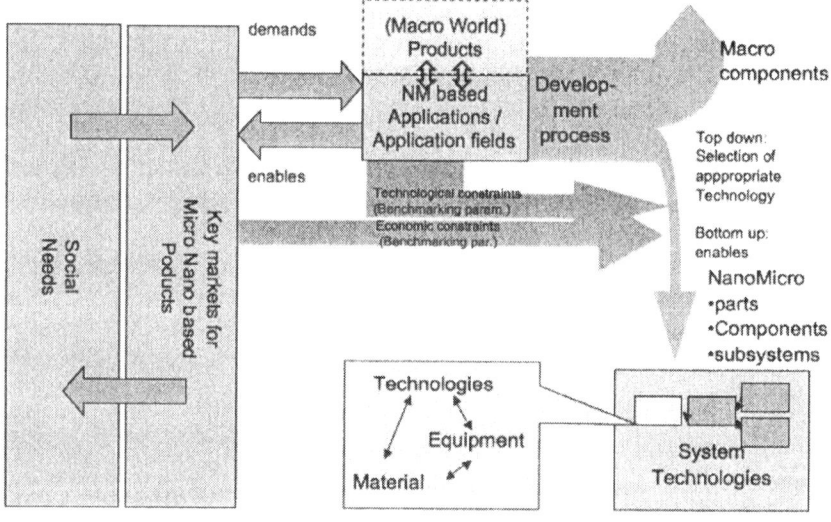

Fig. 2. Bottom up – Top down approach of Product development in MST

In the following chapter the focus will be on the description of a methodology for technological aspects in the above mentioned manner (see also lower right section of Figure 2).

1.3 Knowledge management of technical aspects in MST

Today the classification parameters of MST technologies are often mixed, consisting of production related parameters, design rules and sometimes (estimation of expected) product properties:

- Process Parameters describe typical technical factors that can be achieved in a reliable manner (width, length, material parameters, milling tool size, feed, milling velocity, etc)
- Design Rules (or product parameters) describe the resulting capabilities, e.g. resulting minimum grooves, surface roughness, which again correlate with application specific requirements coming from physical effects often utilised in Sensor or actuator applications.
- As it is not possible to directly match technological parameters with *application requirements* in a product independent manner as long as these are not defined with a specific approach in mind , we defined an additional "physical effects" layer, allowing for the linkage of process capabilities with application requirements via an application independent description of "physical properties".

An analysis of the most influencing production factor resulted in a set of up to six factors of influence, in ProWiDa called *aspects,* for the characterisation of *process (step) related parameters:*

- Materials (substrates and layers)
- Procedure
- Geometry
- Machines
- Tools

A process step as the smallest modelling element consists of n-tuple of parameters sets attached to the aspects mentioned before. Additional parameter sets can be attached to the competence itself (e.g. product properties related to a specific combination of material, design and technology). These combinations (competences) can be additionally attached to a specific group or company, if the knowledge shall be shared as a knowledge map in a distributed environment. The approach describing the prototype which has been realised so far is outlined in the next chapter. A schematic overview is presented in Figure 3.

- Material, procedure and geometry are almost sufficient for a clear definition of parameter sets that can be approximately compared with the Design Rules, we already know out of the silicon MST. These parameter sets are of relevance for a design engineer or the marketing department to check out the feasibility of a customer' request.
- The remaining aspects are important for the clear and reproducible definition of fabrication parameters you need at the shop floor level. Remark: Not every aspect needs to be of relevance for a proper description of a competence.

114   Bridging the Gap – From Process Related Documentation to
        an Integrated Process and Application Knowledge Management

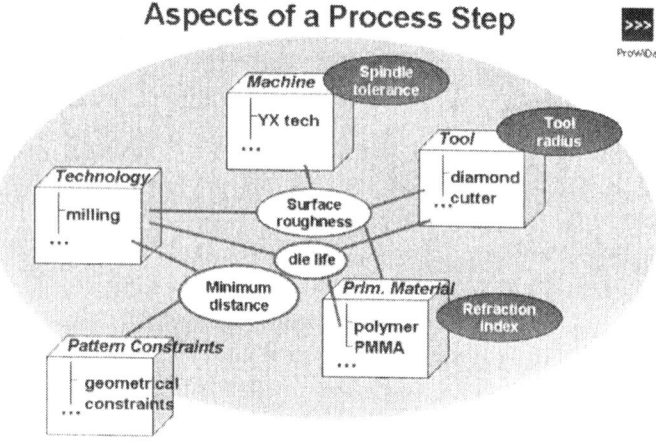

Fig. 3. Technology-oriented "Aspects" of a process step

1.4 Knowledge about relations and constraints between process steps.

State of the art process documentation is typically realised along the requirements of a quality management system, which usually just describes the relevant business process and - as a subset - the product related production chain. We aimed to find a way to show the relations between the process steps in a more flexible and transferable manner, which also allows the user to store information about "Meta" parameters. These Meta-parameters are additional properties or differing properties resulting out of the combination of single process steps to a process sequence or process chain.

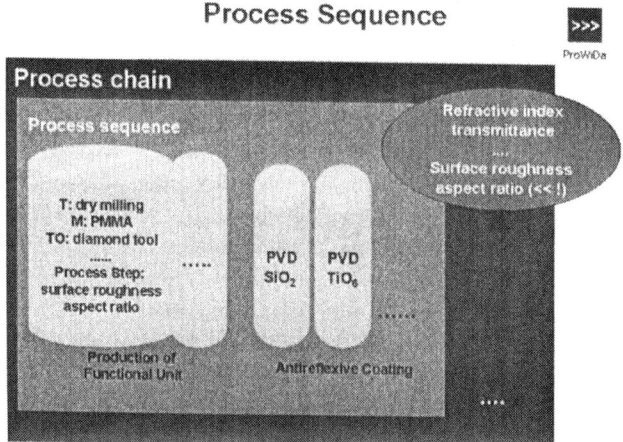

Fig 4. Example of a production process for an optical surface using generic modelling elements

Most work in this field suffers from a lack of real world process parameters which are imperative to serve as a basis for the analysis of processes and process relations, e.g. with AI methods.

2 The ProWiDa System

For acquisition and order-independent filing of process parameters (esp. during the development phase) a database approach with a web based user interface for the augmentation of the feasibility studies and product development cycle in a distributed SME network was developed and successfully tested [06]. The system today consists of five modules:

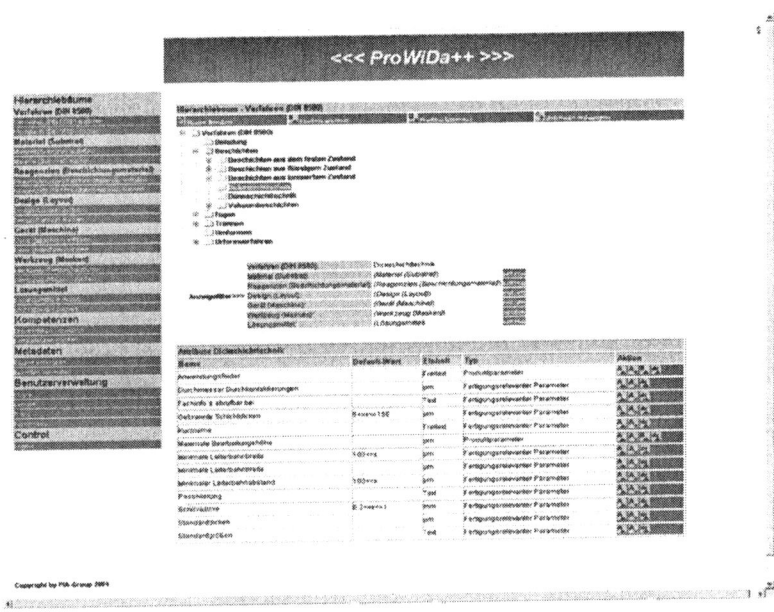

Fig 5. Aspects and parameter sets attached to a competence

2.1 Parameter classification module

The *parameter* module (Fig. 5) is one of the core elements of the ProWiDa System. Like many database driven onthologies it is based on generalisation hierarchies for the description and structuring of the major influencing aspects of a technology. These hierarchies can be compared with the trees structure of the file system on a PC. The place where one branch is attached to the next is called a node. At each node level information can be stored, inherited or generalised to the next levels above or below. This allows for the definition of template structures and parameters that can be e.g. refined on the next level of detailing. In this way, technologies, materials and other aspect can be structured in a standardised manner.

In ProWiDa - instead of information elements or files - process attributes and values are assigned to each node and leaf of the hierarchy tree. Each tree represents an aspect as defined in Chapter 3. Summation of parameter sets out the different hierarchy trees then yields a list of parameters for the "competence" on the lowest levels of the tree.

The values of specific parameter sets attached to the process step in a specific sequence describe an order related process chain. If one only would look at a specific combination of these parameter sets, there would be at first sight no difference to the specification of a work list being stored in a database system.

Filters allow for a retrieval of the competences e.g. one can mask out the material parameter dependency. This is useful as an indication for a first feasibility estimation if there exists no parameter set including the specified material.

The parameter module allows also for the integration of competence- or process sequence specific additional information, documentation (e.g. pictures, documents, hyperlinks).

Customized features like integration of related FMEA-templates and links to work list have also been realized.

2.2 Modeling Module

The *modelling component* allows for a simplified modelling of MST processes (sequential and parallel). Three major elements allow for the description of a complete production process.

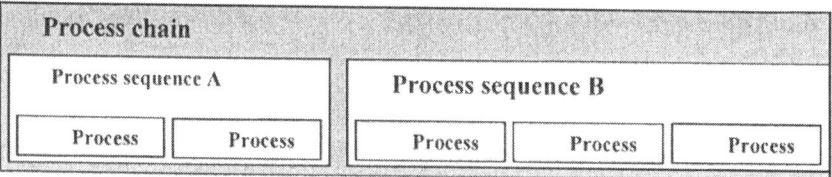

Fig. 6. Scheme of the major modelling elements in ProWiDa

The *process chain* element represents an order/product specific set of process sequences. It is equivalent to the model of a product or customer related production process. The *process sequence* represents a characteristic of a basic technology. The technology is represented by a set of *process steps*. The process step itself is the smallest modelling element, which represents a subtask that has to be executed when processing a technology. For further detailing the use of more sophisticated modelling methods like Petri Nets is more adequate [07].

2.3 Interfaces for integration in Enterprise Resource Planning and other Business management systems

In addition to the nearly "product independent" acquisition and storage of parameters, the ProWiDa system allows the collection of test and measurement data associated with of the iterative execution of a production or development process from feasibility studies up to the industrialisation. The data are stored in both ways, according to the product oriented work lists and as another view to the data pool in the way described above. Administrative features such as the definition of number ranges, batch tracing and statistical process control are also a part of this module.

For the manual acquisition of test and measurement data the ProWiDa System allows the automatic generation of process data sheets. The relevant process parameters for each process step can be derived from highlighted parameters of the equivalent competences in the parameter definition module. A "History" function supports the documentation of test series during the development phase.

Since the ProWiDa System is not intended to become an isolated tool in the enterprise software environment, connectors to ERP-systems had been defined. These connectors will allow for the exchange of accompanying information about process chains and their process steps in a technological and economical context (e.g. test and measurement data, process sheets and parts lists) to the ERP System and vice versa. Workflow functionality enables the model execution especially in the development phase, if there is no support by the ERP system. These functions are of extreme importance for user acceptance. For an easier integration the ProWiDa Integrating Infrastructure comparable to a service oriented architecture (SoA) approach had been developed allowing for an easier connectivity to other standard ERP software with XML interfaces (Fig 7).

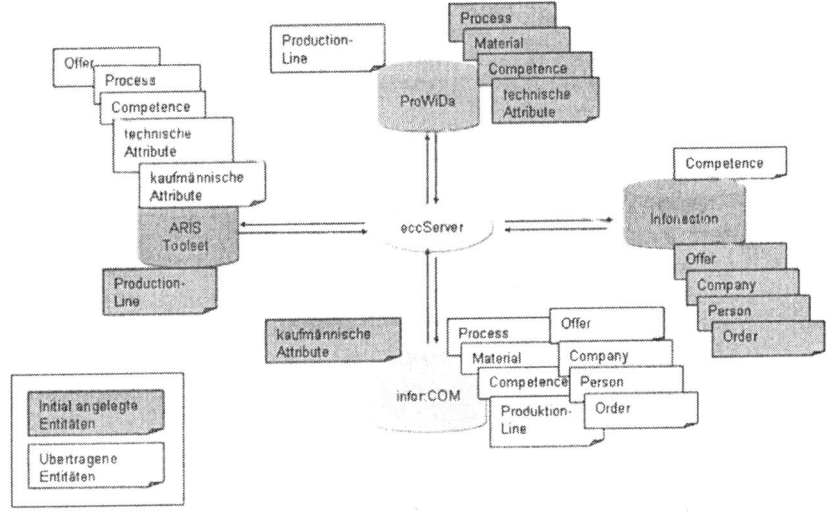

Fig. 7. The ProWiDa "Integrating Infrastructure"

2.4 Action planners as an approach for recursive detection of the "ideal" process chain

Most tools available for process analysis and control focus on the aspects of "Total quality management" offering features like statistical process control etc., which are also covered by the batch tracing and test data module of the ProWiDa System.

In addition the ProWiDa System consists of a "process chain analyser". This prototype approach for detection of the most appropriate process chain bases on ideas coming originally from action planner approach This approach is quite common in the track planning of robots. The prototype approach was developed, to allow for an identification of potential solution paths, coming from a product requirement.

Pre-, runtime, and post- conditions had been defined for description of the "behaviour" of the competences. A software routine allows the comparison preconditions of a process step with post conditions of process steps stored in the central ProWiDa database [07].

The basis for this approach is the development of a MST description language that needs to be established and further developed. First ideas for a MST markup language, also coming from the field of Si-MST attest the feasibility of such an approach [08]. The ProWiDa process chain detection system in its current state demonstrates that such an approach, based on the modelling of MST process step behaviour, could be appropriate for future detection and benchmarking of MST technologies compared to a specific product requirement.

3 Summary and Outlook

The paper basically describes a knowledge management approach for MST production technologies allowing for a product independent definition, storage and retrieval of design rules, business rules and process parameters for the development of micro system products according to customers' needs. The approach is based on a generic database concept for the flexible combination of parameter sets attached to product and fabrication relevant aspects (e.g. technology, design, material, machines tools).

First prototype interfaces to enterprise information systems and components for analysis of data collected in this specific manner have been realised or are under development. They will allow for the import of measuring data from the production for a continuous optimisation of business rules/design rules based on "real world" process data. A very first prototype of a process chain analyser –allowing for an identification of the most appropriate solution path related to a specific task given– has been developed and will be enhanced in the next few years.

References

[01] Nonaka, I.; Takeuchi, H.: The Knowledge-Creating Company: How Japanes companies create the dynamics of innovation. Oxford University Press, New York; 1995.

[02] Heisig, P. in Report Wissensmanagement Hrsg.: C. H. Antoni, T. Sommerlatte, 4. Auflage 2001, Symposion Verlag

[03] Bouwstra, S., Da Silva, M.; Schroepfer, G.; White paper: Towards Standardization of MEMS Materials Characterization; Conventor Inc.; 2003

[04] Wagner, A. Hahn, K.; Considerations for MEMS physical design states, Sophia Antipolis Microelectronics Forum, France; 2003

[05] Brueck, R.; Dickerhof, M.; Hahn, K.; Langbein, I.; Towards an integrated design approach in Si+non Si MEMS Methodology; Presentation at the Patent-Dfmm Workshop in Berlin, Germany, 2007,

[06] Dickerhof, M.; Gengenbach, U.; Kooperationen flexibel und einfach gestalten, Hanser Verlag Munich, Germany, 2006 (only in German language)

[07] Dickerhof, M., Didic, M., Mampel, U.; Workflow and Cimosa –background and case study; Computers in Industry 40 P. 197-205, Elsevier Science, 1999

[08] Niedermaier, M., Entwicklung und prototypische Implementierung eines Konzepts zur Modellierung von prozessschritt-übergreifenden Aspekten der Mikrosystemtechnik auf der Basis von Technologieontologien; Diploma thesis; University of Karlsruhe, Faculty for Computer Science, Germany, 2007 (only in German language)

[09] Wagener, A.; Popp, J.; Hahn, K.; Brück, R.; PDML - A XML Based Process Description Language. In: Proceedings of the 9th European Concurrent Engineering Conference, Modena, 2002. – ECEC 2002

DISTRIBUTED SIMULATION IN MANUFACTURING USING HIGH LEVEL ARCHITECTURE

J. Rodríguez Alvarado*, R. Vélez Osuna**, R. Tuokko*

*Tampere University of Technology, Institute of Production Engineering, Automatic Manufacturing and Assembly Laboratory, P.O. BOX 589, FIN-33101 Tampere, Finland.
Tel +3583 3115 4487, Fax +358 3 3115 2753
jose.rodriguezalvarado@tut.fi, and reijo.tuokko@tut.fi.

**Visual Components Oy, Korppaanmäentie 17 CL6,
00300 Helsinki, Finland, Tel +358 9 323 2250
ricardo.velez@visualcomponents.com

Abstract Distributed simulation has the potential to become widely applicable for geographically-dispersed manufacturing environments, as is the case with desktop manufacturing or rapidly deployable micro-assembly stations. The work presented here discusses in detail the theory of the distributed manufacturing simulation infrastructure based on the IEEE HLA standard and clarifies with practical examples the potential of this approach in modeling and simulation of assembly systems in complex and distributed manufacturing environments.

Keywords distributed simulation in manufacturing, HLA

1 Introduction

Manufacturing is a critical industry for all major economies. Every individual and industry depends on manufactured goods which makes manufacturing crucial to the national economies. Competition is increasingly hard and globalisation is leading to worldwide distribution of production, products and services, affecting all countries and economical regions. At the same time markets are changing. Customers call for faster product changes and demand products which are increasingly targeted to individual needs. Mass production is therefore replaced by customised and personalised production of individual products.

Manufacturing companies are now living in a time where fast prototyping is the rule and products should be assembled fast and accurately. Moreover, the life cycle of products is decreasing constantly on the market and costs have to be cut everywhere. One way to keep up with all those changes has been shrinking of the value chain with the main objective of producing high quality and fast deployable customised products.

Constructing a prototype may be costly, unfeasible, and/or dangerous [1]. Using 3-D models, designers can study and refine assembly sequences for ease of execution, and identify problems that otherwise might not be detected until significant resources were already committed to production [2].

Consequently, 3-D software had become a fundamental part of manufacturing when designing and simulating virtual environments. 3-D models are not only a visual representation of the real components but also can be a portrayal of performance.

2 Distributed Simulation

One of the problems that realistic simulation usually faces is the need of large resources for execution of simulation models with increased complexity. A normal solution to the simulation model complexity problem is to decrease the level of details. This solution allows home-office machines to perform complicated simulations without any radical slowdown in the machine performance.

Distributed simulation has been widely used in numerical simulations and made its path for fifteen years in the virtual environments for military applications. The advantage of using distributed simulations is that each calculation can be done in a single ordinary computer and when the partial result is ready, it is sent to the computer that uses those results for its own calculations, improving the speed and power of the computations. Another advantage is redundancy; in case of a failure of one machine, other computers can handle over the data that the failing computer should have processed.

In the middle of the past decade the US Department of Defense (DoD) came up with the need to standardise the way simulators and simulations could communicate between each other. The result of this standardisation was the High Level Architecture (HLA), with a mandate that simulations still in use and all future developed simulations should be compliant with the HLA standard [3]. HLA was a successful answer to the need of interaction between simulations and simulators from different manufacturers, which later on was defined as Standard 1516 by the IEEE for the open community.

The HLA environment includes certain definitions and components. The main ones are the *RTI*, the *federation object model*, the *federation* and the *federate*. The Runtime Infrastructure (RTI) is the core of the HLA where many federations can reside. A federation is a named set of federate application and a common Federation Object Data that are used as a whole to achieve some specific objective. A federate is an application that may be coupled with other software applications under a Federation Object Model Document Data and a RTI [4]. A federate can be a simulation or a human interaction like a pilot in a cockpit simulator.

Manufacturing simulations, including those of microfactories, have been until now centralised simulations. This has made it imperative to use powerful computers for extensive or high-detailed simulations.

Utilising a distributed approach, every part of the simulation can be displayed in high detail, running the simulations on ordinary computers.

3 The MS2Value project

The MS2Value project (Modeling and Simulation of Manufacturing Systems for Value Networks) aims at developing a distributed simulation environment for manufacturing operations, building it on top of open source or commercially available components [5]. By using this approach, manufacturing operations could be simulated across the world, where each federate would be running a small, but rich in detail, portion of the simulation in each machine and all the statistics of the simulation can be retrieved via the RTI. These simulations can run in different geographical locations as shown in Figure 1.

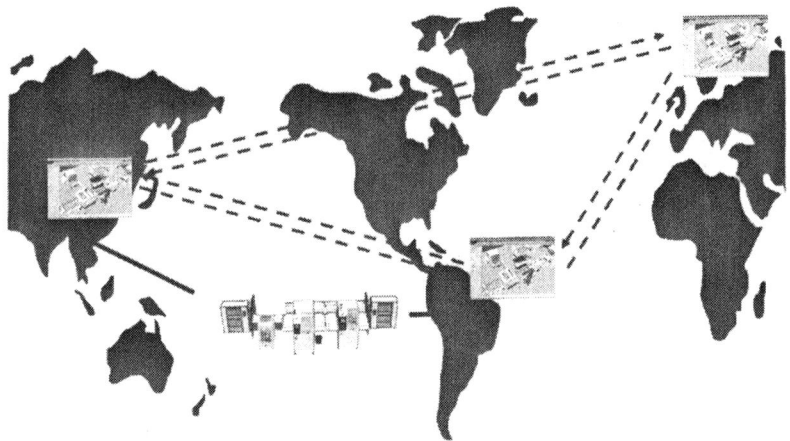

Fig. 1. Distributed and real time simulation is possible using HLA and SOA

In the MS2Value project, the *DS-HLA (Distributed Simulation based on High Level Architecture)* application was created to interconnect Visual Components' 3DCreate [6] simulations through an open source RTI (jaRTI) [7]. In each simulation, it is possible to implement one or more federates. These federates would be handling only a part of the microfactory operations and after the product has been processed, the federate would give an update to the federation related with that operation.

Figure 2 shows the architecture developed for the MS2Value project. As is shown, different federates can connect to the RTI via sockets or web services and share information between the 3DCreate applications. It is a HLA rule that all the interactions must be made through the RTI. It is also possible to monitor the federation through the console and verify the state of the federation.

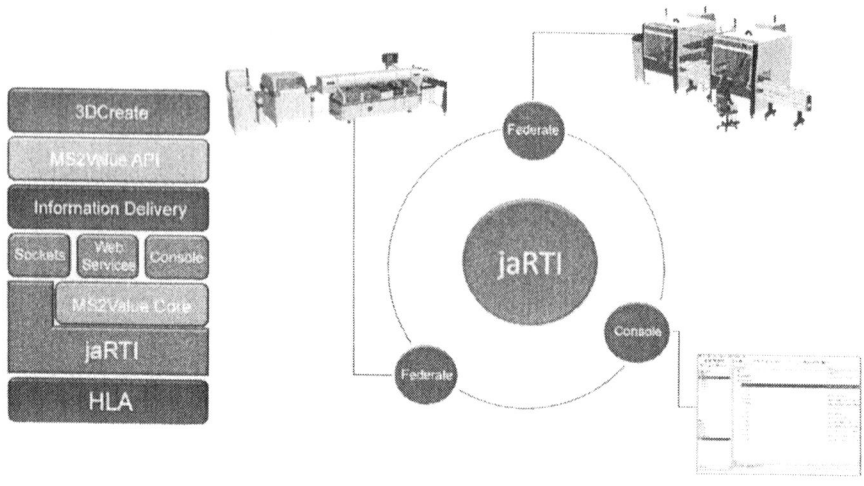

Fig. 2. Architecture of the MS2Value Project

The architecture is designed not only to work with Visual Components' software; any application can be designed to monitor a federate, a parameter or a federation and provide statistical data about the simulation.

4 Applications of the DS-HLA

The MS2Value architecture can be used for multiple applications concerning the simulation field. In addition to complex macro-scale assembly systems, simulation of micro and desktop factories can also be challenging, if the models should be accurate in size and details so that the model can give a realistic representation.

By using the DS-HLA application one can simulate the complete process of the micro-assembly station or a micro-factory represented in different machines with high detailed models which would be the main pattern at the time of doing the real implementation. All of these simulations can be done in lightweight processes without overloading the computer's processors.

When modeling an assembly system, some modifications can be done in the process using this approach that would allow comparing an end product going through different processes. For example, a simulation federate assembles an accelerometer and other simulation federate handles the packaging. By connecting more federates to the federation, it would be possible to make a parallel packaging process simulation where we can compare different types of machines, assembly techniques, cycle times, etc.

Figure 4 shows a graphical representation of the example mentioned above.

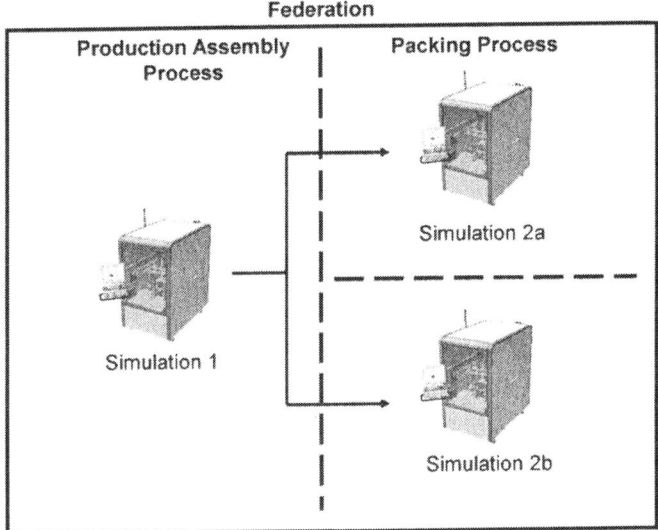

Fig. 4. Diverging of the same product into several processes to weigh their performance

These simulations can grow as the needs of the process designers evolve, models can communicate back and forward and information can be grabbed and stored by other federates in databases or files.

Another theoretical example can be that several micro or desktop factories are simulated in different parts of the globe. Those simulations would offer certain processes that could be used in the production of goods. Using a similar approach to the DS-HLA application, users could design an assembly sequence based on the availability of the services, run the simulation from where it could be possible to monitor the production, get the statistics and validate different scenarios.

5 Conclusions

Distributed simulation has the potential to become an important tool for widely distributed manufacturing organisations. Global competition forces more efficient manufacturing and assembly operations which are more and more often distributed over a value network. This kind of networked manufacturing increases the complexity of overall capacity calculations and overall system simulation, and micro and desktop factories are also on this development track. The DS-HLA application developed as part of the MS2Value serves as a theoretical proof of distributed simulation for manufacturing systems. A pilot case study has been developed, but industrial applications are still to be waited for. Development on the DS-HLA will continue still as an open source project, partly funded and developed by Visual Components and research partners, such as Tampere University of Technology.

References

1. R. Fujimoto, Parallel and Distributed Simulation Systems, p.4.
2. Integrated Manufacturing Technology Initiative; Integrated Manufacturing Technology Roadmapping Project, Modeling and Simulation (2000) p.3-6.
3. IEEE Std 1516-2000, IEEE Standard for Modeling and Simulation (M&S) High Level Architecture (HLA) Framework and Rules.
4. Defense Modeling and Simulation Office, United States Department of Defense, High Level Architecture. (August 20, 2007); https://www.dmso.mil/ https://www.dmso.mil/public/transition/hla/
5. R.Velez Osuna, R. Tuokko, Modeling and Simulation as a Tool for Decision Making in Adaptive Value Networks, (International Conference of Concurrent Enterprising ICE, 2006)
6. Visual Components Oy, http://www.visualcomponents.com
7. jaRTI (now renamed poRTIco), http://www.porticoproject.org

Chapter 4

High Precision Packaging and Assembly Processes

ADAPTIVE PACKAGING SOLUTION FOR A MICROLENS ARRAY PLACED OVER A MICRO-UV-LED ARRAY

Markus Luetzelschwab, Dominik Weiland, Marc P. Y. Desmulliez

Microsystems Engineering Centre, Heriot-Watt University
School of Engineering & Physical Sciences, Earl Mountbatten Building
Edinburgh, EH14 4AS, Scotland, UK

Abstract In this article a versatile packaging solution is presented that allows the static and active alignment of a microlens array that is to be placed over a micro-UV-LED array. A modified UV-LIGA process is applied for building up the structure. For the static approach, the microlens array rests on four posts with the aim of reducing the contact area between the two parts, hence reducing the probability of vertical misalignment. The fine height adjustment is done by electroplating a certain thickness to the electrodes where the posts are being placed. Since the electrodes can be individually addressed, a possible tilt, caused by uneven post heights, can be compensated. With minimal modifications, the structure can be rendered into a dynamic alignment system, featuring actuators for vertical and lateral movement. Even though the microlens array is part of the actuator itself, it is not connected to any potential or energy sources. A magnetic actuator is proposed and partly tested that is capable of simultaneously perform a lateral movement while the vertical actuation is in progress. As a restoring means, a gel material is used as a precursor for a photo-patternable PDMS structure.

1 Introduction

Surface active devices (SAD) require precise positioning of their micro-optical components. This has long been being a critical issue, since it is an important factor for the interaction between optical devices and the environment [1, 2]. Different alignment solutions have been developed to resolve some of the inherent difficulties associated with this objective [3, 4]. The majority of these solutions make use of bulky and expensive external alignment set-up configurations [5, 6]. They also only offer the option of static alignment, meaning that the device will be permanently fixed after having reached its specified position. Herein is presented a versatile packaging concept that allows a microlens array (µLens) to be aligned either statically or dynamically over a micro-UV-LED array (µLED). As shown in Fig.1 both the µLens and the µLED are delivered to us in chip size which complicates the packaging process significantly.

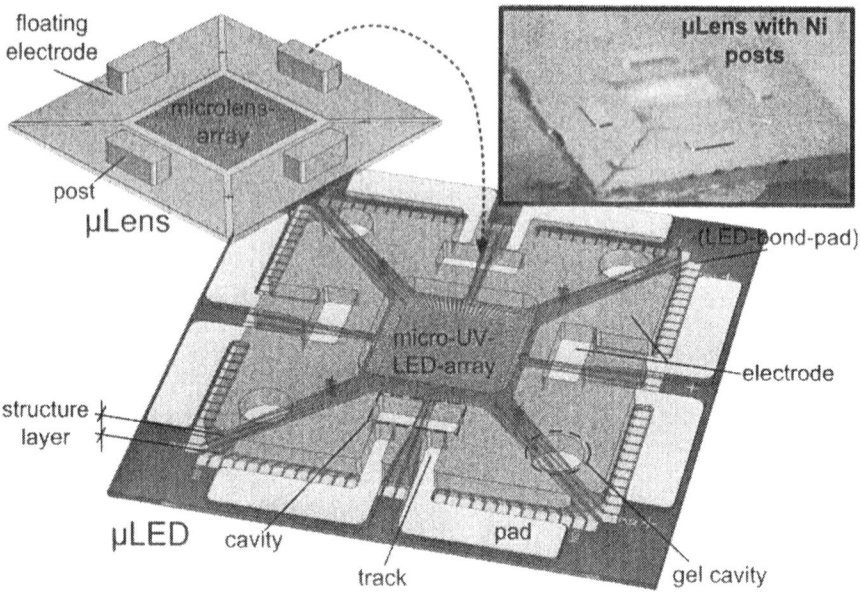

Fig. 1. The microlens array is flipped and placed on top of the micro-UV-LED array. For lateral magnetic actuation, the posts are made of a ferromagnetic material such as nickel.

The μLens is flipped and placed over the μLED in order for the post to fit into the cavities. Since the posts are slightly higher than the depth of the cavities, the contact area is reduced between the two devices. This means that the probability of vertical misalignment of the whole device is reduced. With this method it is possible to "bridge" potential particulates and surface imperfections situated on the structure layer of the μLED or the μLens. Each of the cavities features a pair of electrodes which can be accessed via the corresponding pads. By electroplating the electrodes to a certain thickness, vertical fine adjustment can be performed. The posts and cavities are designed to have a reduced contact area also for the lateral direction. This reduces the over determinations of the alignment process. After placing the μLens, the assembly can be fixed using for example UV curable adhesive. The assembly can be sealed afterwards.

With few modifications of the existing structure, the actuation of the μLens for dynamic alignment is possible. Electrostatic actuation in the vertical direction can be achieved as well as simultaneous lateral magnetically induced movement. Four floating electrodes are evaporated onto the μLens as shown in Fig.1. The purpose of the floating electrodes is to reflect the electrostatic field generated by the pair of electrodes situated underneath on the μLED. No electrical potential needs to be applied to the floating electrode. No wire, which could cause mechanical disturbance, needs to be attached in consequence. Each of the four sides features an electrostatic actuator; therefore vertical movement as well as tilting of the μLens can be accomplished.

2 Static approach

The fabrication of the posts (Fig.1) revealed that the spin-coating of SU8 provides the specified thickness within a tolerance of around ± 10 %, giving thereby ± 4 μm of thickness non-uniformity for an intended thickness of 40 μm. A method was found by fine adjusting the distance between μLED and μLens caused by this tolerance by electroplating a well controlled thickness of the electrodes that are situated within the cavities of the μLED. As shown in Fig. (top right), the overall distance between μLens and μLED includes the thickness of the electrode (t_{elec}) and the thickness of the post (t_{post}). Since the electrodes are individually addressable, uneven post heights of μLens posts can be equalised as shown in Fig. 2 (left). This is particularly important, since the posts of the μLens might vary over the whole device. For instance, if post C of the μLens was found to be too small, then the electrode C needs to be electroplated in order to compensate for that mismatch.

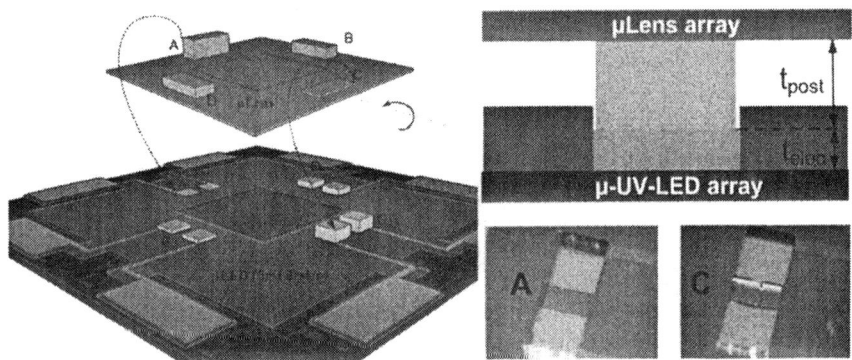

Fig. 2. Left) Different post heights on the μLens can be compensated. *Top right)* The distance between μLens and μLED is comprised through the thickness of the post, t_{post}, and the thickness of the electrode, t_{elec}.

Several electroplating tests were conducted as exampled in Fig. 3. An overall targeted thickness of 15 μm is requested to level the μLens. The left picture in the figure is the height uniformity of the evaporated titanium electrode, taken prior to the electroplating process. The peaks are artefacts from previous process steps and are not considered in the following discussion. It is however assumed that due to the small lateral dimensions, the peaks would not contribute to a displacement with their full height, since they would probably partly deform or penetrate the SU8 post during the assembly process. The right hand side of Fig. 3 displays the situation after electroplating a nickel layer of approximately 5 μm. The resulting surface yields a thickness between 15.1 and 15.6 µm and is therefore within the sub-micron range specification as far as the variation of thickness is concerned. The calculation for the required thickness was conducted by firstly determine the grow rate for a given geometry and parameters and then extrapolating the required time to achieve the intended thickness in a following electroplating process.

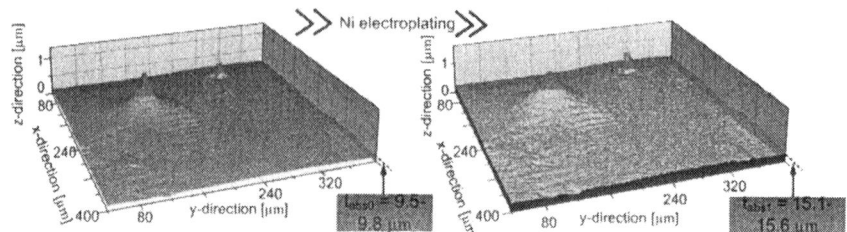

Fig. 3. *Left)* Height distribution of the surface of the titanium electrode with the z-axis highly exaggerated. The two peaks that are seen are due to impurities of the sample. *Right)* Surface height distribution of the electrode after electroplating approximately 5 µm of nickel.

For the measurements shown in Fig. 3, the thickness was achieved with only one measurement for determining the grow rate. Higher accuracy could be achieved by carrying out several measurements, to approximate the target thickness very slowly at the end of the plating process.

3 Dynamic Approach

With minor modifications, the positioning of the µLens can also be dynamic and be monitored in real-time in the lateral and vertical directions. The electrodes shown in Fig. 1 then serve as a means to generate an electrostatic field [7]. The floating electrode, which is attached to the µLens, reflects the electrostatic field from one electrode to the other so that the assembly forms a capacitor (Fig. 4, top right). The floating electrode and hence the µLens does not need to be connected by any wire since no potential is required. This eases the fabrication and assembly process and frees the µLens from any possible mechanical disturbance caused by a wire during its movement. The FEM simulation of the force and analytical calculation curves agree with an error of less than 17% as shown in Fig. 4 on the left hand side.

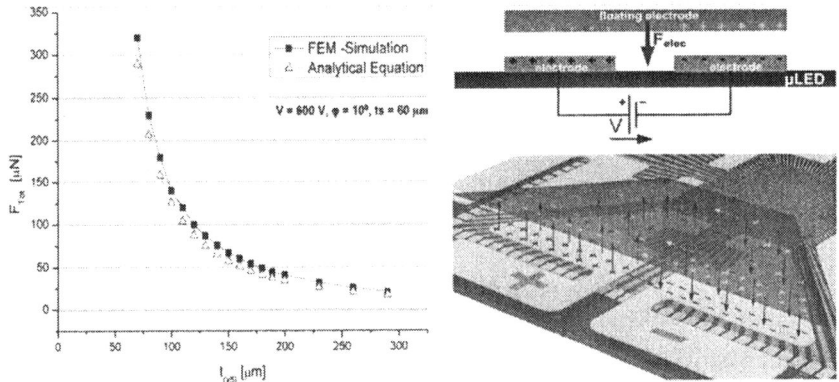

Fig. 4. *Left)* Analytical and FEM (COMSOL®) simulation solutions of the force generated as a function of the displacement of the electrode. *Top right)* Schematic of the µLens floating electrode forms a capacitance with the µLED electrodes. *Bottom right)* 3-D representation of the figure above.

As a way to restore the µLens back to its initial position, gel bumps are placed within the circular cavities (Fig. 1) as shown in Fig. 5 (left) and the µLens is then flip-chipped on top. In the future the gel material will be replaced by photo-patternable PDMS structures as PDMS offers a greater control of the shape and height [8, 9]. The characterisation of the bumps in terms of robustness and reproducibility of their response was investigated with long term step voltage measurements using a setup similar as shown in Fig. 5 (right), without using the PI-controller. One sample exhibited displacements of around 25 µm at a voltage of 350V. Long term measurements revealed that the repeatability with 12 step responses, next to other different step voltages, taken over a time period of over 15 hours was good with a standard deviation of 0.43 µm at a mean value of 24.75 µm. By reducing the size of the gel bumps, the distance between floating electrode and the electrodes is reduced as well. Due to the nonlinear relationship between electrode-to-electrode distance and the force exerted, the voltage can then be drastically reduced to achieve the same magnitude of displacement. An advantage of using gel is the possibility to have not only elastic behaviour in the vertical but also in the lateral directions. This will be of benefit for the lateral magnetic actuator as described later. A negative side effect of using gel bumps is the difficulties that are involved when trying to reproduce similar bumps in shape and volume. Each device does exhibit different displacements as a function of the voltage due to the lack of control of depositing the gel.

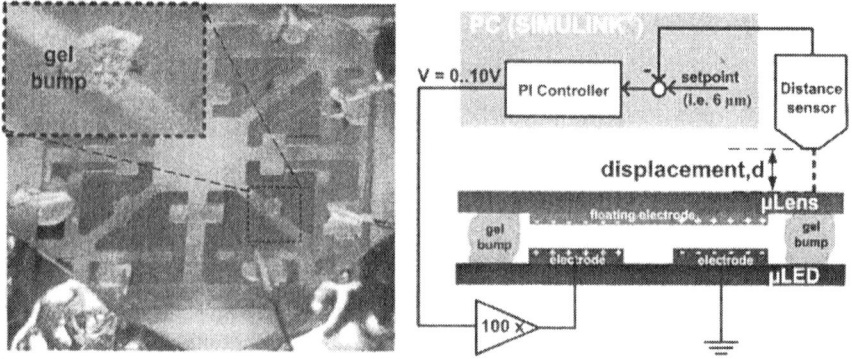

Fig.5. Left) Gel bumps, placed within circular cavities, provide the restoring force for the actuation. *Right)* Step response measurement setup (without PI controller) and closed control loop for accurate displacements (with PI controller).

To see whether it is possible to control the μLens for a certain displacement, the step response for voltages between 0 V and 600 V were recorded, as shown in Fig.6 with a voltage on time of 2 minutes and a voltage off time of 6 minutes. The progression of the plateau of the step response, as seen in Fig.6 at time slot 1 (ts1) is not flat but with an angle β. This indicates that the restoring force does not reach an equilibrium but instead slowly continuous to increase. Furthermore, the response to the rising edge of the step is much faster compared to the falling edge. It is assumed that the stickiness of the gel and its memory properties with regard to stress and strain are the main causes of this behaviour. The mathematical description is not trivial as well as the physical effects within the system. In a first approach, the step response of the system at 400 V was treated as a PT2 controlled path. The slope with the angle β, was neglected. By applying a tangent to the positive slope, the dwell time, t_u, and compensation time, t_g, could be determined to be 0.25s and 0.65s respectively. The proportional gain (Ks) was calculated to be around 0.03. The characteristic times and gains for the different voltages steps differ from each other so a compromise had to be taken between the obtained values. However, the dwell time, t_u, seemed to be quite consistent for the different voltage steps and even samples with a value of around 0.25 s. From this information, the corresponding parameters were calculated with the CHR (Chien, Hrones and Reswick) method. After an initial rest run, the values needed to be changed however, since the CHR method only provides a rough approximation. The controller was chosen to be a PI controller in favor of a PID controller, since the noise of the overall system, most likely caused by the vibration of the building, would have had a disturbing effect on the differential parts of the controller. The controller was set up by using SIMULINK® and the displacement was measured by using a confocal chromatic displacement sensor (Micro-epsilon, optoNCDT 2400) as shown in Fig. 5 (right).

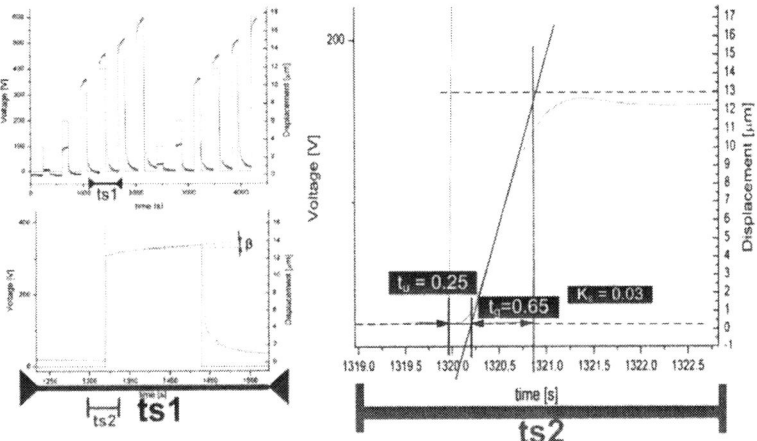

Fig.6. Top left) Step responses were recorded for voltages between 0 and 600V. *Bottom left)* Zoomed out picture of times slot 1 (ts1) for the 400 V step response. *Right)* Time zoom for the time slot, ts2, where different parameters were obtained.

Figure 7 depicts the approximation of the effective displacement, after changing the set point (intended displacement) discrete steps. The setpoint is achieved in approximately 25 seconds for the positive step response and around 30 to 35 seconds for the negative step response.

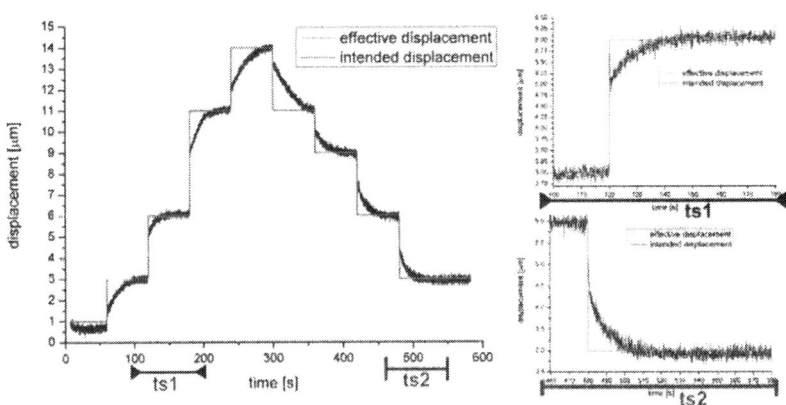

Fig. 7. Left) The setpoint of the closed control loop is changed in discrete set and the response of the system recorded. *Right top and bottom)* Zoomed out time slots (ts1 and ts2) from the left hand picture.

The noise caused by the setup and the vibration of the building makes it difficult to accurately determine the control difference that remains due to the measurement itself as well as the feedback to the system and its disturbing consequences. It is feasible to use an adaptive control loop feedback, which changes the parameters of the controller depending in which voltage range the device is operated, hence making the controlled system more responsive. The setup should be installed on top of a vibration-controlled table.

A magnetic actuator is also proposed and has been partially tested. This actuation mechanism is responsible for the lateral movement and uses the posts of the μLens that need to be made of a ferromagnetic material such as nickel. Each side features a magnetic actuator, thereby allowing lateral alignment of the μLens. The magnetic actuator does not possess a physical connection with the μLens which leaves the μLens, from a mechanical point of view, almost undisturbed. By attaching an external driving device (EDD) as shown in Fig. 8 (left), simultaneous vertical electrostatic movement and magnetic lateral movement can be performed, since the two fields can coexist within the actuator without influencing each other. The electrical insulator of the EDD needs to be robust at high voltages, but must let pass a magnetic flux that is generated by the coil with the current, i. The magnetic flux coming from and into the tracks (Fig.8, right bottom) attracts the Ni post with a force, F_m, making the distance, d_m, as small as possible to achieve the most flux for a given current i.

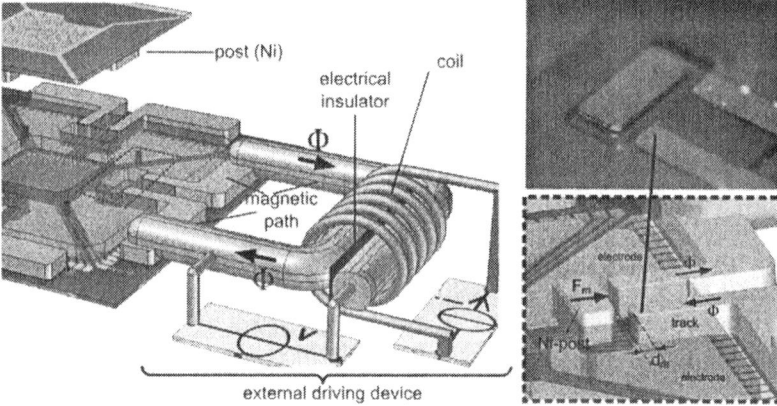

Fig. 8. Left) External driving device (EDD) attached to the ferromagnetic pads of the μLED. The EDD allows a magnetic flux going through the magnetic path, but also provides a voltage for the electrostatic actuation. *Right top)* Photograph of electroplated Ni tracks leading towards the cavity where a Ni post is placed for experimental purposes. *Right bottom)* close up of the magnetic actuator.

Preliminary test were conducted, and a movement of around 50 μm was observed, which is well above the intended displacement range. However, a high current of 400 mA is necessary for this to happen and further improvement of the magnetic path is necessary. Fig. 9 displays a magnetic circuit with lumped elements where, with the exception of the magnetic resistance of air, all materials involved exhibit a nonlinear and hysteresis behaviour. In particular the latter is mathematically difficult to grasp since during operation, many different hysteresis loops are "created" while starting from different initial points. Fig. 9 (right) only depicts qualitatively the different hysteresis loops for magnetic materials that have not yet been exposed to any magnetic field. The objective is to find a MEMS process able material that exhibits high permeability but low magnetic hysteresis. With that it might be possible to represent the single magnetic resistors with lookup tables and using SIMULINK® for numerically solving the given problem within a certain degree of accuracy.

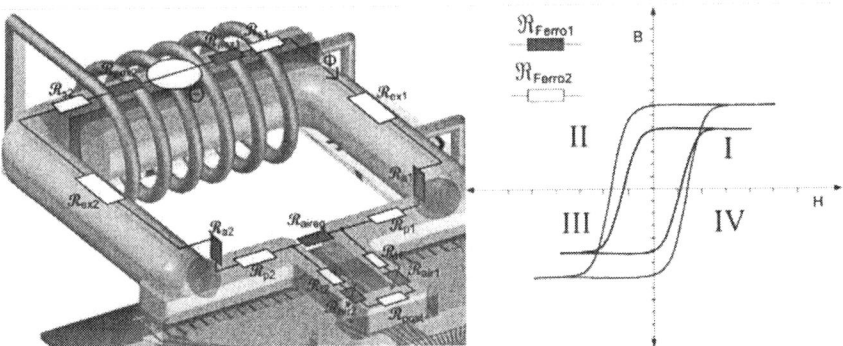

Fig. 9. Left) For illustrating purposes only, a schematic is drawn with the highly nonlinear (green and white) magnetic resistors which are supposed to have a hysteresis which is not to be neglected. *Right)* Schematic drawing of the magnetisation curve of the different magnetic material used.

The proposed solution with its dual actuator has the potential to have a positive impact on the production cost of such a device, in particular for the static approach. It is for instance feasible that the external driving devices are only temporarily attached to the assembly during the process of alignment and fixing in position. By doing this, the most expensive device within a production line would be 3 small intensity detectors placed closely above the μLens which detects the most power output, if the individual microlens is to be perfectly aligned, in vertical as well as lateral direction towards the μLED. Three detectors would be necessary to control the μLens with regard to tilting and angular alignment. No expensive alignment tools are therefore necessary. It is estimated that the cost reduction of the alignment devices involved might be in the order of 10 times smaller than using a conventional alignment system.

4 Conclusions

In conclusion, a device was presented that allows the accurate static alignment as well as dynamic vertical movement of a microlens array on top of a micro-UV-LED array. Both methods are capable of achieving sub micron accuracies. The use of gel bumps showed a simple and robust means for generating the necessary restoring force which will be further investigated and refined by using photo-patternable PDMS structures. Furthermore, a concept of simultaneous vertical electrostatic movement and lateral magnetic movement was proposed and partially tested.

Acknowledgments

The authors would like to express their gratitude for the financial support from the UK Engineering and Physical Science Research Council (EPSRC) through its Basic Technology Programme. The work was carried out under the project entitled "A thousand Micro-emitters per square millimeters" referenced GR/S85764.

References

[1] C. Pusarla and A. Christou, "Solder bonding alignment of microlens in hybrid receiver for free space optical interconnections," presented at Electronic Components and Technology Conference, 1996. Proceedings., 46th, 1996.

[2] S. S. Lee, L. Y. Lin, K. S. J. Pister, M. C. Wu, H. C. Lee, and P. Grodzinski, "Passively aligned hybrid integration of 8 × 1 micromachined micro-Fresnel lens arrays and 8 × 1 vertical-cavity surface-emitting laser arrays for free-space optical interconnect," *IEEE Photonics Technology Letters*, vol. 7, pp. 1031-1033, 1995.

[3] G. C. Boisset, B. Robertson, W. S. Hsiao, M. R. Taghizadeh, J. Simmons, K. Song, M. Matin, D. A. Thompson, and D. V. Plant, "On-die diffractive alignment structures for packaging of microlens arrays with 2-D optoelectronic device arrays," *Photonics Technology Letters, IEEE*, vol. 8, pp. 918-920, 1996.

[4] M. S. Cohen, M. J. DeFranza, F. J. Canora, M. F. Cina, R. A. Rand, and P. D. Hoh, "Improvements in index alignment method for laser-fiber array packaging," *IEEE Transactions on Components, Packaging, and Manufacturing Technology Part B: Advanced Packaging*, vol. 17, pp. 402-411, 1994.

[5] M. T. Gale, J. Pedersen, H. Schutz, H. Povel, A. Gandorfer, P. Steiner, and P. N. Bernasconi, "Active alignment of replicated microlens arrays on a charge-coupled device imager," *Optical Engineering*, vol. 36, pp. 1510-1517, 1997.

[6] S. Eitel, S. J. Fancey, H. P. Gauggel, K. H. Gulden, W. Bachtold, and M. R. Taghizadeh, "Highly uniform vertical-cavity surface-emitting lasers integrated with microlens arrays," *Photonics Technology Letters, IEEE*, vol. 12, pp. 459-461, 2000.

[7] S. J. Woo, J. U. Jeon, T. Higuchi, and J. Jin, "Electrostatic force analysis of electrostatic levitation system," Hokkaido, Jpn, 1995.

[8] K. M. Choi, "Photopatternable Silicon Elastomers with Enhanced Mechanical Properties for High-Fidelity Nanoresolution Soft Lithography," *J. Phys. Chem. B*, vol. 109, pp. 21525-21531, 2005.

[9] W. O. J.C. Loetters, P H Veltink, P Bergveld, "The mechanical properties of the rubber elastic polymer polydimethylsiloxane for sensor " *Journal of Micromechanics and Microengineering*, vol. 3, pp. 145-147, 1997.

SOLDER BUMPING – A FLEXIBLE JOINING APPROACH FOR THE PRECISION ASSEMBLY OF OPTOELECTRONICAL SYSTEMS

Erik Beckert[*], Thomas Burkhardt, Ramona Eberhardt, Andreas Tünnermann

Fraunhofer Institute for Applied Optics and Precision Engineering
Dr. Erik Beckert, Fraunhofer IOF, Albert-Einstein-Strasse 7, 07745 Jena, Germany
Phone: +493641807338, Fax: +493641807604
email: erik.beckert@iof.fraunhofer.de

Abstract For microoptics and microsystem assembly, solder bumping is introduced as a flexible and high precise joining alternative to adhesive bonding, especially if adhesives limit the system performance in terms of long term, high temperature, vacuum or UV stability. The bumping technology, which is based on jetting the liquid solder onto wetting surfaces, can be parameter optimised to work in complex environments as well as with submicron accuracy.

Keywords optics, microassembly, laser beam soldering, bumping, jetting

1 Motivation

Joining of components is, next to an accurate alignment, the most challenging process in the precise microassembly of optical and optoelectronical systems, since it is the last step of the process chain to securely fixate the position of sensitive and functional relevant parts in the system and to connect them to respective thermal and electrical interfaces. Bonding with UV-curing adhesives has been the standard technology for this task, but is limited in terms of long term stability, radiation robustness and outgassing in vacuum environments. Force-based clamping technologies are not an appropriate alternative since miniaturised components and brittle materials often not allow for the adaption of clamping mounts.

Soldering, being a metal-based adhesion joining technology, can be such an alternative. With a wide variety of available, lead-free solder alloys that exhibit excellent and variable mechanical parameters solder joints provide creep-minimised, electrically and thermally conducting connections between a mounted part and its system environment. These joints are high temperature and cryogenic stable, but are non-transparent and thus cannot be used, in contrast to polymeric adhesives, within any optical beam propagation path. But of more importance is that soldering in microassembly requires different reflow approaches than that known from electronics manufacturing. Due to sensitive parts and materials the reflow energy should be applied thermally and locally restricted, e.g. by laser. Second, fluxless processing is required to prevent any contamination of optical surfaces. Finally, solder application, its reflow and resolidification must not influence the alignment state of the component to be assembled outbound of tolerable limits that often are in the micron and submicron range. Solder bumping is a technology that provides all these features and incorporates the necessary equipment in one singular and flexible device.

2 Solder Bumping – the process

2.1 Technology basics

Laser based Solder Bumping [1], shown schematically in Fig. 1, was developed to individually place bumps for the flip chip assembly of microelectronic circuits [2]. The patented process places bumps made of various soft solder alloys with melting temperatures ranging up to 280° C on different substrates usually without preheating the metallic wetting surface.

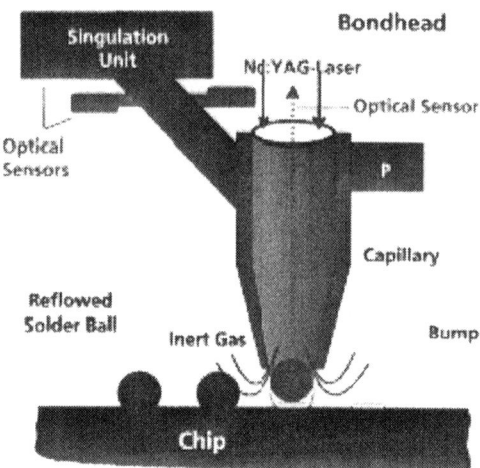

Fig. 1. Solder Bumping [1], Source: Pac Tech GmbH

In a reservoir the solder alloy is fed to the bumping device in the form of solid spheres with a uniform, narrowly (±3 μm) tolerated diameter, ranging from 100 μm to 760 μm. A singulation unit transfers a sphere into the bumping capillary at the beginning of the bumping process. The inner diameter of the capillary is at its end smaller than the diameter of the sphere, so by gravity force the sphere gets locked at the outlet of the capillary. This locking state is further secured by applying nitrogen pressure from within the capillary, which now can be flexibly moved towards the desired bumping position, e.g. by a robot. It is noticeable that this positioning desired bumping position, e.g. by a robot. It is noticeable that this positioning movement, after locking the solder sphere at the outlet of the capillary, can incorporate all six degrees of freedom, especially tip and tilt (Fig. 2), enabling for the positioning of the capillary in even complex, miniaturised and three dimensional integrated system environments.

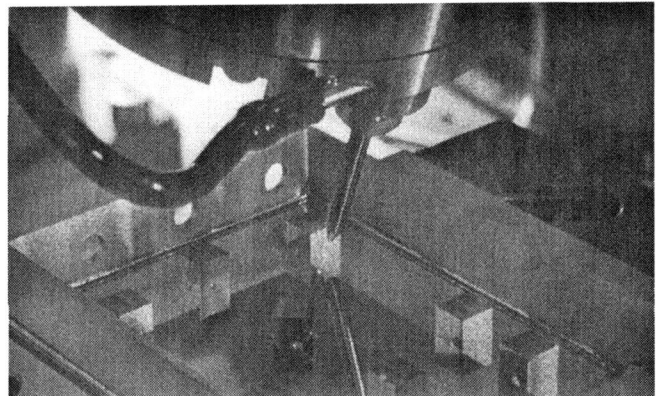

Fig. 2. Tilted capillary in a complex assembly environment

After the desired position is reached, a reflow laser is applied through the capillary. In the current setup a fibre coupled, pulsed Nd:YAG-laser with pulse widths ranging from 15 ms to 25 ms and pulse energies up to 4 J is used, but can be adapted to the actually chosen solder material and volume. The laser pulse liquidifies the solder volume and stores thermal energy within it. Now being liquid the solder gets pressed out of the capillary by the nitrogen pressure that also forms an inert atmosphere around it, and flies towards the substrate, while the laser beam during pulse time pre-heats the wetting area by exiting the now empty capillary outlet. 0.5 mm up to 6 mm free flow bumping distances for the solder spheres where demonstrated.

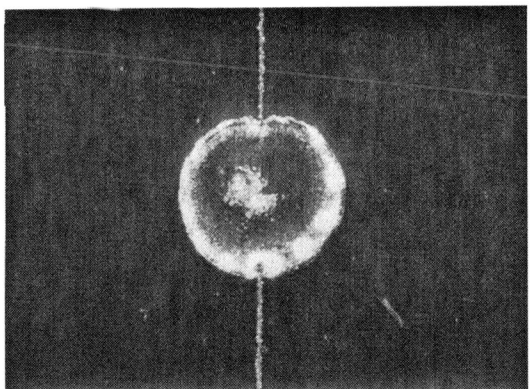

Fig. 3. Bump applied to vertically orientated wetting surfaces

When the still liquid solder hits the wetting surface, it locally heats the area by giving its stored thermal energy away, thus realising a stable joint after resolidification by creating the necessary intermetallic phases. The wetting area does not necessarily have to be a horizontal or planar surface, but can be shaped in any geometry as long as the solder volume can reach the wetting surfaces on both parts to be joined. Bump application was demonstrated even on vertical surfaces (Fig. 3) and with joining gaps up to 100 µm present.

2.2 Parameter analysis

Applying the bumping technology to the various circumstances during precision assembly of optoelectronic microsystems requires not only a fundamental understanding of the wetting physics of the liquid solder hitting a cold or only locally preheated metallic joining surface with different shape, but also an analysis of the various parameters that influence the bumping results. In terms of optics, the results are characterised by:

- dealigment during bumping (required usually to be in the micron and submicron range) due to solder resolidification and long term creep of the solder resulting from intermetallic phase changes and diffusion processes,
- mechanical stability of the joint including repeatability of the wetting behaviour under fluxless processing conditions,
- introduced mechanical stress and micro cracks, thus reduction of mechanical stability, due to thermal shock when liquid solder hits the wetting surface.

A robustness analysis according to Taguchi rules [3] of the bumping process under various conditions is currently carried out. In a more general investigation on the influencing factors a parameter field shown in Table 1 was generated. First experiments indicated, that bump distance and pulse energy, coupled with the deposited energy within the solder volume, are of the most sensitive parameters to be optimised in order to reach a minimised dealignment while securing wettability on different substrates without damaging the bulk material of the components to be joined [4].

Parameter	Range	Influence
Laser reflow parameters		
Pulse energy	0 J to 4 J	Thermal shock, wettability,
Pulse width	15 ms to 25 ms	Pre-heating of wetting surface
Bump parameters		
Solder sphere diameter	100 µm to 760 µm	Pulse energy, dealignment
Bump distance	0.5 mm to 6 mm	Bump formation, Dealignment
Bump direction	0° to approx 45°	Bump formation
Wetting area geometry	planar, complex	Bump formation
Joining gap	up to approx. 100 µm	Bump formation, Dealignment
Nitrogen pressure		Bump formation, Wettability, Delignment
Material parameters solder and components to be joined		
Solder melting temperature	140° C to 280° C	Pulse energy
Surface metallization	Au, AgPd, others	Wettability
Substrate thermal conductivity	up to 400 W m^{-1}K^{-1}	Pulse energy, Wettability

Table 1. Parameter field for the bumping process

Fig. 4 demonstrates the result of an accuracy test with respect to the bumping distance. Motivation for this experiment was that in complex assembly environments it is often difficult to reach the desired joining area even with the small and up to 15 mm long bumping capillary. Thus it would be feasible to increase the bump distance as much as possible, so that only the flying solder volume after reflow would need a path through the environment. But investigations, where bumps were set into a target grid, led to the result that accuracy of setting the bump at the respective target decreases with increasing bumping distance. For a 400 µm solder sphere a bump distance above approx. 5 mm becomes thus useless.

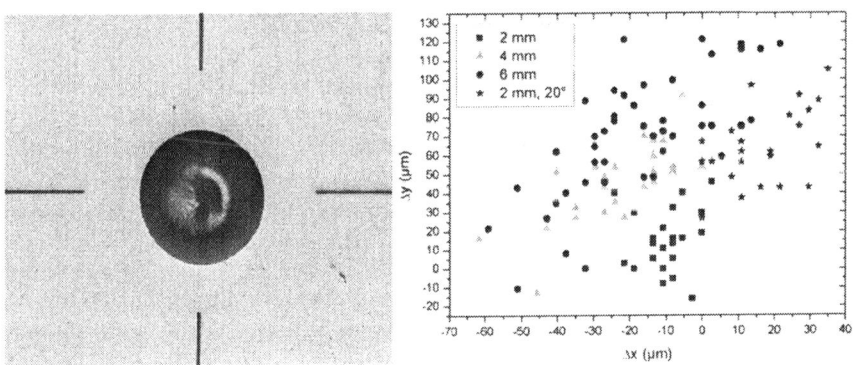

Fig. 4. Bump applied to a vertically orientated wetting target, accuracy with respect to different bumping distances

Fig. 5 shows the link between pulse energy, wettability and damage to the components to be joined. Wettability becomes especially difficult if thermal conductivity of the bulk material is high, thus requiring a large amount of energy deposited in the solder volume to make wetting possible. In contrast, high pulse energy leads to an increased thermal shock when the liquid solder hits the wetting surface, which is not feasible for brittle materials like glass. Consequently, the optimisation goal is to find a pulse energy high enough to secure wettability and low enough to prevent damage to the components. This process window narrows if completely different materials have to be joined (e.g. a lens made of glass and a mount made of metal) and widens for smaller solder volume. In the current setup thus the solder sphere diameter was changed from initial 760 µm to currently 400 µm. It was found that even on difficult wetting surfaces with underlying copper bulk material joints could be created with the solder fracture strength being the limiting factor for mechanical stability.

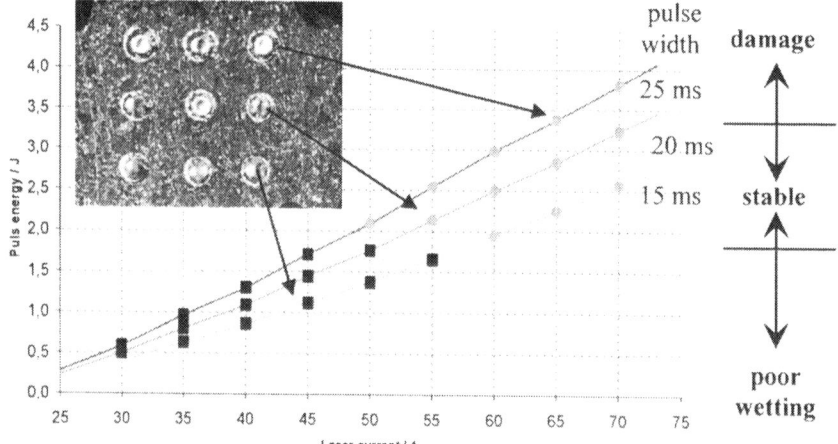

Fig. 5. Pulse energy process window for a copper substrate

3 Application example

A very common task in the precise assembly of optoelectronical systems is the accurate alignment and fixation of collimating lenses in front of laser source. So-called aspherical Fast Axis Collimators (FAC) create a parallel beam from the divergent propagating light of solid state lasers (SSL). While having the advantage that collimation is integrated in on singular optical element, the alignment of this component is very sensitive. With focusing lengths down to 300 µm these FAC have to be aligned with an accuracy down to 0.5 µm and less. While alignment itself is not as challenging with precise positioning devices available, the stable fixation of the alignment state is more difficult. When polymeric adhesives are used the shrinkage of these adhesives during curing (1% to 10%) is one source of dealignment, others are aging due to the continuous radiation load and large thermal expansion in the often high power environment of the laser. Replacing the polymeric adhesive by a solder not only increases stability, but also provides better heat transfer away from the component.

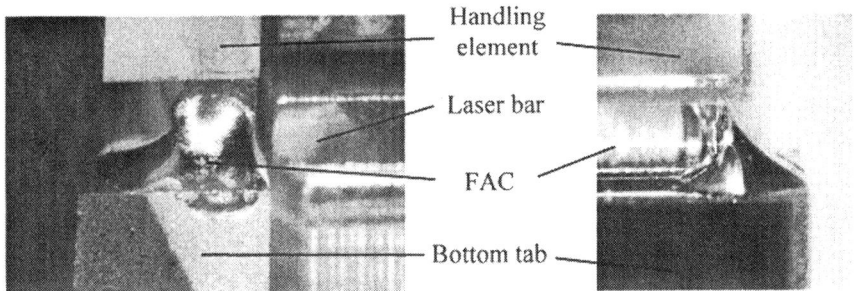

Fig. 6. FAC in front of a laser bar (left: side view, right: front view)

Fig. 6 illustrates how such a FAC can be assembled in front of a laserdiode bar. The FAC in this case is a 11 mm long aspherical cylinder lens with a diameter of approx. 0.5 mm and a focal length of 300 µm. With a working distance of 90 µm it is very sensitive in terms of axial and lateral dealignment. This allows for a very precise detection of the FAC behaviour during solder bumping, since in the far field. There micron lateral movements of the FAC are translated into millimetre movements of the collimated beam projected onto a screen, where the beam profile can easily be viewed by a camera (fig. 7). Axial movement (defocus) results in sharpening and desharpening of the beam profile.

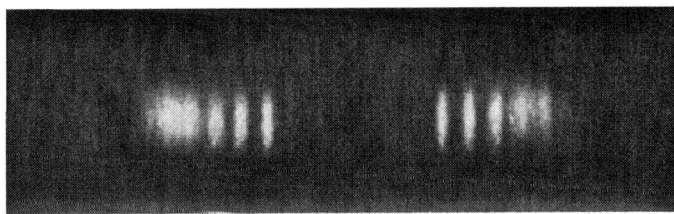

Fig. 7. Camera view of the collimated beam in the far field

To ensure a minimised joining gap between FAC and heat sink of the laser while maintaining 6DOF alignment capability, an additional mounting element ("bottom tab") was introduced. The FAC is joined at its end faces to the bottom tab, while the bottom tab has a planar contact face with respect to the heat sink. This overall leads to a complex sequential joining procedure in which the initially realised alignment state of the FAC gets degraded. This is shown in Fig. 7. The process steps include pre-processing (handling and alignment, steps 1 to 3) and post processing (releasing handling equipment, steps 12 to 15). Thus between steps 3 and 12 the solder bumping process chain itself takes place and results in a dealignment of approx. -0.6 µm.

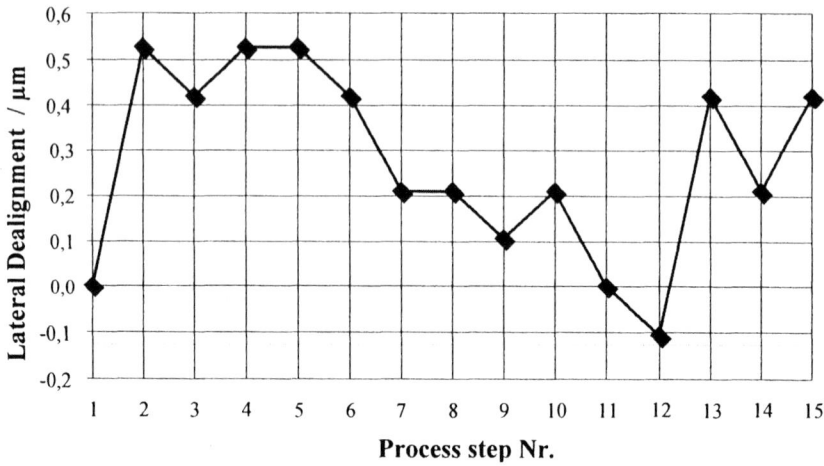

Fig. 8. Dealignment of a FAC during a complex bumping sequence

In a series of six complete FAC processing the dealignment was measured to reproducible in a range of ±0.2 µm. This allows for pre-compensation of the expected dealignment and further increases joining accuracy

4 Outlook

While first experiments proved that solder bumping is a sufficient method to increase repeatability and accuracy compared to conventional flux-free laser beam soldering methods [5], the technology still has to mature for complex precision assembly tasks. Placing accuracy needs to be enhanced in order to use larger bumping distance to reach into complex assembly environments. Long term stability is an issue, since intermetallic diffusion and phase changes are different to what is known from reflow soldering in electronics due to the short laser beam reflow. Most important is a better understanding of the wetting and resolidification process in order to optimise and minimise dealignment during solder bumping, which is still at the limit for some micro-optical applications.

Nevertheless solder bumping technology has a substantial potential for fast varying, high precise assembly processes in microoptics and microsystem technology, where multi-functional and high accuracy packaging techniques are required.

Acknowledgement

The work presented in this paper was partially funded by the german Federal Ministry of Education and Research within the "Optische Technologien" framework programme (Funding call BrioLas, subproject BriMo, FKZ 13N8612).

References

[1] Patent US020040069758A1, "Method and device for applying a solder to a substrate".
[2] E. Zakel, Th. Oppert, G. Azdasht: Laser assisted Wafer Level Packaging for MEMS. In: Proceedings of the Semicon Europa (Munich, Germany), April 19, 2004
[3] S. H. Park, "Robust design and analysis for quality engineering", Chapman& Hall, London, 1996.
[4] K. J. Puttlitz (editor), "Handbook of lead-free solder technology for microelectronic assemblies", Dekker, New York, 2004.
[5] H. Banse, R. Eberhardt, E. Beckert and W. Stöckl, "Laser Beam Soldering – a New Assembly Technology for Microoptical Systems", Microsystems Technologies 11, 2005, pp.186-193.

FLUIDASSEM - A NEW METHOD OF FLUIDIC-BASED ASSEMBLY WITH SURFACE TENSION

N. Boufercha[1], J. Sägebarth[1], M. Burgard[1], N. Othman[2], D. Schlenker[2], W. Schäfer[2], H. Sandmaier[1]
[1] Universität Stuttgart, IFF / MST, Nobelstr. 12, 70569 Stuttgart, Germany
www.iff.uni.stuttgart.de
[2] Fraunhofer, IPA, Nobelstr. 12, 70569 Stuttgart, Germany
www.ipa.fraunhofer.de

1 Introduction

Regarding electronic components the reduction of costs will be a challenging goal in the next few years. The use of polymer electronics will not help to minimise this problem. Cheap products made of silicon are here to stay for a while.

The latest roadmap of the VDI/VDE-IT (see MINAM-Roadmap: www.micronanomanufacturing.eu) indicates that the size of assembled silicon chips must dramatically be reduced. The size of these chips will be less than 500 µm x 500 µm and will have a thickness lower than 50 to 70 µm. The reduction of the silicon material must be in parallel to the reduction of the costs for the manufacturing process.

An indicator for this trend is also shown in Figure 1. For example the expected world market share for RFID in 2010 will be about 11.7 billion USD (Frost & Sullivan 2004). This can only happen if the overall costs (microchip and manufacturing) decrease. The conventional manufacturing methods for microchips with conventional die-bonders have also reached their limit concerning accuracy of positioning, throughput and handling of smallest microchips and micro components. The expected requirements effected by downscaling make it necessary to invent new innovative technologies for production like **FluidAssem**.

The main topic in the next few years will be the development of processes which have the capability to use self assembly methods ([3], [10], [6]). The presented paper investigates methods for processes allowing self assembly of microchips. Additionally the vision is traced to establish an assembly process without using pick and place methods as used in die-bonder processes today. A new concept **FluidAssem** was developed to establish methods for reaching this challenging goal of a safe assembly process for the future.

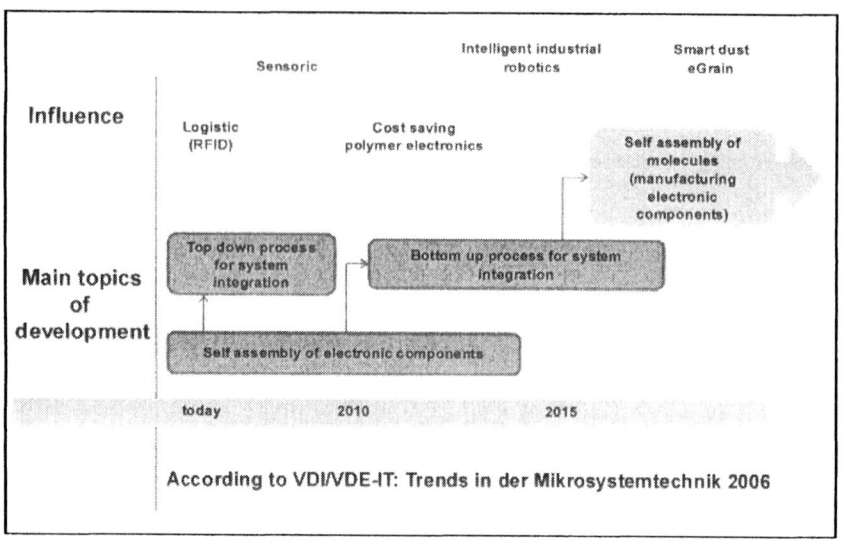

Fig. 1. Roadmap microsystems

2 The Micro Cosmos

It is well known that in the micro cosmos surface forces prevail against volume forces. Why not use these forces for transporting and assembling small chips and components? The surface forces are high enough to do all the things needed for the assembly process of micro components!

To move droplets on a substrate, various methods can be used. The most famous technique is the use of electro wetting. An electric field partially changes the contact angle on one side of the droplet. The droplet moves in the direction of the smaller contact angle because the surface tension of the droplet wants to minimise the surface energy. This motion stops if the droplet has reached a position with a local minimum of energy. In this case the movement of the droplet depends directly on the contact angles realised by an electric field.

Positioning of droplets is also possible by changing only the surface properties ([14], [13], [11], [5], [2]). Therefore, the surface of a substrate must be structured with hydrophilic and hydrophobic areas [7]. These structures are usually a combination of coating with additional micro- and nanostructures to get super hydrophilic or super hydrophobic areas. An area structured this way allows guiding droplets along a specified way.

If a single droplet can move from one to another position, this can also be done with small chips "swimming" on those droplets. With a skillfully structured surface on the substrate and an intelligent use of the surface tension of the fluid, a self alignment process for micro components can be achieved.

In Figure 2 a principle method for fluidic assembly is given. In a first step a large droplet (diameter about 200μm) is placed in the centre of the substrate. This droplet has the job of catching the falling microchip and placing this microchip in the correct position. Smaller droplets are used to contact the microchip in the final position. Special micro and nano structured regions are used to hold the drops in the starting position on the substrate.

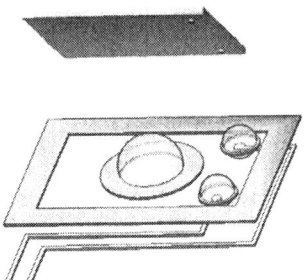

Fig. 2. Principles of fluidic assembly

To achieve a fluidic assembly system, the realised surface properties ([9]), the structures on the substrate and the whole process should first be simulated and optimised.

The whole process can be analysed via decomposition. The steps in detail are as follows:

- placing the droplets for catching and contacting the microchip on the correct starting position on the substrate
- putting the micro chip on the top of the catching droplet (middle one) without mismatching the position (vertical alignment)
- adding a mismatch between starting and end position of the chip (lateral
- alignment)
- pinning effect (contact angle hysteresis)
- checking the accurateness of the assembly process

This decomposition of the whole process allows an easy and effective estimation of the assembly process. It can also be used to calculate an upper limit of the assembly process.

3 Simulation Process

The simulation of the dynamic behaviour of a droplet on a structured substrate surrounded by a gaseous environment is done with multiphase models which take wetted walls into account. These models must have the capability to track the fluid-fluid interface and to describe changes in physical properties such as density, viscosity, contact angle etc. The non-conservative level set method [4] is used as the computation method for solving these problems. The level set method is based on continuum approach in order to represent surface tension and local curvature at the interface as a body force. This allows capturing any topological changes due to changes in surface tensions. In this method the interface between two phases is represented by a smooth function, called level set function $\Phi(r, t)$. The level set function is always positive in the continuous phase and is always negative in the dispersed phase. The interface is implicitly represented by the points where the level set function is zero. From such a representation of an interface we can calculate the motion of the free surface by advection of the level set function:

$$\frac{\partial \Phi}{\partial t} + u\nabla \Phi = \gamma \nabla \left[\epsilon \nabla \Phi - \Phi(1 - \Phi) \bullet n \right] \qquad (1)$$

The governing equations of motion for the incompressible isothermal flow can be written in terms of the Navier-Stokes equation which is the equation for the fluid velocity u and pressure p [12]. The Navier-Stokes equation describes the balance of force densities f_{type} acting on fluid elements.

$$\varrho \left[\frac{\partial u}{\partial t} + (u\nabla)u \right] = f_{press} + f_{frict} + f_{vol} \qquad (2)$$

The body force f_{vol} includes gravitational force and the surface tension term due to the level set treatment of interfacial stresses. The f_{vol} term is represented by the two components

$$f_{volx} = \sigma \cdot \kappa \cdot \frac{\partial \Phi}{\partial x} \cdot \delta(\Phi) \qquad f_{voly} = \sigma \cdot \kappa \cdot \frac{\partial \Phi}{\partial y} \cdot \delta(\Phi) + \rho \cdot g \qquad (3)$$

The surface tension term at the interface which is determined by the position of the zero level set is treated by the delta function $\delta(\Phi)$. The curvature κ of the fluidic interface is represented in terms of level set function

$$\kappa = \nabla n \qquad (4)$$

The unit vector n on the interface points from dispersed phase to continuous phase. In terms of the level set function the unit vector can be described as

$$n = \frac{\nabla \Phi}{|\nabla \Phi|} \qquad (5)$$

The change in physical properties is described by the Heavyside function which is represented in terms of level set function

$$H(\Phi < 0) = 0 \qquad H(\Phi = 0) = \frac{1}{2} \qquad H(\Phi > 0) = 1 \qquad (6)$$

Density and viscosity which are constant in each fluid are represented in terms of Heavyside function as

$$\rho = H(\Phi) + \frac{\rho_{discret}}{\rho_{continous}}(1 - H(\Phi)) \qquad \eta = H(\Phi) + \frac{\eta_{discret}}{\eta_{continous}}(1 - H(\Phi)) \quad (7)$$

The finite element method is used for modeling and simulation. The set of equations are solved by using "COMSOL Multiphysics" which allows also the coupling of different physical models [15].

4 Self assembly Process

For the analysis of the assembly process the model will be simplified. In the following the droplets for contacting the microchip will not be taken into account. So the first process step will be the placement of a droplet (diameter about 200 µm on the substrate. This droplet has the job of catching the microchip.

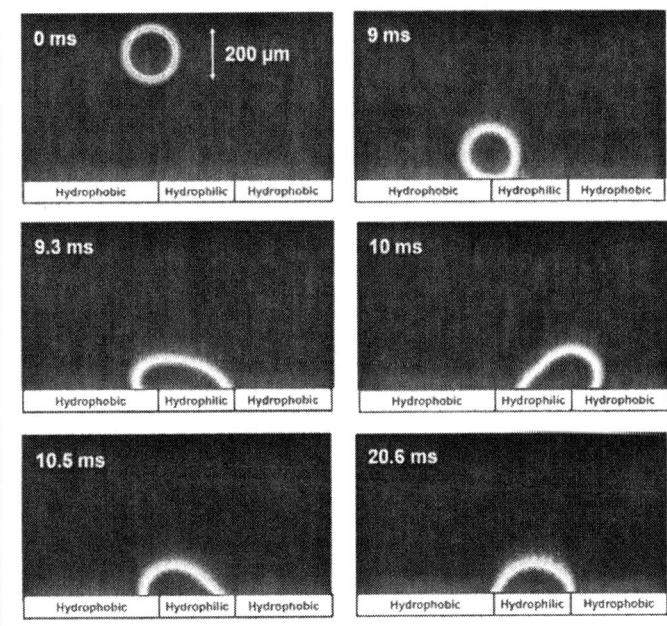

Fig.3. Simulation of the self alignment of a fluid

The droplet in Figure 3 will fall on the functionalised surface. In the example shown in Figure 3 the droplet hits the surface at the interface between the hydrophilic and hydrophobic area. This is to simulate the limited exactness by dispensing the droplets for catching the microchips. After the droplet hits the substrate, it moves in the direction of the hydrophilic surface (picture of the simulation at 9 ms in Figure 3). Reaching the right hydrophobic area, the droplet will overshoot the hydrophobic area because of its mass inertia (time = 10 ms). The acceleration of the droplet is again in direction of the hydrophilic array. The droplet will slow down and move back into the other direction. After some oscillations the droplet will reach a stable position. In about 20 ms the droplet is placed fully on the hydrophilic area and is ready for catching a microchip from a dispenser.

Beside the time needed to get a stable drop, it is important to know the size of the hydrophobic ring around the hydrophilic area to fix the droplet in the correct starting position (see figure 2). Some simple considerations with undamped oscillations show that the overshoot of the droplet will be in the same range as the deviation from the centre where the drop is dispensed.

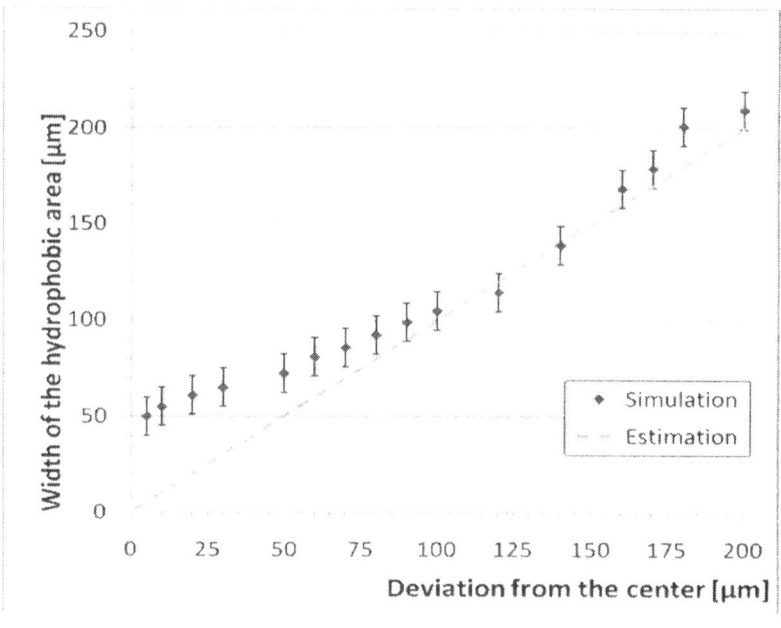

Fig. 4. Size of the hydrophobic barrier

In Figure 4 this estimation is presented as a dotted line. With a detailed and very time consuming simulation of a two phase fluidic system, this simple consideration is relatively good for a large deviation from the centre, but for small deviations the size of the hydrophobic ring will not become zero. The minimum size of this ring will be about 50 µm for the parameter used in the simulation (contact angle hydrophilic: 60°, contact angle hydrophobic: 120°, surface tension 72.5 mN/m, viscosity, ...). With an estimated accuracy of 50 µm by dispensing the droplet, the hydrophobic barrier must be about 80 µm (see figure 4).

After the droplet is placed on the substrate, the microchip is set on the top of the droplet.

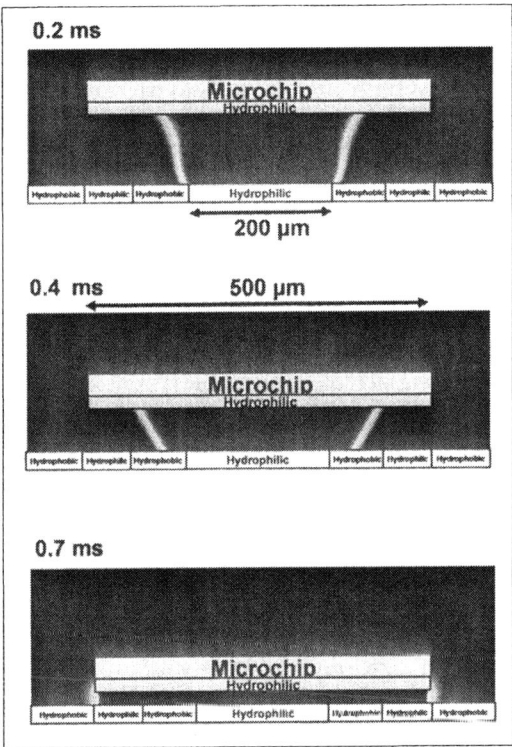

Fig. 5. Simulation of the self alignment capability of a microchip

As described before this phase of the assembly process is split into two parts. In the first part only the vertical lowering of the microchip is investigated. The horizontal alignment of the microchip is considered in the second part.

In the first part there is only a vertical and no horizontal mismatching of the chip. Figure 5 at t=0.2 ms shows the situation when the microchip reaches the top of the droplet placed on the substrate.

If the hydrophilic microchip (yellow) touches the fluidic interface of the droplet, the fluid creeps along the surface of the microchip. The high surface tension of the changed shape of the droplet overrides the hydrophobic area (white). A new fluidic interface between the chip and the substrate is set up (figure 5 after 0.7 ms).

In the second part the lateral (horizontal) alignment function is investigated. For this alignment the surface tension on the right and left side of the chip tries to minimize the surface energy. The fluidic segment shell allows an easy horizontal movement and the microchip will be automatically horizontally aligned on the correct final position for assembly.

An analytical estimation of the time for this horizontal alignment is

$$t \approx l \cdot \sqrt{\frac{\rho \cdot d}{\sigma} \cdot y} \qquad (8)$$

l is the edge length and d is the thickness of the microchip, y is the relative mismatching of positioning of the microchip.

Figure 6 shows the relation between the time needed for the horizontal adjustment and the deviation from the final position based on the above mentioned estimation for the alignment of a microchip.

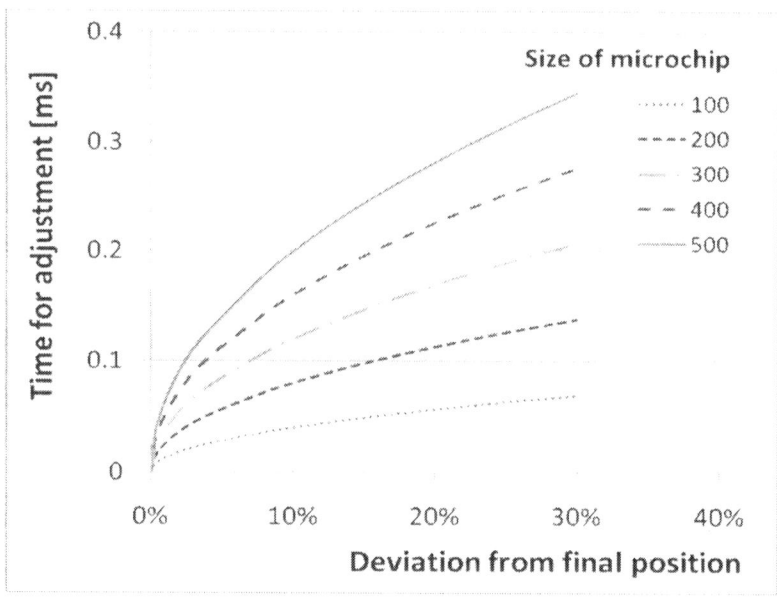

Fig.6. Alignment of the microchip

The analytical estimation in Figure 6 shows that a microchip with a size of 500 μm and a deviation of up to 30 percent from the final position needs less than 0.35 ms for horizontal adjustment. The catching of the microchip and filling of the area between microchip and substrate with the fluid is calculated by the simulation. The numerically calculated time for the vertical adjustment is about 0.7 ms (see figure 5).

The horizontal and vertical adjustments are taking place simultaneously. The total time for the alignment process will be less than one millisecond.

After the microchip has reached the final position, it can be fixed and contacted on the substrate.

5 Accuracy of Assembly

The whole assembly process is driven by the surface tension. In principle this surface tension builds up an interface with the lowest possible energy. To get the best results the hysteresis of the contact angle should be zero. The improvement of the hydrophilic and hydrophobic areas on the surface of the substrate needs a nanostructured interface. Unfortunately the nanostructuring leads to a stronger pinning effect and also to a bigger hysteresis γ of the contact angel. According to Figure 7 this pinning effect prevents a proper alignment of the microchip.

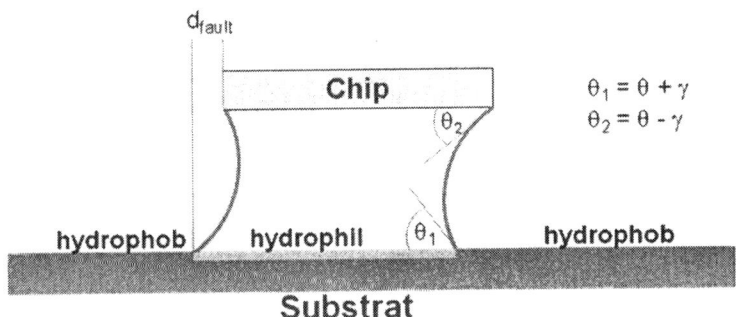

Fig. 7. Exactness of positioning (model)

Assuming that the vertical size (thickness) of the fluid between the substrate and the microchip will be very small, an approximation of the form of the interface by a sphere structure is appropriate enough. The biggest error of the alignment will come up if on the top (side of microchip) the contact angle θ will be reduced by the hysteresis angle γ and on the bottom (side of substrate) the contact angle θ will be increased by the hysteresis angle. In this case an approximation of the position fault d_{fault} depending of the drop size d_{drop} and the diameter of the microchips l_{chip} can be given

$$d_{fault} = \frac{1}{6} \cdot \pi \cdot \frac{d_{drop}^3}{l_{chip}^2} \cdot \tan \gamma \qquad (9)$$

The simulation (figure 5) shows a microchip with a diameter of about 500 μm which is caught by a droplet with a size of about 200 μm. The hysteresis γ of the contact angle is about 20°. In this case the error of positioning of the microchip is smaller than 10 μm.

6 Summary

This paper shows a new assembly method called **FluidAssem** for the assembly of small chips and other micro components. This new assembly process uses the downscaling effect where surface forces prevail over volume forces. The basis of this process is a special substrate with hydrophobic and hydrophilic structured areas. These structures allow the guidance of micro components via microfluidic forces to the exact assembly positions.

In the assembly process the micro components can be supplied to the substrate via a microfluidic dispensing process ([1], [8]). The fluidic system aligns the micro components on the correct positioning by the use of surface tensions. In this final position the micro component will be fixed and connected. The predicted exactness of the positioning of the microchips will fulfill the requested requirements (exacter than 10 μm) for the assembly process. The simulation shows that the time for placing a droplet on a substrate will be about 20 ms and the time for aligning the micro component will be about 1 ms. Compared to die-bonders used today, a higher throughput of assembled chips per hour can be expected.

With the resign of kinematic components the assembly process will have the capability to handle smallest microchips and micro components.

References

1. H. Sandmaier, R. Zengerle, B. de Heij, C. Steinert. A Tunable and Highly-Parallel Picoliter-Dispenser Based on Direct Liquid Displacement. In *MEMS 2002, Las Vegas, USA*, 2002.
2. J. Berthier and P. Silberzahn. *Microfluidics for Biotechnology*. Artech House, 2006.
3. J. Fang and K. F. Böhringer. High yield batch packaging of Micro Devices with uniquely orienting Self-Assembly. In *IEEE International Electronic Devices Meeting*, 2005.
4. M. Fietz. *Numerische Simulation von Oberflächeninstabilitäten in Zweiphasenströmungen mit Hilfe einer Level-Set Methode*. PhD thesis, RWTH Aachen, 2003.
5. T. Franke and A. Wixforth. Das Labor auf dem Chip. *Physik unserer Zeit*, 38:88-94, 2007.
6. G. Engelmann, H. Reichl, M. Hutter, H. Oppermann. High Precision Passive Alignment Flip Chip Assembly Using Self-alignment and Micro-mechanical Stops. In *Electronics Packaging technology Conference*, 2004.

7. F. Mugele. Liquids *in Contact with Solids: Nanotribology and Micro-fluidics*. PhD thesis, Universität Ulm – Abteilung Angewandte Physik, 2004.
8. C. P. Steinert, H. Sandmaier, S. Messner, B. de Heij, M. Daup, R. Zengerle, O. Gutmann, R. Niekrawietz. Droplet Release in a Highly Parallel Pressure Driven Nanoliter Dispenser. In *Transducers '03*, Boston, 2003
9. S. Pal, D. Roccatano, H. Weiss, H. Keller and F. Müller-Plathe. Molecular Dynamics Simulation of water near nanostructured hydrophobic surfaces: interfacial energies. *ChemPhysChem*, 6:1641-1649, 2005
10. J. S. Smith. High density, low parasitic direct integration by Fluidic Self Assembly (FSA). In *IEEE International Electronic Devices Meeting*, 2000.
11. A. Torkkeli. *Droplet microfluidic on a planar surface*. PhD thesis, VTT Publications 504, 2003.
12. E. Truckenbrodt. *Fluidmechanik, Band 1, Grundlagen und elementare Strömungsvorgänge dichtebeständiger Fluide, 4. Auflage*. Springer-Verlag, Berlin, 1996.
13. F. Exl und J. Kindersberger. Messung von Tropfenrandwinkeln auf Isolierstoffoberflächen. *In ETG-Fachbericht 97, VDE-Verlag GmbH Berlin Offenbach, S. 67-72*, 2004.
14. L. Zhu, Y. Feng, X. Ye and Z. Zhou. Tuning wettability and getting superhydrophobic surfaces by controlling surface roughness with well-designed microstructures. *Sensors and Actuators*, A 130-131:595-600, 2006.
15. W. B. J. Zimmerman. *Multiphysics Modelling with finite Element Methods*. World Scientific Publishing Co.Pte.Ltd., 2006.

CONCEPTS FOR HYBRID MICRO ASSEMBLY USING HOT MELT JOINING

Sven Rathmann, Annika Raatz, Jürgen Hesselbach

Institute of Machine Tools and Production Engineering
Technical University Braunschweig
Langer Kamp 19b, 38106 Braunschweig, Germany

Abstract Nowadays, the production of 3D MEMS and MOEMS is carried out by using hybrid integration of single components, for which batch production is normally preferred. In this field, adhesive technology is one of the major joining techniques. At the Collaboration Research Center 516, a batch process based on a joining technique which uses hot melt adhesives was developed. This technique allows the coating of micro components with hot melt in a batch. The coating process is followed by the joining process. Due to this, the time between coating and joining can be designed variably. Because of the short set times of hot melt adhesives, short joining times are possible. For this assembly process adapted heat management is necessary. This paper presents adapted heating management concepts and gripping systems which allow a fast and accurate assembly of hybrid micro systems with hot melt coated components. Therefore, the chosen gripping system depends on the process and heat management concept as well as the thermal properties of the components. Furthermore, the simulative and experimental results of the heat management concepts will be discussed.

Keywords precision joining operations, hot melt adhesives, heat management concept

1 Introduction

The continuous miniaturisation of hybrid micro systems requires joining processes which are suitable for batch production [1]. The most important technology is the adhesive joining technique [2]. Mostly, viscous adhesives are used. Disadvantages of viscous adhesives are long set times and the low suitability for batch processes. A new approach is the usage of hot melt adhesives. The main advantages are extremely short set cycles, the possibility of pre-applying the adhesive and the time-delayed joining procedure [3]. Therefore, the use of hot melt adhesives can be an interesting alternative for the assembly of hybrid micro systems. To establish this technology, adhesive and application technologies as well as the assembly process must be developed, and the process parameters must be determined. A very important aspect for the process design is the kind of heating technology. Therefore, special process components, such as fixtures and grippers as well as the heat management concepts should be developed. One project at the Collaborative Research Centre 516 "Design and Manufacturing of Active Microsystems" is concerned with the development and modification of hot melt adhesives and coating concepts and also with the development of the special assembly process with hot melt coated micro components.

2 Joining Technology

The listed limitations of viscous adhesives in the micro system engineering are to be compensated by using hot melts. These hot melts are thermoplastics, physical setting adhesives, which are single-component, non-viscous and non-solvent at room temperature. One of the most important advantages of hot melts in comparison to viscous adhesive systems is the possibility of pre-applying hot melt systems in a batch process, e.g. as powder or adhesive spheres, as dispersion or as an adhesive foil. The joining procedure does not have to take place directly after the adhesive has been applied to the substrate; this can happen at any time later on, i.e. hot melts possess no pot life time [4]. The adhesive is only melted during the bonding process by a thermal impulse and moistens the surface of the other substrate. The heating can be accomplished directly by heating the adhesive itself or indirectly by heating the substrate. The adhesive sets, once the temperature of the adhesive has fallen below the melting temperature. If an appropriate heat management concept is used, the hot melts set very fast, which means that the handling strength (usually the ultimate strength) can be achieved in less than one second as experiments have shown.

3 Assembly Process

This section discusses the design of an assembly process which makes use of a joining technique based on the application of hot melt adhesives described above. The design of the joining process is the main focus when developing the assembly process [5]. However, in order to put the concept into practice, further requirements have to be imposed with regard to the design of the process. First, the clamping and gripping units should not come into contact with the adhesive covered surfaces of the components, once the hot melt has been melted. Second, the coating should not distort or cover any marks and parts of the component needed for the assembly. Third, possible ways of how the necessary heat can be transmitted into the components must be considered, which is the most important of all requirements. The different heat management concepts which are usable will be discussed on the basis of a specific micro system, which is presented below. Figure 1 shows the assembly scenario of a linear micro stepping motor. This motor consists of a stator, two guide blocks, a traveller, a traveller block as well as guiding balls. Moreover, a distance foil is needed for the assembly.

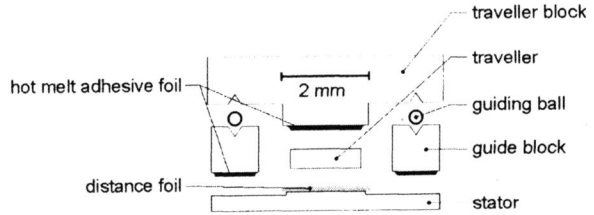

Fig. 1. Example product for micro assembly using hot melt adhesives

The assembly steps are: joining the guide blocks, inserting the distance foil, placing the traveller, inserting the guiding balls and joining the traveller block with the traveller. In a pre-assembly step, a thin film of hot melt adhesive (about 50 μm) is applied onto the guide blocks and the traveller block. The melting temperature of the hot melt ranges from 100 °C to 110 °C. In the following, the assembly process will be discussed on the basis of the joining procedure of a guide block, which consists of six steps:

1. measuring the position of the stator
2. measuring the position of the guide block
3. gripping the guide block
4. heating up the hot melt coated guide block
5. aligning the guide block with the joining position
6. joining the guide block with the stator

The heating of the hot melt adhesive can be carried out in various ways. Not only does the kind of heat source play an important role for the process design, but also the way the heat should be transmitted into the component as well as the starting time of the heat input must be taken into account. These factors exert influence also on the load of the components during the heating phase and the assembly uncertainty. In the next two sections, the different heat management concepts and experimental results will be presented and their advantages and disadvantages analyzed.

4 Adaptive Heat Management Concepts

Heat management is an integral part for the selection and the design of an assembly process using hot melt adhesives. Since the volume of the hot melt is quite small, the thermal capacity is rather low, which is why heating the hot melt itself is not very practical. Contrarily, the thermal capacity of the component and the gripping system is much higher. Hence, the characteristics of the grippers and the components such as their volume, their thermal capacity as well as their thermal conductivity must be taken into consideration when designing the assembly process. Besides, the heating source is another crucial factor for the process design. The following heat sources can be used to heat up the gripping unit: heating plates, infrared heaters, lasers, heating foils and Peltier elements as well as a combination of these heat sources. In general, there are two different kinds of heat management concepts – a passive and an active one. The passive heat management concept makes use of the principle of heat storage to supply the energy for the joining process.

The solid line in Figure 2 shows the typical temperature history of a component according to the passive heat management concept. Before the actual joining process, the gripper and the component inside it are heated up by a heat source until the working temperature T_{Hp} has been reached. The time needed until T_{Hp} has been reached determines the process time to a large extent. If neither handling nor measuring operations have to take place right after the gripper has picked up the component, the gripper should be heated up before the gripping process. Besides, the working temperature T_V must be much higher than the melting temperature of the adhesive to provide enough time to position and join the components until the temperature falls below the working temperature, which is marked by t_{Kp}. The time span between t_{Hp} and t_{Kp} is referred to as the processing time. Contrary to the setting, the process time marks the time span during which the position of the components can still be measured and adjusted. The processing time depends mostly on the material of the components and the gripper. Once the joining component comes into contact with its joining partner, the temperature drops substantially, whereupon the hot melt sets and reaches its final strength. In general, the assembly time is less than a second.

As Figure 2 shows, the working temperature is much higher than the melting temperature, as a result of which the components are exposed to very high thermal load. Under certain circumstances, this load will lead to a deterioration of the accuracy of the assembly process. Nonetheless, the fact that the passive heat management concept can be easily integrated into already existing assembly systems is a major advantage of this concept. Disadvantages are the long heating times, the inflexibility of the process design, the high thermal load of the components as well as the immense efforts to monitor the assembly process. In particular, the controlling process of the temperature variation is rather complex.

In contrast to the passive heat management concept, the heat source of the active heat management concept is integrated into the gripping or clamping unit. Therefore, heat can be continuously transmitted into the component during the handling process. As another result, the temperature- and material-dependent processing time of the passive heat management concept can be omitted, which is why the processing temperature can be set at a lower level. In most cases the processing temperature can be set right above the melting point. This results in a lower thermal load of the components. Due to the continuous heat input, the processing time of the active heat management concept can be variably chosen. Thus, the assembly process can be designed very flexibly. By actively cooling the components, the joining time can be further shortened. The typical temperature profile of the active heat management concept is illustrated by the dashed line in Figure 2.

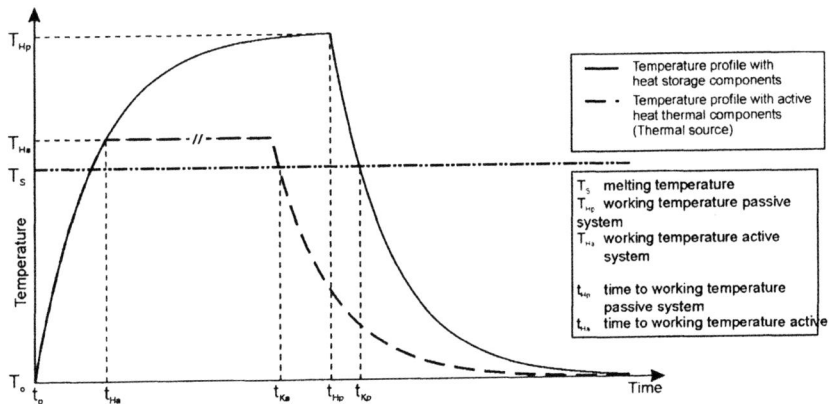

Fig. 2. Temperature profiles on different heat managements

The advantage of the active heat management concept is the independence of the assembly process from the heat capacity of components which leads to a flexible process design. The complex design and the integration of active heating components is a disadvantage of the active management concept, mainly, if small and accurate assembly systems are used. In contrast, the integration of heat storage components is easier to handle. However, the control and monitoring of the passive concept, especially the control and monitoring of the temperature, are considerably more difficult than with the active heat management concept. The described heat management concepts allow a variable design of the assembly processes. However, not only the active or the passive heat management concept can be used, but also a combination of both is possible.

5 Experimental Setup

For the experimental validation, a micro assembly system was used. Figure 3 shows the micro assembly robot AUTOPLACE 411 with different fixture and heating systems. The robot has 4 DOF. The uncertainty of the robot system is 1 µm in the x-, y-, and z- direction. An adapted gripper system is mounted onto the robot system. For the several heating concepts, a precision heating plate and an infra red heater are used. The IR heater has a surface power of 64 kW/m^2 with a typical working temperature of 860 °C.

To investigate the passive and the active heat management concepts a passive and an active gripper system were designed. Figure 4 shows the passive gripper concept for the assembly of the guide blocks in Figure 1. It consists of a base plate, isolation ceramic and a vacuum gripper with an integrated heat accumulator. The material of the heat accumulator is copper. The heating of the guide block and the gripper is done by an infra red heater. The emitted wave length ranges from 2 to 10 µm. Due to the high reflection and transmission index, the heat input in the guide block into makes up only 6 % of the emitted heat radiation [6]. Hence, the guide block is heated by conduction between the gripper system and the guide block. The heat that is needed for this process is brought into the gripping system through heat radiation.

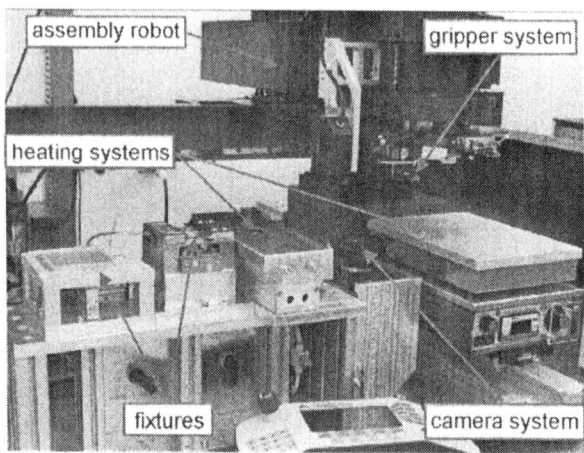

Fig. 3. Assembly robot AUTOPLACE 411 (SYSMELEC)

Figure 5 shows an active gripper concept. The heat source of the gripper is a peltier element. The gripper consists of a heat sink, isolation between heat sink and the heat flux elements and the peltier element in the middle of the active gripper. The gripping concept has to guarantee a good contact surface for the heat flux from the gripper to the guide block.

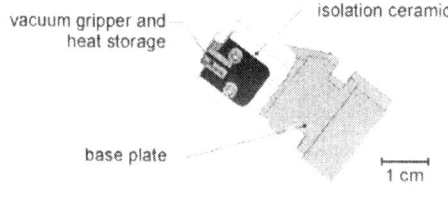 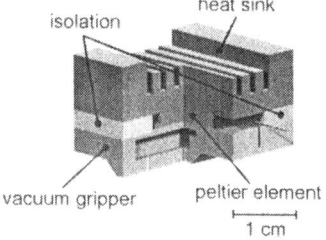

Fig. 4. Passive gripper concept   *Fig. 5.* Active gripper concept

6 Simulation and Experimental Results

As a first step the process parameters have to be defined. Parameters of interest are: the warming time, the cooling time, the components' temperature and the heat flux to the ambiance. A simulation model was designed to determine these parameters. Figure 6 shows the passive gripper simulation model and the result at the time of 50 s of a transient temperature simulation. The gripper system is heated with an infra red heater. The infra red radiation is simulated by a heat flux to the gripper system. For the calculation of the heat flux an effective area of 144 mm² was assumed. Based on the heater properties, the heat flux to the gripper system, which affects to the bottom of it, was estimated at 9 W.

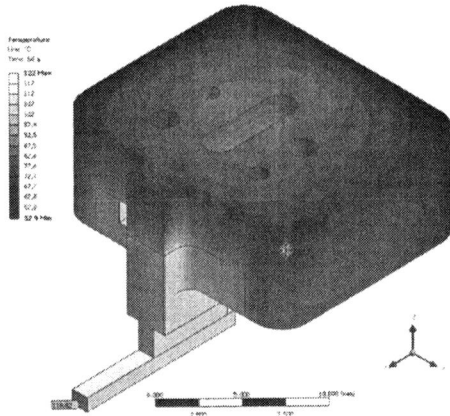

Fig. 6. Simulation model and results

The transient simulation shows the heat distribution in the gripper and the picked guiding block during and after the warming phase of 50 s (Figure 7). The maximum gripper temperature is 122 °C. The hot melt coated area of the guide block, the bottom side, has a mean temperature T_{Hp} of 121 °C at this time.

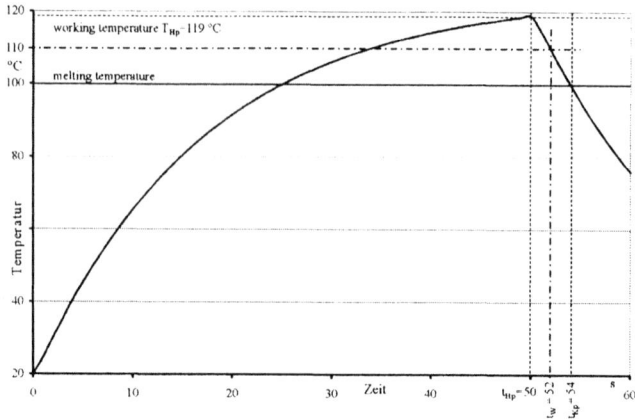

Fig. 7. Temperature profile of the hot melt coated side of the guide block

For the movement to the assembly position a maximum time of 2 s is scheduled. After this time the mean temperature on the coated side of the guide block is 110 °C, which is adequate for the hot melt joining process. The critical time t_{Kp}, i.e. the time that elapses until the temperature has fallen right below the melting point, is reached 4 s after the warming phase. To control and validate the simulation parameters and results, the passive heat management concept was implemented for the described micro assembly system. As the temperature is difficult to measure, only samples of the gripper temperature can be taken. But the results from the simulation are in accordance with the results of the experiments that were carried out.

7 Conclusion and Outlook

In this paper an alternative joining process for hybrid micro systems based on hot melt adhesives was presented. In particular, the process design of the joining process in combination with different heat management concepts was described. In this context, an active and a passive heat management concept were discussed. The objective of this work is to determine the expedient parameters to design the assembly process. As a first step, the simulation results and the experimental validation show that the passive heat management concept is suitable for joining with hot melt adhesives. The disadvantage of the passive management concept, i.e. the long heating time, can be reduced by using the active concept. The first steps for an active heat management concept have been done by developing an active gripper system. Next steps will be the integration of the assembly system and experimental investigations. In the active management concept has to be investigated. In addition, the combination of the active and passive heat management concept seems to be expedient for a joining technology in micro assembly using hot melt adhesives.

Acknowledgments

The authors gratefully acknowledge the funding of the reported work by the German Research Centre (Collaborative Research Centre 516 "Design and Manufacturing of Active Micro Systems")

References

1. H. Van Brussel, et. Al., Assembly of micro systems. In: Annals of the CIRP, 49(2), pp. 451-472 (2000).
2. M.F. Zäh, M. Schilp and D. Jacob, Kapsel und Tropfen - Fluidauftrag für Mikrosysteme. Evolutionäre und revolutionäre Verfahren in der Dispenstechnik. In: Wt-Werkstattstechnik online 92(9), pp. 428-431 (2002).
3. S. Böhm, et Al., Micro Bonding with non-viscous adhesives. In: Microsystem Technologies 12(7), pp. 676-679 (2006).
4. S. Böhm, et Al., Micro Bonding using hot melt adhesives. Journal of Adhesion and Interface. The Society of Adhesion & Interface 7(4), pp. 28-31 (2006).
5. B. Lotter and H.-P.Wiendahl, (editors), Montage in der indusriellen Produktion - Ein Handbuch für die Praxis. Springer Verlag Berlin Heidelberg, ISBN 3-540-21413-5 (2006).
6. R. Hull, Properties of Crystalline Silicon. Emis DataReviews Series No. 20, INSPEC, IEE, London, UK (1999).

APPLICATION OF MICROSTEREOLITHOGRAPHY TECHNOLOGY IN MICROMANUFACTURING

Hongyi Yang, Gregory Tsiklos, Ronaldo Ronaldo, Svetan Ratchev
School of Mechanical, Materials and Manufacturing Engineering,
University of Nottingham University Park Nottingham, NG7 2RD

Abstract Stereolithography (SL) technology is the mostly widely used Rapid Prototyping (RP) process in which complex structures are fabricated in a layer-by-layer fashion. Currently Microstereolithography (MSL) is undergoing fast development. In this paper, the development of MSL and its application in micromanufacturing are reviewed. One promising direction for the development of MSL is the assembly-free process, through which a complex structure can be manufactured in a single process thus avoids the extra assembly procedure. With the assembly-free process, the manufacturing cost can be greatly reduced and the manufacturing reliability dramatically improved. The assembly-free process may play an important role in the future fabrication practice.

Key words microstereolithography, micromanufacturing, assembly free process

1 Introduction

The sterolithgraphy (SL) process is the first rapid prototyping technique which was patented in 1984. Currently it is widely used in various industrial and technological fields that require the direct manufacturing of three dimensional prototype parts [1, 2]. In SL process, small-size, high-resolution three-dimensional objects are built by superimposing a certain number of layers obtained by a light-induced and space-resolved polymerization of a liquid resin into a solid polymer. The laser beam is focused and polymerisation occurs locally, which allows creating the shape of one layer of the object. When a layer is finished, fresh resin is spread on top of the already manufactured part of the object, and the light-induced solidification of the next layer is started. SL technique is regarded as a milestone in the manufacturing practice, because the layer-by-layer building process enables applying adding material process as compared to the removing of materials used by traditional techniques.

Microstereolithography (MSL) is the general designation of various microfabrication technologies with high-resolution based on the principle used in SL, which was proposed by Ikuta et al. in 1993[3]. Ikuta et al. also named MSL as

Micro-Photoforming, IH process, Spatial Forming, 3D Optical Modelling etc., depending on the different designs of the building apparatuses used in the manufacturing process. Compared with SL, MSL has a resolution of submicron scale for the x-y-z translational stages, also the UV laser spot in MSL is finely focused to submicron scale, and thus MSL allows for polymerizations of layers with 1-10um thickness [4].

Modern micro-manufacturing industries are facing increasing requirements of miniaturization, customisation and integration in the fabrication of micro devices such as Micro-electro-mechanical systems, microfluidic systems. The MSL technology provides an ideal solution to such challenges due to its unique characters of high resolution, high liability and lower cost. Currently MSL is being exploited in different fields to cater for various new applications.

2 State of the art

The MSL system mainly consists of four parts: a laser source, a beam delivery system, a CAD design tool and computer-controlled precision x-y-z stages, and a vat containing UV curable resin. After designing the 3D solid model with CAD software, the 3D model is sliced into a series of 2D layers with uniform thickness. The NC codes generated from each sliced 2D file are then executed to control the UV-beam scanning.

Based on the different beam delivery system used, two major beam scanning techniques have been developed: scanning MSL (or vector-by-vector MSL) and projection MSL (or integral MSL). Figure 1 schematically shows these two kinds of systems.

The Scanning MSL machines build the solid micro parts in a point-by-point and line-by-line fashion. A classical MSL apparatus is shown in Figure 1 (a), in which the laser beam is focused by a dynamic lens and deflected by two low-inertia-galvanometric on the surface of a resin system containing UV photoinitator, monomer, and other additives. An acousticoptical shutter cuts out the laser beam when a switch between the last polymerised segment and the next segment to be polymerised must be made without polymerization. To get a better resolution, the beam is focused more precisely in order to reduce the spot size to a few micrometers in diameter. Additional technological developments in the designed SL machines may be required. Constrained surface technique, Free surface technique are normally used in the commercial MSL systems. A main problem of the Conventional MSL is the limitations in terms of the minimum thickness of the resin layers during the layer preparation because of viscosity and surface tension. This problem was solved by two-photon MSL process recently, with this process the resin does not need to be layered. With two-photon MSL process, micro and even nano structure can be fabricated precisely [5].

Projection MSL builds 3D microstructure in a layer-by-layer fashion. Each layer is built by exposure through mask once, which significantly saves the time compared to the scanning MSL. A dynamic pattern generator is used to generate pattern projection for exposure curing (Figure 1 (b). When a beam passes through the projector, it contains the pattern of the layer and focused on the resin surface for the polymerization of the exposed areas. The time required to build a 3D structure only depends on the number of layers of the structure. Nowadays, two kinds of computer driven dynamic pattern generators, liquid crystal display (LCD) and Digital Light Processing (DLP) are used. For the LCD projector, an addressed LCD light valves array (made of liquid-crystal material) is used to control the on/off of the light, acting as a projector. The LCD technique has some intrinsic drawbacks, including large pixel sizes, low filling ratio, low switching speed (~20ms), low optical density of the refractive elements during the OFF mode and the higher light absorption during the ON mode [6]. Such drawbacks limited further improvement of the LCD projection MSL system. In DLP projectors, the image is created by microscopically small mirrors laid out in a matrix on a semiconductor chip, known as a Digital Micromirror Device (DMD), which is a competing technology in the field of digital display. Each mirror represents one pixel in the projected image. The number of mirrors corresponds to the resolution of the projected image. High resolution DMD like 2800x2100 has been used in MSL. These mirrors can be repositioned rapidly, which is essential for switching between 'on' and 'off'. Using DMD, Sun etc. fabricated complex 3D microstructures with the smallest feature of 0.6μm [6]. Because of its capability in batch fabrication of 3D Microsystems, the dynamic projection MSL is a promising technology in micro manufacturing.

Currently efforts are also being made to incorporate a broad spectrum of materials into MSL fabrication for the manufacturing of micro-electro-mechanical systems with special functions. Metal, ceramic as well as polymer micro parts have been fabricated [7, 8]. Similar to the SL of ceramic and metals, micro or even nanoceramic and metallic powders are mixed with photocurable resin for the MSL of ceramics and metals. Researches on functional photocurable polymers such as biodegradable polymers also have found applications in Tissue Engineering, drug delivery systems and so on [9,10].

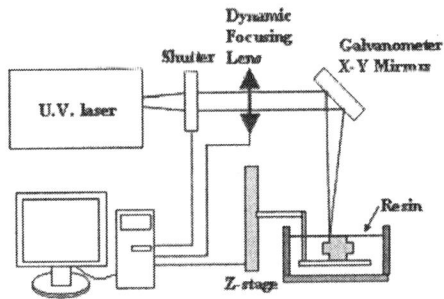

Fig. 1(a). Schematic view of MSL system: Scanning MSL

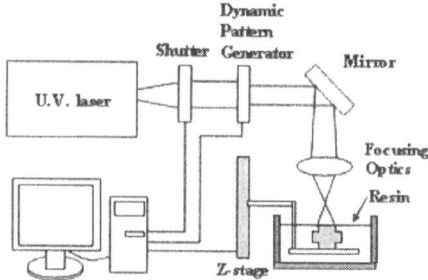

Fig. 1(b). Schematic view of MSL system: Projection MSL

3 Case study

In this study, the samples are fabricated by Perfactory® MSL system of EnvisionTec, Germany. This projection MSL system has a DLP projector and has a high resolution of 7.5μm and visible light ($\lambda = \sim 475$nm) was used as energy source. Figure 2 shows some micro structures fabricated with this system. From these pictures, we can see that with this technique complex structure with fine figures can be realized. Porous scaffolds (Fig 2a, Fig 2e and 2f) and designed micro structures (Fig 2b and Fig 2d) can be used for tissue engineering scaffold, drug delivery system, micro fluidic and other micro scale applications.

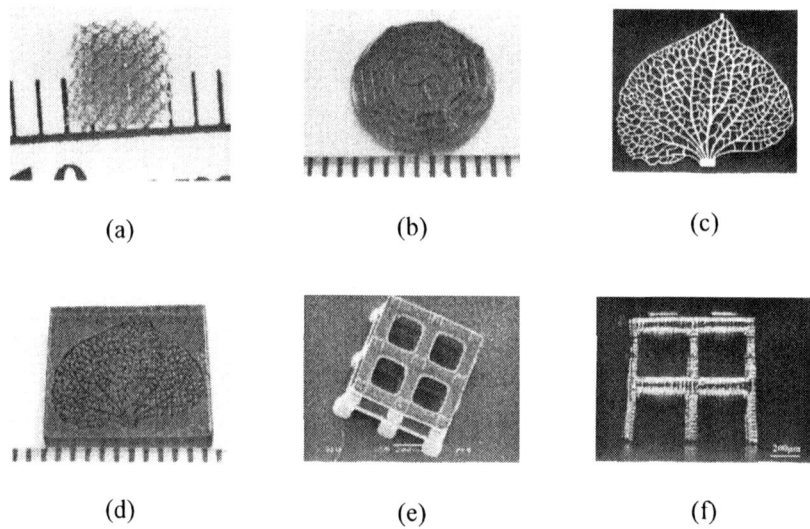

Fig. 2. Microstructures made by Perfactory® MSL

Microsystems are usually complex, and composed of a number of unit devices to be fabricated separately and then assembled together. However, with the miniaturisation of the ☐icrosystems, the assembly process becomes more and more difficult, time-consuming, and expensive. It is estimated that assembling process takes 80% of the whole manufacturing costs in the Microsystems. Thus the assembly-free design represents a new and promising direction in the manufacturing technology.

With MSL it is possible to fabricate complex free-forming three-dimensional microstructures in a single process, and the assembly process can be avoided. Typical example of assembly-free manufacturing has been published [11], such as the micro-fluidic systems which have seen wide applications in the area of microchemistry and microbiology. Generally, micro-fluidic systems have several components, such as the micro-channels, micro-pumps and micro-mixers. The traditional technologies used for micro-fluidic systems manufacturing are lithography, etching and micromachining [12]. Assembly process is needed to fabricate a closed system because of the micro fluidic structure can be only made on a 2D layer. Using MSL, 3D microstructure can be manufactured directly. Fig 3 shows a multi-analysis micro fluidic system with a size of 2x10x14mm and 300μm fluidic channel inside. For each analysis part, there are two inlets, mixing channel, testing chamber and one outlet. The total manufacturing process for this structure was only taken less than 2 hours. The whole process was greatly simplified by reducing assembly process and other assistant processes compared with other traditional methods.

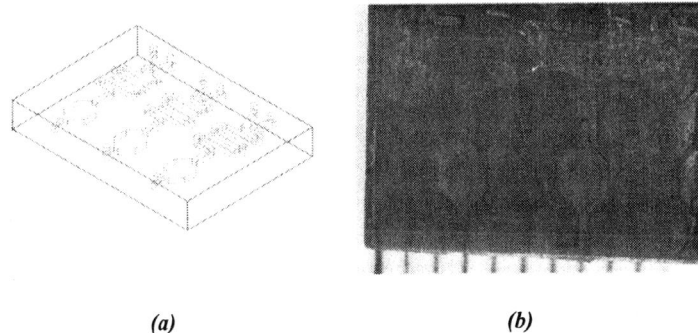

Fig. 3. Micro fluidic device made by Perfactory® MSL: *(a)* designed structure and *(b)* Microscope photo.

4 Conclusion and future works

With MSL, complex micro structure can be realised with high resolution. Assembly-free process of micro systems can also be realized with the improvement of MSL, which is promising because of the reducing of microassembly cost. Especially, with the development of the projection MSL system with DLP projector, mass production can be realised.

Current commercial SL or MSL techniques still have many technical limitations, such as low reliability and limited applicability of materials. Currently researches are working on several key issues to improve the MSL technology. The first is to improve the resolution of the fabrication process. The resolution is affected by many factors including machine, software, process, and resin related factors. The second is to search for new photo curable materials. Integration with other microfabrication technologies is also a promising direction for assembly-free process of multimaterials.

Acknowledgements

The authors are grateful to the Centre of Excellence in Customised Assembly (CECA) for supporting this work.

References

[1] C. Hull, Method for production of three-dimensional objects by stereolithography, US patent No. 4575330, 1984.

[2] J.C. Andre, M.A. Le, and O. de Wittee, French Patent, 8411241, 1984.

[3] K. Ikyta and K. Hirowatari, Real three dimensional microfabrication using stereo lithography and metal molding, Proc. IEEE MEMS, 1993, p 42-47.

[4] V. K. Varadan, X. N. Jiang, V. V. Varadan, Microstereolithography and other fabrication techniques for 3D MEMS, Wiley, 2001, p105.

[5] S. Wu, J. Serbin, M. Gu, Two-photon polymerisation for three-dimensional microfabrication, Journal of Photochemistry and Photobiology A: Chemistry, 191, 1-11, (2006).

[6] C. Sun, N. Fang, D. Wu, X. Zhang, Projection micro-stereolithography using digtal micro-mirror dynamic mask. Sensors and Actuators A121,113-120, (2005).

[7] A. Bertsch, S. Jiguet and P. Renaud, Microfabrication of ceramic components by microstereolithography, J. Micromech. Microeng. 14, 197-203, (2004).

[8] K.F. O Connor, D.C. Nohns, W.A. Chattin, Method of combining metal and ceramic inserts into stereolithography components, United States Patent 5705117, 1998

[9] K.S. Anseth, A.T. Metters, S.J. Bryant, P.J. Martens, J.H. Elisseeff, C.N. Bowman, In situ forming degradable networks and their application in tissue engineering and drug delivery. Journal of Controlled Release, 78, p199-209, (2002).

[10] J.W. Lee, B. Kim, G. Lim and D.W. Cho, Scaffold Fabrication with Biodegradable Poly(propylene fumarate) Using Microstereolithography. Key Engineering Materials, 342-343, p141-144, (2007).

[11] H.W. Kang, I.H. Lee, D.W. Cho, Development of an assembly-free process based on virtual environment for fabricating 3D microfluidic systems using microstereolithography technology. Journal of Manufacturing Science and Engineering, 126, p766-771, (2004).

[12] P. Abgrall and A.M. Gue, Lab-on-chip technologies: making a microfluidic network and coupling it into a complete microsystem- a review, J. Micromech, Microeng, 17, R15-R49, (2007).

IN SITU MICROASSEMBLY

Ronaldo Ronaldo, Thomas Papastathis, Hongyi Yang, Carsten Tietje, Michele Turitto, Svetan Ratchev

The University of Nottingham, Precision Manufacturing Centre,
School of Mechanical, Materials and Manufacturing Engineering,
Nottingham, NG7 2RD, UK

Abstract The 21st century sees significant breakthroughs in fabricating micro devices in the quest of miniaturising. Most micro parts have been manufactured in the range of less than 1mm. However, they are built based on material that is process dependant, resulting in monolithic parts. For example the Integrated Circuits, Micro Electro Mechanical System (MEMS) are silicon based, and on their own do not constitute a complex system that requires various functions. To pursue fully functional and miniaturised complex devices, microassembly is therefore necessary. However, microassembly processes differ from the assembly processes in the macro world. Microassembly encounters sticking effects in parts handling, adhesive forces from electrostatic attraction, van der Walls forces and surface tension [1, 2]. This paper envisions microassembly processes by using an innovative approach. It departs from the traditional assembly process by utilising the Projection Micro Stero Lithography, with a positioning algorithm to assemble micro parts without traditional handling and joining, named in situ microassembly process.

Keywords microassembly, in situ microassembly, projection micro stereolithography, rapid prototyping.

1 Problems in Microassembly

Micro products are defined as products whose functional features or at least one dimension are in the order of micrometers. The products are usually characterised by a high degree of integration of functionalities and components [3].

The continuous miniaturisation of parts has led to difficulties in handling, transporting, gripping and positioning whilst assembling them. Furthermore, these difficulties have naturally led to optimisation in vision, precision positioning and miniaturisation of assembling tools. Vision systems help the assembly process, however the limitations lie on the short depth of focus which creates difficulties during the manipulation process [2]. For the assembly of micro components, this restriction is a major barrier especially when two micro parts are required to be monitored during the whole process.

Miniaturisation of assembly tools, such as micro grippers, has adverse effects associated with contact to micro parts, such as the electrostatic attraction, and van der Walls forces, and contamination. These effects have been avoided by a number of ways of non contact handling utilising magnetic, electric, optical, aerodynamic, ultrasonic methods, as reviewed by Vandaele (2005) [4], or by pneumatics as used for a contactless feeder reported in [5].

Böhringer et al (1998) having studied extensively the difficulties of assembling micro parts serially, envisioned the idea for parallel microassembly and introduced the self assembly concept based on electrostatic force fields as a solution [1]. Despite many successful publications on self-assembly, the assembly process of single parts providing true three dimensional structures should not be ignored. Thus, a direct control and in situ microassembly with the utilisation of photo-projection micro stereolithography is one of the solutions for non self-assembled material. The system can be extended further with contactless feeders and handling systems, to form a holistic approach encompassing the whole production cycle.

2 In situ assembly methodology

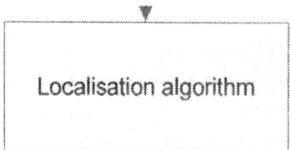

The part to be assembled can be transported with an ordinary vibratory feeder, ultrasonic plate feeder, or contactless feeder. The part is placed with reasonable positioning accuracy on the working table.

A small amount of liquid photopolymer is then added to surround the part. The amount of the liquid polymer must be at least sufficient for a layer of polymerisation. The part must be on its equilibrium state while resting on the working table.

The vision system takes measurements of some points on the parts. These points serve as datum points that will be used later for the localisation algorithm.

Once the positioning of the part is detected, the algorithm, implemented in software, calculates the Euclidean displacement of the part in comparison to the part of the model file. Euclidean displacement and transformation are explained in detail in the workpiece localisation algorithm section.

As long as the part is located within reasonable angle, the Computer Aided Design (CAD) or Stereolithography (STL) model can be translated and rotated as much as the Euclidean displacement to match the placement orientation of the part. The new CAD or STL model orientation and location must be within the built area of the polymerisation process, so the whole product can be built successfully.

Finally, the polymer around the part is solidified by photo projection. For further parts to be assembled, a similar approach can be taken. If it is not possible, then the outline around the part can be solidified to provide the exact imprint for the part to be placed in the correct orientation and place. Once it is placed, the solidifying process can take place layer by layer by the Projection Micro Stereo Lithography process until the whole product is built. This process is described in the following section.

2.1 Projection Micro Stereo Lithography

Projection Micro Stero Lithography (PµSL) – utilises a UV light source and a digital micro mirror display to project and solidify a layer at once. 3D microstructures with 120nm resolutions are reported in [6].

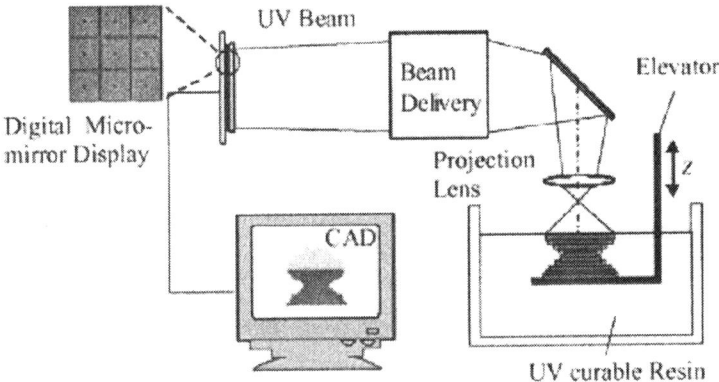

Fig. 1. Schematic diagram of PµSL using digital micro mirror display [7]

Fig. 1 shows the principle of of PµSL in a schematic way. The system contains a state of the art product Digital Light Processor (DLP®) from Texas Instruments. The Digital Light Processor (DLP®) contains a 1.3 million Digital Micromirror Device (DMD) and can reflect 1024 shades of gray to create a grayscale image. When the micromirror is tilted towards the photopolymer liquid, it projects a photon that is absorbed by the photo-initiator and generates radicals. These radicals then react with the rest of the monomers until a stable polymer chain is formed. These chains lead to the solidification of the polymer [7].

The advantage of this concept in comparison to the more commonly known stereolithography process is that in PµSL the polymerisation across the whole surface of the layer is done at once due to the planar projection. Whereas the ordinary stereolithography has a laser beam that traces a cross section pattern on the surface of the photopolymer. Therefore PµSL is faster and for the case of in situ microassembly process, PµSL is superior due to the capability to polymerise the surrounding area of the part at once, and hence a more accurate geometry can be achieved.

180   *In Situ Micro-Assembly*

As with any other Rapid Prototyping System, PμSL has no measurement and feedback to control the process.

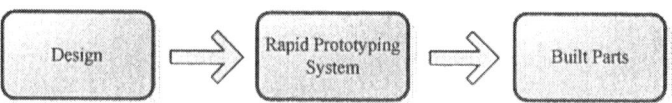

Fig. 2. Ordinary Rapid Prototyping System

2.2 Measurement system

In in situ microassembly, a measurement system is used to define the position of the parts to be assembled on the working table. A non contact measurement technique is preferred as it does not exert any force onto the object, and hence the part will always stay in its equilibrium state. Non contact measurement techniques such as laser interferometry, ultrasound or optical measurement can be utilised. Fraunhofer IPT Germany reported utilising the Digital Micromirror Device for lateral scanning with resolution of 1μm in 2001[8]. The schematic diagram below is proprietary of GFMesstechnik (patent no. EP0943950A1).

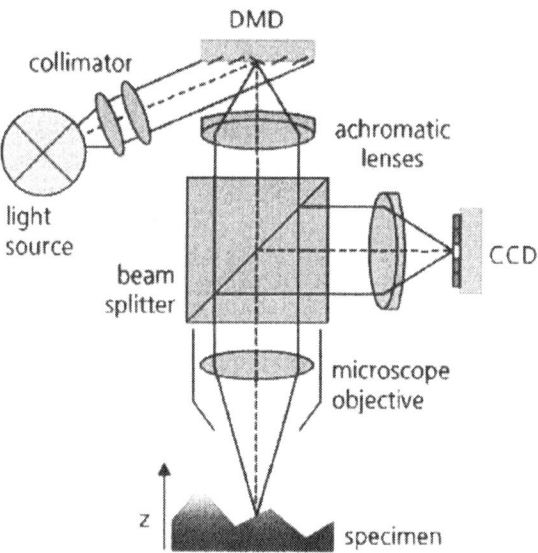

Fig. 3. Schematic diagram of optical metrology using DMD [8]

Integrating the PμSL and the optical metrology, both utilising the same DMD technology, is the most desired solution. Because the whole system uses the same

DMD component there is no need to alter the structure of the PµSL system, and therefore representing the most cost effective solution.

In terms of precision, utilising the DMD as a metrology and fabrication tool will be the most precise solution for the reason that DMD is fixed to the same location and both processes are conducted by the same device, and therefore there is no device performance discrepancy.

For the in situ microassembly, the measurement system is only used to check the position of parts that are going to be assembled, as the light source can polymerise the uncured polymer and alter the intended geometry.

3 Workpiece Localisation Algorithm

Workpiece localisation algorithms have been researched in depth and applied widely for positioning correction in conventional machining. The typical examples of such algorithms include the variational algorithm, tangent algorithm and Menq algorithm. In particular, the localisation of 3D objects has been developed by Menq, Yau and Lai (1992) as a least squares problem with non linear optimisation. Hong-Tan algorithm is a similar iterative technique where the home surface point and the Euclidean transformation are optimised separately. The computational time based on Hong-Tan algorithm with 45 measurement points using a Pentium PC (166MHz) is reported to be under one second [9].

The functional model that represents the 3D template and surface discretely by the least square matching concept had been developed in parallel by Gruen (1984), Ackermann (1984) and Pertl (1984) [10]. The basic model is shown below:

$$f(x,y,z) = g(x,y,z)$$

Where the $f(x,y,z)$ is template model location and $g(x,y,z)$ is the search element location. However the ideal equation above is not true in the real world, as it contains random or stochastic error. The true error vector $e(x,y,z)$ is noted as with the equation below:

$$f(x,y,z) - e(x,y,z) = g(x,y,z)$$

Nevertheless, the aim is always to minimise the error and to converge the search element location as closest to the template model location through iterative process. In conventional machining operations that utilise the workpiece localisation algorithm, it is required to move the object from the search element location to the template model location. The process is represented by transformation matrices and is called Euclidean transformation).

3.1 Application of the localisation algorithm to in situ microassembly

The localisation algorithm can be applied in the in situ microassembly for an accurate assembly process. The process starts by placing a part to be assembled on the working table. Once it is placed, an optical measurement is used to obtain the actual part position.

Applying the localisation algorithm, which compares the datum points of the CAD model with the actual measurement, gives the Euclidean distances. This is the distance between the template model position and the actual part position.

Once the Euclidean distance is acquired, the next action is to perform the Euclidean transformation, that is to translate and rotate either the template model or the part to closest match into the ideal model $f(x,y,z) = g(x,y,z)$.

When performing the physical translation of the part, the actual position and the intended one will not match completely. This is caused by both the positional error of the stage used to re-position the part and the relative displacement of part to the stage surface, since the part is not held rigidly by a fixturing device.

A more accurate concept is introduced in the in situ microassembly process. The positional correction is done through feeding the error values to translate the CAD/STL model. This concept is far more accurate due to the fact that no physical movement is experienced by the parts, and hence the eliminating the transport, handling and orientation errors are eliminated.

Once the CAD/STL model is translated to the match the position of the part, the 'updated' model is then sent to the PμSL system. The surrounding medium of the part is solidified and it is then securely held. The process may be repeated to assemble different parts on different layers.

To summarise, there are four components in the in situ microassembly process, the design, localisation algorithm, the PμSL, and measurement system. These components work mutually to build complex and functional devices.

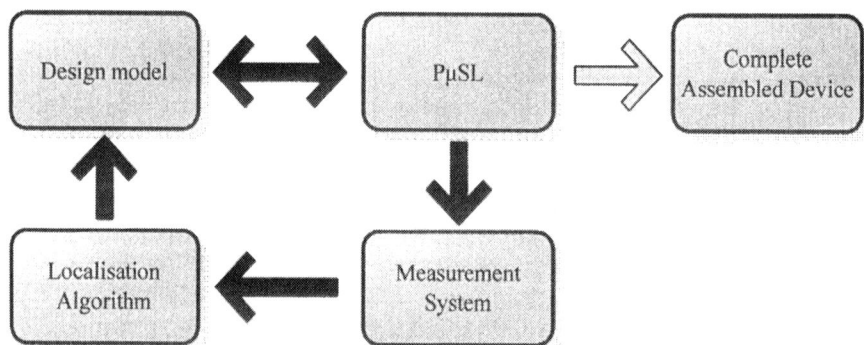

Fig. 4. In Situ Microassembly components

An experimental study has been carried out to validate the concept and applicability of in situ microassembly process. This case study is described in the following section.

3.2 Case Study

A battery and a Light Emitting Diode (LED) are assembled as a mini torch. The width and height of the conductive LED legs are 480x480 µm and the battery has a diameter of 7.7mm and depth of 5.3mm (Figure 5). The mini torch is designed as the model below (Figure 6).

Fig. 5. A battery and an LED

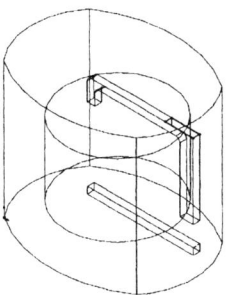

Fig. 6. The CAD model

If the design were to be manufactured conventionally, the casing would have to be split into two as it is not possible to machine the battery cavity and LED leg channels from a single part. Therefore the final assembly process constitutes of handling, orientation, and joining of four parts; the first half case, battery, LED and the other half case to make the final product.

In the case of in situ microassembly, the casing with the complex cavity and micro channels are built around the parts layer by layer. Firstly, the CAD model is converted to STL format. The LED is placed on the working table. To validate the concept, the LED is placed at the angle of 27 degree. The STL model is then rotated by 27 degrees to accommodate the error. The new STL model is then transferred to the PµSL system (Perfactory). Fig. 7 shows the STL model and the rotation process. Stereolithography process then takes place solidifying the polymer around the LED and battery according to the new orientation of the STL model.

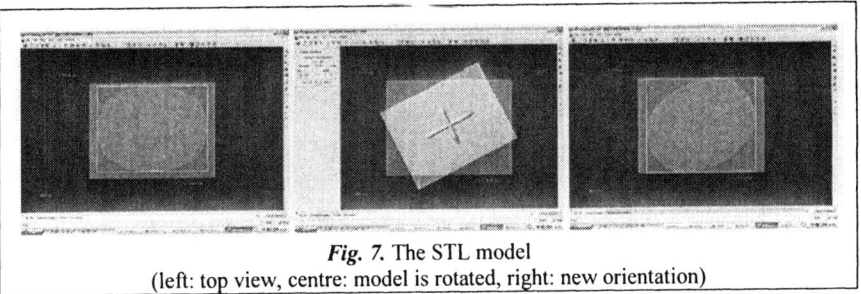

Fig. 7. The STL model
(left: top view, centre: model is rotated, right: new orientation)

The mini torch has been successfully fabricated through layer by layer solidification of polymer with the LED and the battery was being assembled through the in situ microassembly process as shown by the figures below.

Fig. 8. In situ micro assembled mini torch

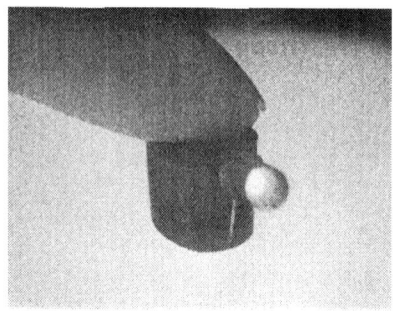

Fig. 9. Fully functional mini torch

4 Conclusions and outlook

The case study has confirmed the applicability of the in situ microassembly process that could be used instead of the conventional microassembly processes, and the same principle can be applied to even smaller devices. The design rule of thumb is to place the largest surface area of the part away from the projection path of the Stereolithography beam to ensure that no geometry of the part blocks the projection of the beam.

Accuracy of the in situ microassembly process depends on the accuracy of the measurement system. Depending on the need, different measurement systems can be used. However, the DMD measurement system is a promising system that can be integrated with the PμSL. The local minimisation algorithm can also be integrated into the design software. Once all the components are integrated together, in situ microassembly process can take place autonomously, being a solution for the creation of complete and functional micro devices. For small quantity production, in situ microassembly process is the most cost effective as it does not require any expensive tooling.

References

1. Böhringer, K.F., R.S. Fearing, and K.Y. Goldberg, *The Handbook of Industrial Robotics: Microassembly*. 2nd ed. 1999: John Wiley & Sons.
2. Van Brussel, H., et al., *Assembly of microsystems*, in *Cirp Annals 2000: Manufacturing Technology, Vol 49/2/2000*. 2000. p. 451-472.
3. Alting, L., et al. *Micro Engineering*. in *CIRP*. 2003.

4. Vandaele, V., P. Lambert, and A. Delchambre, *Non-contact handling in microassembly: Acoustical levitation.* Precision Engineering, 2005. **29**(4): p. 491.
5. Turitto, M., Y.-A. Chapius, and S. Ratchev. *Pneumatic Contactless Feeder for Microassembly.* in *International Precision Assembly Seminar.* 2006. Bad Hofgastein, Austria: Springer.
6. Cox, A., C. Xia, and N. Fang, *Microstereolithography: A Review.* Proceeding of the 1st International Conference on Micromanufacturing, 2006.
7. Sun, C., et al., *Projection micro-stereolithography using digital micro-mirror dynamic mask.* Sensors and Actuators A: Physical, 2005. **121**(1): p. 113.
8. Bitte, F., G. Dussler, and T. Pfeifer, *3D micro-inspection goes DMD.* Optics and Lasers in Engineering, 2001. **36**(2): p. 155.
9. Chu, Y.X., J.B. Gou, and Z.X. Li, *Workpiece localization algorithms: Performance evaluation and reliability analysis.* Journal of Manufacturing Systems, 1999. **18**(2): p. 113.
10. Gruen, A. and D. Akca, *Least squares 3D surface and curve matching.* ISPRS Journal of Photogrammetry and Remote Sensing, 2005. **59**(3): p. 151.

Chapter 5

Micro-Assembly Applications

INTEREST OF THE INERTIAL TOLERANCING METHOD IN THE CASE OF WATCH MAKING MICRO ASSEMBLY

Maurice Pillet, Dimitri Denimal, Pierre-Antoine Adragna, Serge Samper

SYMME Laboratory of Université de Savoie,
B.P.806, 74016 Annecy, cedex
maurice.pillet@univ-savoie.fr, dimitri.denimal@univ-savoie.fr,
pierre-antoine.adragna@univ-savoie.fr, serge.samper@univ-savoie.fr,

Abstract A mechanical part to tolerance is traditionally expressed as a [Min Max] interval which allows the definition of the conformity of the characteristic. Inertial tolerancing offers a new point of view of the conformity based on the mean square deviation to the target. This article demonstrates the efficiency of inertial tolerancing and proposes a comparison with the traditional tolerancing method in the case of watch making micro assembly.

1 Introduction

The functional quality of a mechanical assembly is often a direct function of a functional clearance. In assemblies of watch making, this clearance results from the combination of several elementary components. In micro-assembly, tolerances are very tight with regard to the dimensions, and the yields of these assemblies are often weak. Moreover, in the case of the assembly of a bridge on the main plate, the same assembly must satisfy several clearances (Fig 1). The statistical combination of several clearances having a weak yield leads to an extremely weak global assembly yield. That results in numerous and expensive final improvements. [1]

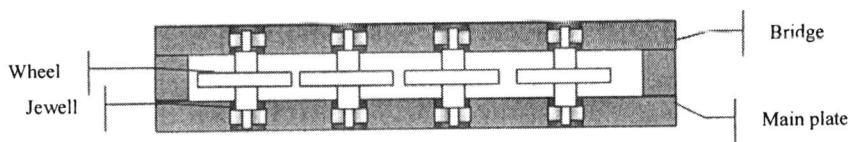

Fig. 1. Typical problem in watch making assembly

To increase the yield (Manufacturing and assembly), different tolerancing methods are available, notably the worst case and statistical tolerancing methods. The worst-case method favours the assembly yield to the detriment of the manufacturing yield. Theoretically by this approach, if tolerances are respected, the assembly yield must be 100%. Unfortunately though such tolerances are not compatible with the production and control means currently available. Statistical tolerancing introduced by Evans [9] through his state of the art could appear to be a solution. The statistical tolerancing method allows an important widening of the tolerances, but does not guarantee the conformity of the assembly. [2][3][13]. Graves[10] warns us against five fails of the statistical tolerancing limit and proposes some precautions at the time of statistical tolerancing design. But even with these precautions, statistical tolerancing remains dangerous.

Inertial tolerancing [4][5][6][7][8] proposes an alternative to the traditional methods, making it possible to obtain good yield at the same time in assembly and production. Tolerance does not define an interval but the maximum mean square deviation compared to the target (Equation 1).

$$I = \sqrt{\delta^2 + \sigma^2} \tag{1}$$

I: Inertia
δ: Off-centring off the mean
σ: Standard deviation

This new expression of the tolerance has many advantages. It makes it possible to increase variability in production while guaranteeing the functional clearance. This characteristic is particularly interesting in the complex case of the assembly of micromechanical parts. We propose to study the application of inertial tolerancing in the case of an assembly of a bridge on a main plate (Fig. 1). This assembly comprises k wheels (k generally lying between 2 and 5), each wheel clearance implying a chain of n components (n generally lying between 3 and 5).

Assumptions:

- The component count c is identical for each wheel
- The functional clearance is identical for each wheel equal to $\pm t$
- In the case of a centred production, we wish to obtain a capability index Cp = 1 that is to say ($\sigma = t/3$)
- Productions are normally distributed
- In the case of traditional tolerancing method, Cp is defined by Equation 2. In the case of inertial tolerancing method Cp is defined by Equation 3

$$Cp = \frac{2t}{6\sigma} \tag{2}$$

$$Cp = \frac{I}{\sigma} \tag{3}$$

In the case of a traditional tolerancing method, *Cpk* is defined by Equation 4.

$$Cpk = \frac{Min(\delta - LSL, USL - \delta)}{3.\sigma} = \frac{t - |\delta|}{3.\sigma} \tag{4}$$

In the case of an inertial tolerancing method, we defined *Cpi* by Equation 3.

$$Cpi = \frac{I_0}{I} \tag{5}$$

> I_0: Inertia tolerance
> I: Inertia of the batch

2 Calculation of inertial tolerance for each component

Several approaches can be considered to calculate the inertial tolerance according to the assumptions of capability, and respect of the centering on the target. Two approaches were proposed which will be tested on this example.

1. Classical inertial tolerancing
2. Adjusted inertial tolerancing

We will compare these both approaches with the worst case and statistical tolerancing method for the yield of assembly and the variations authorizsed in production.

Case #1: Classical inertial tolerancing

To calculate the component inertial tolerances (I_P), the assembly distribution is considered centred, six standard deviations ($\pm 3\sigma$) contained in the functional tolerance interval ($\pm t$).

$$I_C = \sigma_C = \frac{2t}{6} \tag{6}$$

With the precedent assumptions

$$I_P = I_C/\sqrt{n} = t/3\sqrt{n} \tag{7}$$

I_C: Clearance Inertia
I_P: Part inertia

With the assumption of a capability $Cp = 2$ and $Cpi = 1$, the maximum decentring is equal to:

$$\sigma_P = \frac{I_P}{2} \tag{8}$$

$$\delta_P = \pm\sqrt{I_P^2 - \sigma_P^2} = \pm\sqrt{\frac{3I_P^2}{4}} = \pm\sqrt{\frac{3I_C^2}{4n}} \tag{9}$$

The most unfavourable assembly corresponds to the situation where all decentrings are added. With the assumption of independence we can write:

$$\delta_C = n\delta_P$$

$$\sigma_C = \sqrt{n\sigma_P^2} = \sqrt{\frac{nI_P^2}{4}} = \sqrt{\frac{nI_C^2}{4n}} = \frac{I_C}{2} = \frac{t}{6}$$

With a tolerance of $\pm t$ on the clearance, the z index is calculated by the relation (10):

$$z = \frac{t - n\sqrt{\frac{3I_C^2}{4n}}}{t/6} = \frac{t\left(1 - \sqrt{\frac{n}{12}}\right)}{t/6} = 6 - \sqrt{3n} \tag{10}$$

The yield is calculated by $\Phi(z)$ the where Φ represent the standard normal cumulative distribution function.

	N	2	3	4	5
	Worst caseTolérance	± t/2	± t/3	± t/4	± t/5
	Inertial tolerance	0.236t	0.192t	0.167t	0.149t
	percentage of increase in the production dispersion compared to worst case tolerancing (position centrered)	141%	173%	200%	224%
	$Z = 6 - \sqrt{3n}$	3.551	3.000	2.536	2.127
	Yield	1.000	0.999	0.994	0.983
	Global Yield on k assembly				
	2	1.000	0.997	0.989	0.967
K	3	0.999	0.996	0.983	0.951
	4	0.999	0.995	0.978	0.935
	5	0.999	0.993	0.972	0.919

Table 1. Inertial Yield with tolerancing

Case #2: Statistical tolerancing

The same table can be obtained for statistical tolerances under similar conditions: Cp = 2, Cpk = 1:

$$t_P = t / \sqrt{n}$$

$$\sigma_P = \frac{t_P}{6} = \frac{t}{6\sqrt{n}}$$

The maximum decentring is for *Cpk = 1*:

$$\delta_P = t_P - 3\sigma_P = \frac{t}{\sqrt{n}} - \frac{t}{2\sqrt{n}} = \frac{t}{2\sqrt{n}}$$

The most unfavourable assembly corresponds to the situation where all decentrings are added [12]. With the assumption of independence we can write:

$$\delta_C = n\delta_P = \frac{t\sqrt{n}}{2}$$

$$\sigma_C = \sqrt{n\sigma_P^2} = \frac{t}{6}$$

The z index is calculated by the relation (11):

$$z = \frac{t - t\frac{\sqrt{n}}{2}}{t/6} = 6 - 3\sqrt{n} \tag{11}$$

		n	2	3	4	5
		Worst case Tolérance	± t/2	± t/3	± t/4	± t/5
		statistical tolerance	±0.707t	±0.577t	±0.5t	±0.447t
		percentage of increase in the production dispersion compared to worst case tolerancing	141%	173%	200%	224%
		Z= $6 - 3\sqrt{n}$	1.757	0.804	0.000	-0.708
		Yield	0.961	0.789	0.500	0.239
Global Yield on k assembly						
K	2		0.923	0.623	0.250	0.057
	3		0.886	0.492	0.125	0.014
	4		0.851	0.388	0.063	0.003
	5		0.818	0.306	0.031	0.001

Table 2. Yield with statistical tolerancing

Table 2 shows the risks of the statistical tolerancing method. With a $Cp = 2$ on each component and $Cpk = 1$, as soon as there are 4 wheels in the mechanism the yield of the assembly can be close to zero although the individual yield of each component is 0.9973.

In the case of inertial tolerancing the worst yield is equal to 0.919 ($k=5$ and $n = 4$) for the same widening of the production spread in the centred case as statistical tolerances.

Case #3 - Adjusted inertial tolerancing

Adragna [11] showed that it is possible to calculate inertia on each component in order to guarantee an yield of the final assembly.

Adragna proposed an inertial adjusted coefficient (I_C) in order to guarantee that the assembly Cpk index will never be lower that a Cpk_{Min} value. The I_C coefficient is calculated by the simple following relation:

$$I_C = \sqrt{Cpk_{Min}^2 + \frac{n}{9}} \tag{12}$$

This I_C coefficient has the same rule as the inflated coefficient for the traditional inflated statistical tolerancing:

$$I_P = \frac{t}{6.I_C.n^{1/2}} \tag{13}$$

In our application we want to guarantee in all cases the assembly yield greater that 0.997 ($Cpk_{min} = 1$)

$$I_C = \sqrt{1 + \frac{n}{9}}$$

$$I_P = \frac{t}{3.\sqrt{n + \frac{n^2}{9}}} \tag{14}$$

With the assumption of a capability $Cp = 2$ and $Cpi = 1$, the maximum decentring is equal to:

$$\sigma_P = \frac{I_P}{2} = \frac{t}{6\sqrt{n + \frac{n^2}{9}}} \tag{8}$$

$$\delta_P = \pm\sqrt{I_P^2 - \sigma_P^2} = \frac{\pm t}{6} \cdot \frac{\sqrt{3}}{\sqrt{n + \frac{n^2}{9}}} \tag{9}$$

The most unfavourable assembly corresponds to the situation where all decentrings are added. With the assumption of independence, we can write:

$$\delta_C = n\delta_P$$

$$\sigma_C = \sqrt{n\sigma_P^2} = \frac{t}{6\sqrt{1+\dfrac{n}{9}}}$$

With a tolerance of ±t on the clearance, the z index is calculated by the relation [10].

$$z = \frac{t-\delta_C}{\sigma_C} = \frac{t - \dfrac{t}{6}\cdot\dfrac{n\sqrt{3}}{\sqrt{n+\dfrac{n^2}{9}}}}{\dfrac{t}{6\sqrt{1+\dfrac{n}{9}}}} = 6\sqrt{1+\frac{n}{9}} - \sqrt{3n} \qquad (10)$$

This relation gives the following outputs:

		n	2	3	4	5
		I_C	1.106	1.155	1.202	1.247
		I_P	0.2132t	0.1667t	0.1387t	0.1195t
		percentage of increase in the production dispersion compared to worst case tolerancing (position centrered)	128%	150%	166%	179%
		$z = 6\sqrt{1+\dfrac{n}{9}} - \sqrt{3n}$	4.184	3.928	3.747	3.610
		Yield	1.0000	1.0000	0.9999	0.9998
Global Yield on k assembly						
k	2		1.000	1.000	1.000	1.000
	3		1.000	1.000	1.000	1.000
	4		1.000	1.000	1.000	0.999
	5		1.000	1.000	1.000	0.999

Table 3. Yield with adjusted inertial tolerancing

The yield is guarantees for each assembly. It is possible also to guarantee the global yield as we will detail it in the end of the article.

Is the situation of this article, yield is better than 0.997. In fact the situation $Cp = 2$ and $Cpi = 1$ is not the worst case situation

3 Discussion

The assembly of micromechanics components often leads to small yield of production and assembly because of the relative importance of the variability. In this context, it is important to choose a tolerancing method able to guarantee the functional specification for the product, but also to give the greatest possible variability for the production.

Worst case tolerances guaranteed the functional specification but is very restrictive for the production

Statistical tolerances gives freedom to the production but does not guaranteed a good assembly capability.

We show in this paper through generic example of assembly in watch making the great interest to use inertial tolerances. It guarantees at the same time good assembly capability while leaving the most possible freedom to the production.

References

1. S. Koelemeijer Chollet, F. Bourgeois, C. Wulliens, and J. Jacot. *Cost modelling of microassembly.* In Proceedings of the International Precision Assembly Seminar IPAS 2003, Bad Hofgastein, Austria, March 17-19, 2003.
2. Graves S. (2001) – *Tolerance Analysis Tailored to your organization* – Journal of Quality technology – Vol. 33, N°3, 293-303, July 2001
3. K.W. Chase. Tolerance allocation methods for designers. ADCATS Report 99-6, Brigham YoungUniversity, 1999.
4. Pillet M., *Inertial tolerancing in the case of assembled products, Recent advances in integrated design and manufacturing in mechanical engineering,* No. ISBN 1-4020-1163-6, 2003, pp. 85-94,
5. Pillet M., *Inertial Tolerancing,* The Total Quality Management Magazine, Vol. 16, No. Issue 3 - Mai 2004, 2004, pp. 202-209,
6. F. Bourgeois, Y. L. de Meneses, S. Koelemeijer Chollet, and J. Jacot. *Defining assembly specifications from product functional requirements using inertial tolerancing in precision assembly.* In Proceedings of the IEEE International Symposium on Assembly and Task Planning, Montreal, July, 2005.
7. F. Bourgeois. *Vers la maîtrise de la qualité des assemblages de précision.* Thesis from Laboratory LPM1 Ecole Polytechnique de Lausanne(EPFL),Lausanne(Swiss),2007
8. Adragna P.A. *Tolérancement des Systèmes Assemblés, une approche par le Tolérancement Inertiel et Modal.* Thesis from Laboratory Symme Polytech'Savoie, Annecy(France), december 2007.
9. Evans D.H., *Statistical tolerancing the state of the art,* newspaper of quality technology, vol. 7, No. 1, 1975
10. Graves S, Bisgaard S, *Five ways statistical tolerancing can fail, and what to do about them,* Quality Engineering, Vol.13,pp. 85-93.
11. Adragna P.A., Pillet M. Samper S., Formosa F., *Guarantying a maximum of Non-Conformity Rate on the assembly resultant with a statistical tolerancing approach* , Computer Aided Tolerancing (CAT) 2007. Erlangen, Germany.
12. Pillet.M, *Inertial tolerancing,* The TQM Magazine Volume 16 - Number 3 - 2004 - pp. 202-209
13. Nigam S.D and Turner JU *Review of statistical approaches to tolerance analysis,* computer aided Design Volume 27 Number 1 January 1995 pp 6-15

PRECISION ASSEMBLY OF ACTIVE MICROSYSTEMS WITH A SIZE-ADAPTED ASSEMBLY SYSTEM

Kerstin Schöttler, Annika Raatz, Jürgen Hesselbach

Technical University Braunschweig, Institute of Machine Tools and Production Technology (IWF), Langer Kamp 19b, 38100 Braunschweig, Germany

Abstract The paper presents a size-adapted assembly system for the automated precision assembly of active microsystems. The part sizes of the microsystems can reach centimeter range, but they must be assembled with an assembly accuracy of about a few micrometers. The results of a sensor guided assembly process using a 3D vision sensor are shown. This process reaches a positioning uncertainty of 1.2 µm and an assembly uncertainty of 36 µm.

Keywords precision assembly, assembly system, sensor guidance

1 Introduction

Microsystem technology (MST) is a key technology for emerging markets. The third "NEXUS Market Analysis" estimates a projected market growth for 1st level packaged Microsystems and MEMS from US$ 12 billion in 2004 to 25 billion in 2009. Read/write heads, Inkjet heads and micro-displays will account for 70% of the market volume in 2009 [1].

For hybrid microsystems, a high assembly accuracy in the range of a few micrometers is required. In order to reach this accuracy, a size-adapted assembly system for sensor guided precision assembly was developed and will be presented in this paper. A precision assembly process of active microsystems will be described by means of a micro linear stepping motor.

2 Size-adapted assembly system

The size-adapted assembly system (Fig. 1, left) consists of a parallel robot $micabo^{f2}$, which provides a high accuracy due to its structure, and an integrated 3D vision sensor with only one camera. Additional components are an assembly fixture and part trays for the adjustment of parts inside the robot's workspace. High flexibility is reached through product specific part trays and grippers, e.g. vacuum grippers with different nozzles and mechanical grippers [2].

Please use the following format when citing this chapter:

Schöttler, K., Raatz, A., Hesselbach, J., 2008, in IFIP International Federation for Information Processing, Volume 260, *Micro-Assembly Technologies and Applications*, eds. Ratchev, S., Koelemeijer, S., (Boston: Springer), pp. 199-206.

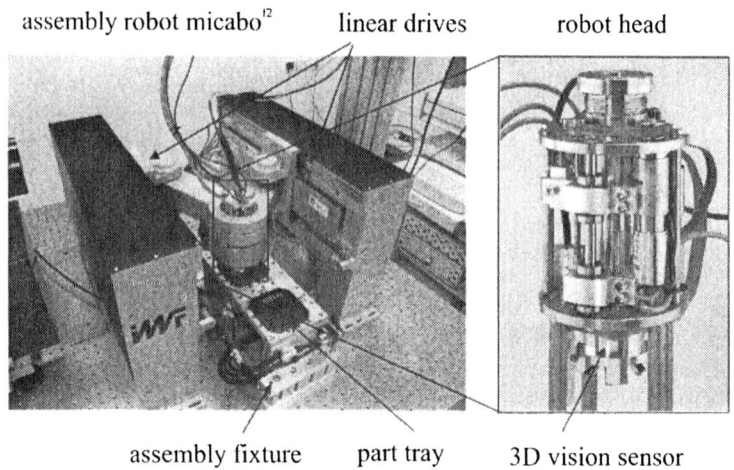

Fig. 1. Size-adapted assembly system

2.1 Parallel robot micabo^{f2}

At the IWF various size-adapted parallel and hybrid robots were developed for precision assembly. First, a functional model of a planar robot *micabo^e* [3] with 3 degrees of freedom (DOF) with a parallel structure and 1 DOF as serial lifting table for movement in z-direction was implemented. Second, a spatial robot *micabo^h* [3] with 6 DOF was designed. Afterwards, a spatial robot *Triglide* [4] based on a parallel structure with 3DOF and one serial rotational axis was realized.

Based on the experiences gained during the development of the above mentioned robot structures, the robot *micabo^f* [5] with 4 DOF for part handling and, as an advanced structure, the *micabo^{f2}* [6] with 4 DOF for part handling and 1 DOF for focusing a vision sensor was developed as a hybrid robot structure. Two parallel linear drives impart motion in the xy-plane. Each of them is connected to a slide, which is coupled to the arms of the structure with rotational bearings. A hollow shaft between the arms takes up the robot head, which is designed like a cartridge and forms the tool center point (TCP). Inside the robot head, two drives are installed. One of them moves a platform with a gripper and the other one moves the 3D vision sensor. The rotation around the z-axis is imparted by a rotational drive that is placed at one robot arm and transmits the movement to the robot head with a gear belt.

The workspace measures 160 mm x 400 mm x 15 mm. The control of the parallel linear drives can hold the desired position with an encoder resolution of 0.1 µm. In accuracy measurements according to EN ISO 9283 [7], the robot *micabo^{f2}* achieved a repeatability of 0.6 µm. This high repeatability is a good precondition for a precision assembly process.

2.2 3D vision sensor

The 3D vision sensor was developed at the Institute of Production Measurement Engineering (IPROM), Technical University Braunschweig [8]. It needs only one camera and a single image for a 3D imaging process as it applies the principle of stereo photogrammetry (see Fig. 2). It is based on a 3-dimensional reconstruction of the objects from a pair of images. The field of vision has a dimension of 11 mm in length and 5.5 mm in width with a resolution of 19 µm/pixel. Repeatability measurements showed standard deviations of $\square_x$ = 0.220 µm and $\square_y$ = 0.290 µm.

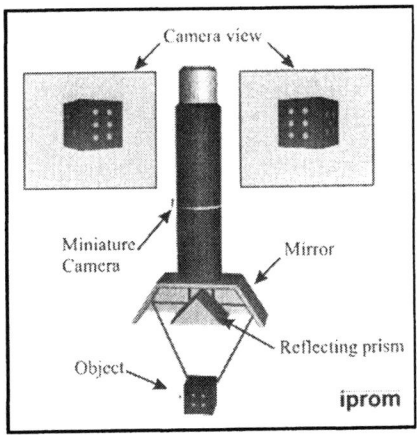

Fig. 2. Functional principle of the 3D vision sensor

2.3 Control system

A real-time system is used to control the assembly system. It features a PowerPC750 digital signal processor (DSP) running at 480 MHz. Two different control loops (Fig. 3) are used for the integration of the sensor information in the control system.

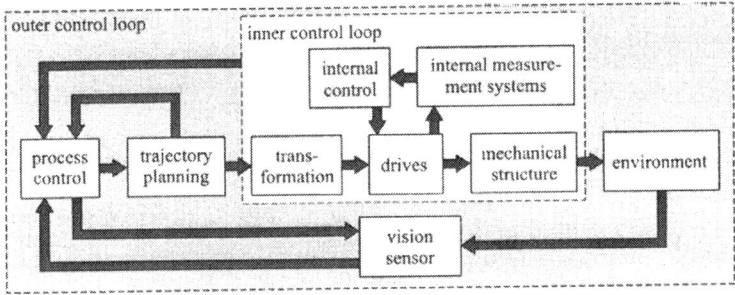

Fig. 3. Control of the assembly system

The process control gives commands to the robot control and demands information from the vision sensor system. The internal control loop works at a system execution time of 0.2 ms. The outer control loop works with 1 ms and contains the 3D vision sensor, which sends information about relative positions of the assembly group to the process control. A resulting vector of the current position from the robot control and the relative position vector from the vision system is calculated inside the process control and is transmitted to the trajectory planning.

At present, the sensor guidance works according to a so called "look-and-move" procedure. This means that the robot's movement stops before a new imaging process starts and a new position correction is executed.

3 Precision assembly of active microsystems

A micro linear stepping motor [9], which works according to the reluctance principle, is assembled with the presented size-adapted assembly system. The motor parts were mainly manufactured with micro technologies developed at the Collaborative Research Center 516. One assembly task is the assembly of guides on the surface of the motor's stator element. In Figure 4 (left) the assembly group of two guides on a stator is shown. The right image shows the view of the 3D vision sensor on the assembly scene.

Circular positioning marks [10, 11], which were manufactured in a photolithographic manufacturing process onto the group of components, are used by the 3D vision sensor for the imaging process. The positioning marks (4 on the stator and 2 on each guide) in both images are measured and a resulting relative position vector is calculated and given to the robot control. Inside the robot control, the relative position vector is separated into a rotation correction, a correction in xy-direction and, finally, the correction in z-direction, which places the guide.

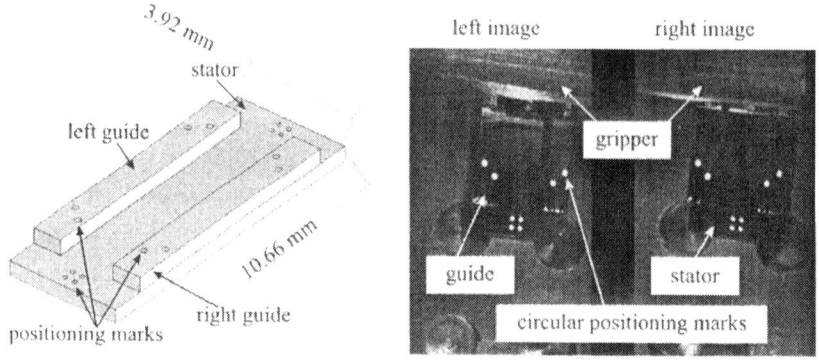

Fig. 4. Assembly group guides on stator (left); view of 3D vision sensor (right)

Figure 5 shows only the assembly steps for the left guide, because the assembly steps for the left and the right guide are equal. In a first step, the marks on the stator are checked. If the part can be recognized, the positioning of guides can be carried out. Otherwise, the stator that has already been checked will be excluded. The guides are checked, too, and are picked up if they are recognised. Gripping the guide and moving it to a pre-defined position above the stator is done without sensor guidance. Afterwards the sensor guidance is started automatically.

If the relative positioning vector reaches the pre-defined limit value of 0.8 µm in x- and y-direction, a final relative position correction is done by moving the robot in z-direction (height). This completes one positioning process. After both guides have been placed, the next assembly process begins until the part trays have to be changed.

At present, the guides are bonded onto the stator element by use of cyanoacrylate which has been pre-applied manually onto the stator element. The joining technology will be improved in future works at the Collaborative Research Centre 516.

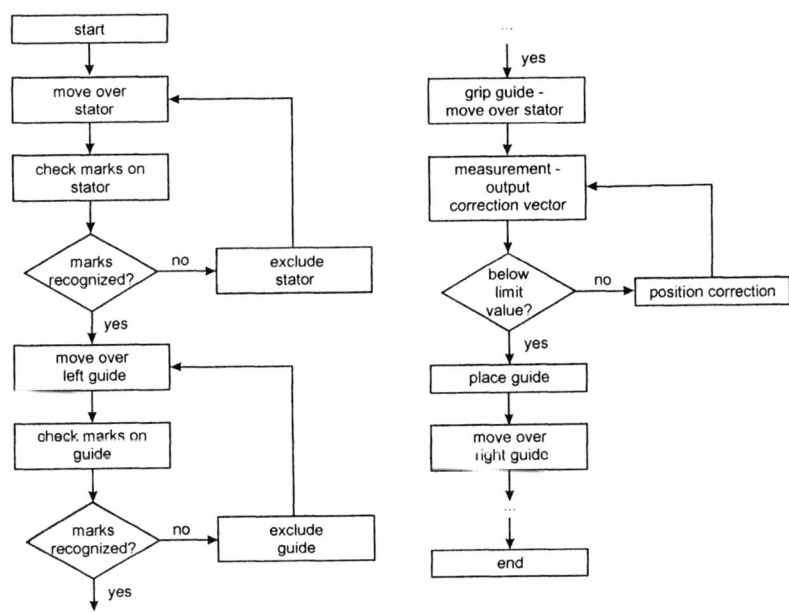

Fig. 5. Assembly of guides on a stator element

4 Results of the precision assembly process

To quantify the precision assembly process, two terms were defined - *positioning uncertainty* and *assembly uncertainty*. According to DIN ISO 230-2 [12], the *positioning uncertainty* is the combination of the mean positioning deviation and the double standard deviation. For precision assembly processes the term *positioning uncertainty* refers to the reached relative position between the assembly parts before the bonding process is carried out (in this case the guide is above the stator and has no contact).

The term *assembly uncertainty* describes the relative position between the assembled parts measured after the assembly process has been completed. This is the combination of the mean assembly deviation and the double standard deviation, too.

Positioning marks are used as an inspection criterion. They are used for quality control of the parts before the process and during the process for the sensor guidance and evaluation of the *positioning uncertainty*. After the process the positioning marks are used for evaluation of the *assembly uncertainty*.

During the process only one end of the assembled parts can be measured, because the gripper covers half of the guide and the stator (see Fig. 4, right). Therefore the relative positioning device observes only the visible sides of the assembled parts. This means, that the measured positioning error and the resulting *positioning uncertainty* is only determined by the visible part side. After the process both ends of the assembly group can be inspected and the overall assembly deviation can be measured. From the deviations the *assembly uncertainty* is calculated. This value is comprised of the overall errors during the assembly of the microsystem.

As expected, the *assembly uncertainty* of 36 µm is much higher than the *positioning uncertainty* of 1.2 µm. This is a result of the relatively long part sizes of 10 mm. A small angular deviation causes a positioning error (in xy-direction) on the side of the part which is invisible during the sensor guided positioning process and which is larger than the error on the visible side. With a greater part length this positioning error will be higher than with smaller parts. Actually the measured assembly deviations are smaller on the end of the assembly group that is visible during the assembly and used for the relative positioning process (average deviation of 7 µm and assembly uncertainty related only to the visible side 16 µm) as on the invisible side (average deviation of 20 µm and assembly uncertainty related only to the invisible side 44 µm).

Furthermore, deviations occurring during the joining process cause an increased positioning error. This leads to the before mentioned gap between *positioning uncertainty* and *assembly uncertainty*. Figure 6 shows the *positioning uncertainty* (left) and the *assembly uncertainty* (right) of 33 assembled groups. The circles in the diagrams show the radius of the uncertainties, which include 95% of the parts.

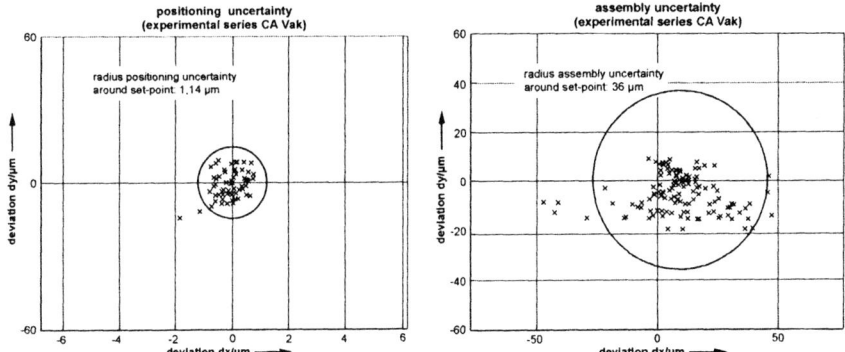

Fig. 6. Reached positioning uncertainty (left) and assembly uncertainty (right)

5 Conclusion

In this paper a size-adapted assembly system consisting of a parallel robot with high accuracy and an integrated 3D vision sensor was presented. Within a workspace of 160 mm x 400 mm x 15 mm a positioning uncertainty of 1.2 µm is reachable. Due to the relatively large part sizes and the joining process of the described assembly process, an assembly accuracy of 36 µm could be reached. However, an assembly accuracy in the range of a few micrometers is needed, as a result of which it should be improved in further research activities concerned with the development of joining and gripping technologies at the Collaborative Research Center 516. Furthermore, another arrangement of the positioning marks with a larger distance between the marks will be studied, which promises an improved positioning uncertainty, especially for angular deviations.

Acknowledgement

The authors gratefully acknowledge the funding of the reported work by the German Research Center (Collaborative Research Center 516).

References

1. H. Wicht, J. Bouchaud, NEXUS Market Analysis for MEMS and Microsystems III 2005-2009, *mst news* (5), 2005, p. 33-34.
2. S. Büttgenbach, J. Hesselbach, R. Tutsch, M. Berndt, K. Heuer, B. Hoxhold, J. Wrege, Sensorgeführte Montage aktiver Mikrosysteme, *Kolloquium*

Mikroproduktion – Fortschritte, Verfahren, Anwendungen, Aachen (März 2005), pp. 263-272.
3. R. Thoben, Parallelroboter für die automatisierte Mikromontage, *PhD-Thesis Technical University Braunschweig*, Fortschritt-Berichte VDI Nr. 758, Reihe 8: Meß -, Steuerungs- und Regelungstechnik, VDI Verlag, Düsseldorf, 1999.
4. J. Hesselbach, S. Dittrich, O. Becker, Sensorgeführte Mikromontage mit Parallelrobotern, *Erstes Internationales Symposium für Mechatronik an der Technical University Chemnitz*, 2002, pp. 229-238.
5. G. Pokar, Untersuchungen zum Einsatz von ebenen Parallelrobotern in der Mikromontage, *PhD-Thesis Technical University Braunschweig*, Vulkan Verlag, 2004.
6. M. Simnofske, K. Schöttler, J. Hesselbach, micaboF2 - Robot for Micro Assembly, *Production Engineering* **XII**(2), 2005, pp. 215-218.
7. EN ISO 9283, Industrieroboter, Leistungskenngrößen und zugehörige Prüfmethoden, Beuth-Verlag, Berlin, 1999.
8. R. Tutsch, M. Berndt, Optischer 3D-Sensor zur räumlichen Positionsbestimmung bei der Mikromontage, *Applied Machine Vision*, VDI-Bericht Nr. 1800, Stuttgart, 2003, pp. 111-118.
9. M. Hahn, R. Gehrking, B. Ponick, H.H. Gatzen, Design Improvements for a Linear Hybrid Step Micro-Actuator, *Microsystem Technologies*, **12**(7), Springer Verlag Berlin Heidelberg New York, 2006, pp. 646-649.
10. M. Berndt, R. Tutsch, Enhancement of image contrast by fluorescence in microtechnology, *Proceedings of SPIE*, Vol. 5856, Optical Measurement Systems for Industrial Inspection IV, München, 2005, pp. 914-921.
11. M. Berndt, R. Tutsch, K. Schöttler, J. Hesselbach, M. Feldmann, S. Büttgenbach, Referenzmarken bei der Montage aktiver Mikrosysteme, *Mikrosystemtechnik Kongress*, Freiburg, 2005, pp. 459-462.
12. DIN ISO 230-2, Prüfregeln für Werkzeugmaschinen, Teil 2: Bestimmung der Positionierunsicherheit und der Wiederholpräzision der Positionierung von numerisch gesteuerten Achsen, Beuth Verlag, Berlin, 2000.

ASSEMBLY OF OSSEOUS FRAGMENTS IN ORTHOPAEDIC SURGERY: THE NEED FOR NEW STANDARDS OF EVALUATION

Olivier Cartiaux[1], Laurent Paul[2], Pierre-Louis Docquier[2], Xavier Banse[2], Benoit Raucent[1]

1Department of Mechanical Engineering, Université catholique de Louvain, 2 Place du Levant, 1348 Louvain-la-Neuve, Belgium

2Department of Orthopaedic Surgery, Cliniques universitaires St-Luc, Université catholique de Louvain, 10 Avenue Hippocrate, 1200 Brussels, Belgium

olivier.cartiaux@uclouvain.be

Abstract In orthopaedic surgery, intra-operative bone machining and assembly of osseous fragments are two very important research areas. One of the most challenging applications is the treatment of malignant osseous tumors within the pelvis due to its complex tri-dimensional geometry. The conventional surgical procedure includes tumoral resection (cutting of the osseous tumor) and reconstruction by allograft (assembly of fragments). Accuracy of bone cutting and osseous assembly has not yet been documented. This paper presents an experimental study on plastic bones, with experienced surgeons working under ideal conditions. The goal was to assess the accuracy by using geometrical parameters resulting from the surgical usual language: surgical margin for tumor cuttings, and maximal gap, gap volume and mean gap between the 2 osseous fragments. Both mean values and correlation between assembly parameters were relatively poor. Experienced surgeons did not manage to consistently perform accurate cuttings and osseous assemblies, even under ideal working conditions. The complex tri-dimensional architecture of the pelvis can mainly explain this inaccuracy. There is a need to adapt computer and robotic assisted technologies for tumor cuttings and osseous assemblies. Finally, our attempt to evaluate accuracy using simple geometrical parameters, was not satisfactory. There is a need to define new evaluation standards for these assemblies. We think that mechanical engineering tools like geometrical tolerances and mechanical fittings are more suitable for this problem of quality evaluation.

1 Introduction

In orthopaedic surgery, intra-operative bone machining and assembly of osseous fragments are two very important research areas. One of the most challenging applications is the treatment of malignant osseous tumors within the pelvis. It is due to the complex tri-dimensional (3D) geometry of the bony pelvis and the proximity of organs and structures difficult to reach, like vessels, bladder, rectum, sciatic nerve [1]. Nowadays, conventional surgical procedure includes tumoral resection (cutting of the tumor), using an oscillating saw, with an adequate margin [2-4].

Most usual available data to deal with this problem are clinical post-operative outcomes. Enneking [5] described a classification of surgical margin as radical, wide, marginal or intralesional. Recurrence is the major post-operative complication. Local recurrence rates from 28 to 35% were reported concerning limb-salvage procedures [6]. However, to our knowledge, geometrical accuracy of bone tumor cutting has not yet been experimentally documented.

Several osseous reconstruction methods exist: autograft (taken on the patient himself), allograft (taken on cadaver), or customised prosthesis. Whatever the used technique, surgeons have to perform assemblies of osseous fragments. Osseous massive allografts present some technical advantages: they can be shaped to perfectly fit the pelvic defect, they allow good reinsertion of soft tissues like tendons and muscles, and they provide true anatomical restoration of the complex 3D architecture of the pelvis [6].

Again, most usual available data are clinical rates, like infection, non-union and fracture of the reconstruction. Many studies have dealt with these post-operative outcomes [7-16]. The main information is the need for an accurate and intimate contact at the host-graft junction (the cutting site) in order to promote and accelerate union [17]. However, ranges of all the reported rates are large. Assessing the quality of a pelvic reconstruction, i.e. an osseous assembly, using clinical outcomes is still a difficult task.

In this study, we designed an experimental model on plastic pelvic bones to simulate the conventional (free-hand) procedure of treatment of osseous tumors. We tried to assess the geometrical accuracy of the surgical gesture, based on geometrical parameters resulting from the surgical usual language.

2 Material and methods

4 experienced senior surgeons (S1, S2, S3, S4), used to pelvic surgery [6], were each asked to perform cutting of 3 different pelvic tumors and their reconstruction. 12 identical plastic pelves and 12 plastic left hemipelves were procured (Sawbones

Worlwide, Pacific Research Laboratories Inc., Vashon, WA) and considered as host bones (host) and allografts (graft) respectively.

2.1 Conventional surgical procedure: design of an experimental model

2.1.1 Virtual tumors

The 12 hosts were scanned (spiral Elscint Twin CT scanner with 2.7 mm slice thickness). 3 sets of tumor were virtually created : tumor 1 in zone I (T1, iliac wing), tumor 2 in zone II (T2, acetabulum), and tumor 3 over zones II and III (T3, acetabulum and obturator foramen) according to Enneking [18]. The tumor was represented by a sphere of fixed diameter and placed on the 3D CT-scan of the pelvis (Figure 1A) using the visualization software Volview, version 2.0.5 (Kitware Inc., New-York, USA).

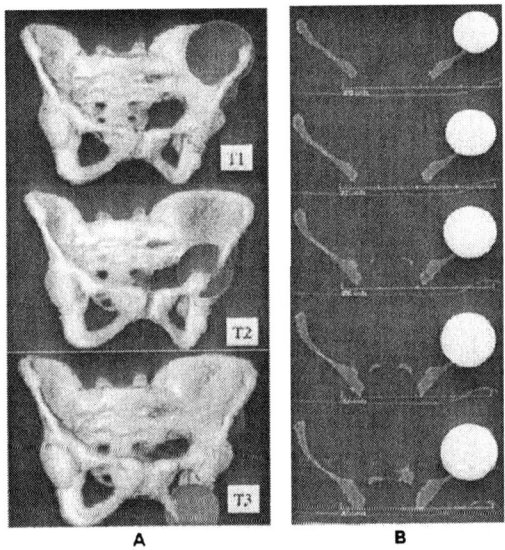

Fig. 1. Virtual tumors. A) 3D views : T1 in zone I of Enneking, T2 in zone II and T3 in zone II-III. B) Examples of transversal slices, given to surgeons, of the virtual tumor in zone I (T1) for cutting planning. A graduated scale of 200mm was given for each slice

2.1.2 Tumor cutting

Each surgeon received a print with 32 sequential coronal slices, 32 transversal slices and 20 sagittal slices for each tumor (Figure 1B). A graduated scale of 200 mm was available on each slice.

The surgeon could plan resection planes and perform cutting of the 3 tumors (T1, T2, T3) without limitation in time. Instruction was given to respect a 10 mm-surgical margin with an accepted tolerance of 5 mm above or below. In other words, surgeon had to perform resection within the interval of tolerance [5 mm; 15 mm]. Based on the 2D slices of the tumor, he could note some landmarks with a skin marker on the host to guide the cutting. After planning and landmarking, he had to perform tumor resection by cutting the host using an oscillating saw (Howmedica Chirodrill 1200 set, Germany). In order to simulate patient positioning on the operating table, the host was rigidly fixed by a steel clamp with a 360°-rotational base (Figure 2).

2.1.3 Host-Graft assembly

Each surgeon had to reconstruct the pelvis, after each tumor cutting. An oversized graft was given to reconstruct T1 and T3 in order to increase the technical difficulties. A size-matched graft was chosen to reconstruct T2 because it was simulating an articular reconstruction. As for tumor resection, surgeon could note some landmarks with a skin marker on the graft to guide the cutting. He could also place the graft into the gap to check its size. Additional cuttings were allowed to each surgeon to optimize the graft cutting. The host-graft reconstruction was temporarily fixed with K-wires (2 mm diameter).

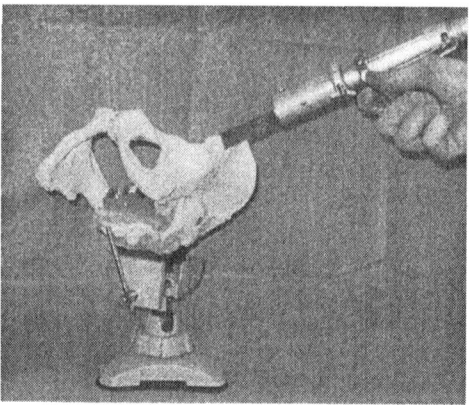

Fig. 2. The sawbone is rigidly fixed by a steel clamp with a 360°-rotational base. The cutting is performed using an oscillating saw.

2.2 Evaluation of tumor cuttings

After tumor cuttings, resected parts were scanned (spiral Elscint Twin CT scanner with 2.7 mm slice thickness) and registered with the 3D CT-scan of the corresponding host. Margin error was calculated as the minimal distance between the 10 mm-surgical margin and each cutting plane, using Volview and its distance measurement tool. Negative values of margin error were given for cutting below the target and positive values for cutting above the target. This allowed verifying if the interval of tolerance [5 mm; 15 mm] had been respected or not.

2.3 Evaluation of host-graft assemblies

Each host-graft junction (HGJ) was evaluated by 4 different methods, each one qualifying a geometrical parameter usually used by surgeons.

3 different observers classified each HGJ according to the degree of contact between host and graft (ordered categorical classification): degree 1 for full contact, 2 for contact > 50%, 3 for contact < 50% or 4 for no contact.

Maximal gap between host and graft of each HGJ was measured with an electronic caliper (CD-15CP, Mitutoyo Inc., Aurora, IL) by the 3 same observers.

Gap volume between host and graft was measured for all the HGJ. An epoxy paste with scanner-high-density (pc-7 Heavy Duty Paste Epoxy, PC-products, Allentown, USA) was used to completely fill the gap of the HGJ. All the constructs were scanned (spiral Elscint Twin CT scanner with 2.7 mm slice thickness). As the density of the paste was high and very different from the Sawbone, a threshold segmentation permitted the elimination of voxels corresponding to the Sawbone (Figure 3). The number of voxels corresponding to the paste was calculated, giving the HGJ gap volume.

Finally, contact surface available on the host was measured for each HGJ. Mean gap between host and graft was then calculated for each HGJ as the ratio of gap volume to the corresponding contact surface.

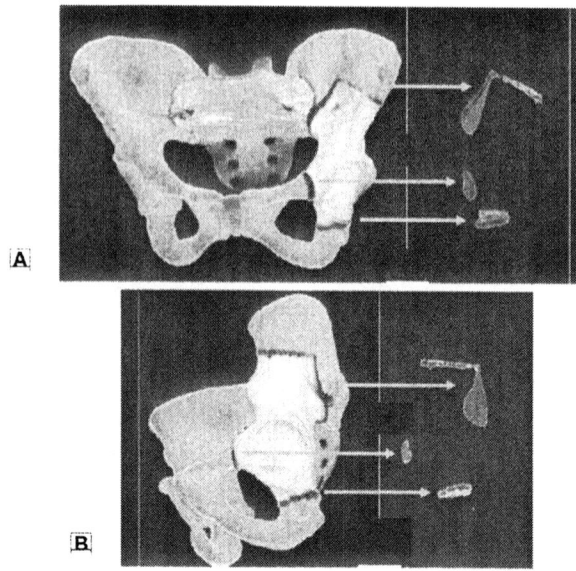

Fig. 3. 3D CT-scan of a host-graft reconstruction with the scanner-high-density epoxy paste filling gaps. A threshold segmentation permitted to eliminate voxels corresponding to the Sawbone and to calculate gap volume of the host-graft junction. A) Antero-posterior view before and after segmentation of the paste. B) Lateral view.

2.4 Statistics

All analyses were performed using statistical software SPSS®, version 12.0 (SPSS Inc., Chicago, IL. The inter-observer agreement of the variable "degree of contact" was calculated using weighted kappa statistics [19,20]. Non parametric Mann-Whitney tests or non parametric Wilcoxon tests were performed to compare each pair of observers for the variable "maximal gap" and each pair of surgeons for the 4 variables "margin error", "maximal gap", "gap volume" and "mean gap". Differences were considered statistically significant when $p < 0.05$. To assess linear relationship between the 4 HGJ variables ("degree of contact", "maximal gap", "gap volume" and "mean gap"), square of Pearson correlation coefficient was measured.

3 Results

24 cutting planes and 24 HGJ were available for evaluation: 6 cutting planes and HGJ for each surgeon (1 cutting plane and HGJ for each T1, 3 for each T2 and 2 for each T3).

3.1 Tumor cuttings

Mean margin error was 5.27 (SD 4.42) mm. Margin errors of the 4 surgeons were not statistically significantly different (Table 1). Among the 24 cutting planes, 11 did not respect the accepted 5 mm-tolerance above or below the target (Figure 4). 2 cutting planes were intralesional: error of 15.02 mm and 15.14 mm below the target respectively.

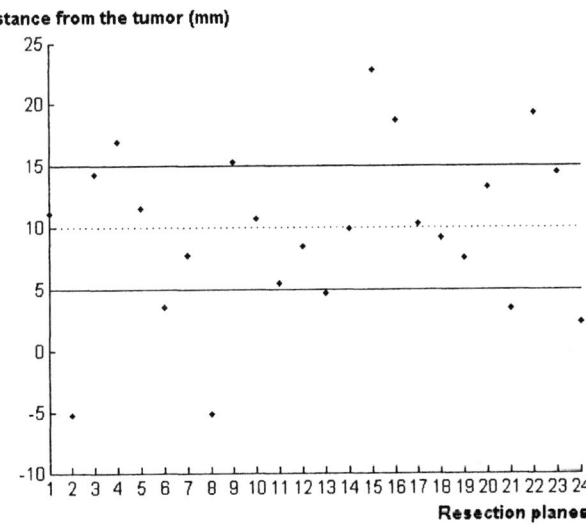

Fig. 4. Distances between the 24 cutting planes and the tumor. The dashed line represents the recommended 10 mm-safe margin, and the lines above and below it represent the accepted tolerance interval.

3.2 Host-Graft assemblies

Of the 24 HGJ, 22 were identically classified by the 3 observers (Obs1, Obs2 and Obs3). Most of HGJ were classified "<50%" contact (Figure 5c). 2 HGJ were classified differently: the first was classified ">50%" contact by 1 observer and "<50%" contact by the 2 others, and the second was classified ">50%" by 2 observers and "<50%" by one observer. Weighted Kappa correlation coefficients between observers were always very good: $\kappa = 0.91$ for (Obs1, Obs2) and (Obs1, Obs3) and $\kappa = 1.0$ for (Obs2, Obs3).

Mean maximal gap between host and graft for the 24 HGJ was 3.31 (SD 1.93) mm. Differences between surgeons were not statistically significant (Table 1). Measurements performed by the 3 observers were highly correlated: Spearman correlation coefficient was 0.95 for (Obs1, Obs2), 0.96 for (Obs1, Obs3) and 0.97 for (Obs2, Obs3). Differences between observers were statistically significant for (Obs1, Obs3) (p = 0.03) and for (Obs2, Obs3) (p = 0.02).

Mean gap volume between host and graft was 2.60 (SD 2.18) cm^3. There was no statistically significant difference between surgeons (Table 1).

Gap volume correlated strongly with contact surface (Figure 5a). Mean gap between host and graft was 3.22 (SD 2.08) mm. There was no statistically significant difference between surgeons, except between Surg1 and Surg2 (Table 1).

Correlation between the 4 HGJ variables was poor (Figure 5). Squared Pearson correlation coefficient r^2 was 0.26 between maximal gap and gap volume, 0.19 between maximal gap and mean gap and 0.05 between gap volume and mean gap.

Table 1. Margin error and host-graft junction parameters according to the 4 surgeons (mean ± SD)

	Tumor cutting	Host-Graft assembly		
	Margin error (mm)	Maximal gap (mm)	Gap volume (cm^3)	Mean gap (mm)
Surgeon				
S1	5.90 ± 5.09	4.22 ± 2.63	2.58 ± 2.84	2.39 ± 0.71[a]
S2	4.89 ± 5.32	2.98 ± 0.94	2.98 ± 2.09	3.92 ± 0.79[a]
S3	4.67 ± 5.24	2.43 ± 1.42	2.17 ± 2.33	2.51 ± 1.43
S4	5.63 ± 2.65	3.63 ± 2.25	2.67 ± 1.88	4.08 ± 3.72

[a] Mann-Whitney U test p = 0.016

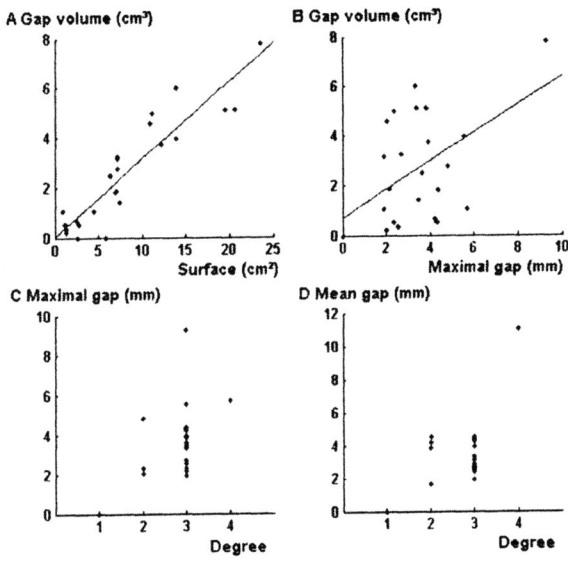

Fig. 5. Correlation between host-graft junction parameters : maximal gap, gap volume, contact surface, mean gap and degree of contact.

4 Discussion

This study was performed on plastic pelves under ideal working conditions: complete visualisation of the bone surface, complete accessibility, absence of muscles, nerves and bleeding. Figure 4 showed that 4 experienced surgeons did not manage to consistently respect a 5 mm-tolerance above or below a requested 10 mm-surgical margin. Moreover, there was a great variation among results of each surgeon, demonstrating the lack of accuracy.

According to the classification of surgical margin made by Enneking [5], Delloye [6] in their series of 24 patients, reported that surgical resection was wide in 19 cases, marginal in 6 and intralesional in 1. Our data, obtained under ideal working conditions, are consistent with those findings.

Inaccuracy during tumor resection can thus be explained by the anatomy of the pelvis. Due to its complex 3D geometry, delineating tumor extension, planning cutting planes on 2D CT slices, manually transferring this planning on the bone and performing the resection are difficult tasks even for experienced surgeons.

We believe that the accuracy of tumor cutting could be improved by using computer and/or robotically assisted (CA) technologies. Optical navigation systems using preoperative CT-scan data already exist for tumor resection in the pelvis [21,22] and also for sacroiliac screw insertion [23], pelvic osteotomies [24,25] or pelvic ring fracture reduction [26,27]. However, to our knowledge, a comparison of accuracy of osseous pelvic tumor resection between conventional and CA procedures has not yet been performed. Data presented in this study could be useful for further evaluation of CA technologies.

We explored several methods to quantify specific geometrical properties of a HGJ. Classification of HGJ according to the degree of contact and measurement of maximal gap between host and graft offered good repeatability. For maximal gap, Obs3 had tendency to measure smaller gaps than Obs1 and Obs2, resulting in a significant difference. However, mean difference was about 10% of the mean maximal gap.

Enneking [17] emphasised the need for an accurate and intimate contact at the osteotomy site in order to promote and accelerate union. Delloye [6] stated gaps and narrow surfaces were both concerns of non-union at the junctions. Consequently, we assumed a good reconstruction, i.e. a good osseous assembly, has a good degree of contact, a small maximal gap, a small gap volume and a small mean gap. However, correlation between our HGJ measures was relatively poor.

Gap volume has some clinical meaning. It quantifies osteogenesis required to completely fill the gap between host and graft. Gap volume was poorly correlated to maximal gap (Figure 5b) and degree of contact. Junctions with smallest gap volumes had not necessarily smallest maximal gaps nor highest degrees of contact.

In conclusion, the lack of correlation and the great variation among each result show that assessing the quality of an osseous assembly by our HGJ geometrical parameters is not efficient. It is also impossible to point the different sources of error that can occur during reconstruction: manual transfer of cutting planes on the allograft, cutting of the allograft, positioning and orientation of the cutting tool on the bone, etc. As for tumor cutting, we believe that using CA technologies could improve the accuracy of allograft cutting.

Even for experienced surgeons working under ideal conditions, our attempt to define what a good pelvic reconstruction is was not satisfactory. Again, this can mainly be explained by the complex 3D geometry of the pelvis and consequently by the great diversity of encountered HGJ. We believe that there is a need to define new standards of evaluation that would be independent of the bony structure.

We think that mechanical engineering tools, like geometrical tolerances and mechanical fittings, could be well adapted to this problem of evaluation of an osseous assembly. Well-known concepts like flatness, roughness, parallelism, angularity, etc, or new concepts like modal tolerancing [28] seem to be interesting tools. New research works are about to be launched to know how these tools could be used to evaluate the geometrical accuracy of bone machining and osseous assembly and how they could be integrated in the usual surgical language.

Acknowledgments

The authors would like to thank Pr Christian Delloye and Dr Olivier Cornu of the Department of Orthopaedic Surgery (Cliniques universitaires St-Luc, Université catholique de Louvain, Belgium) for performing the simulated surgeries.

References

1. D. Donati, A. El Ghoneimy, F. Bertoni, C. Di Bella and M. Mercuri, Surgical treatment and outcome of conventional pelvic chondrosarcoma, *J Bone Joint Surg (Br)* **87**, 1527-1530 (2005).
2. N. Kawaguchi, S. Matumoto and J. Manabe, New method of evaluating the surgical margin and safety margin for musculoskeletal sarcoma, analysed on the basis of 457 surgical cases, *J Cancer Res Clin Oncol* **121**, 555-563 (1995).
3. M.E. Pring, K.L. Weber, K. Unni and F.H. Sim, Chondrosarcoma of the pelvis : A review of sixty-four cases, *J Bone Joint Surg (Am)* **83**, 1630-1642 (2001).
4. K.L. Weber, M.E. Pring and F.H. Sim, Treatment and outcome of recurrent pelvic chondrosarcoma, *Clin Orthop* **397**, 19-28 (2002).
5. W.F. Enneking, S.S. Spanier and M.A. Goodman, Current concepts review. The surgical staging of musculoskeletal sarcoma, *J Bone Joint Surg (Am)* **62**, 1027-1030 (1980).
6. C. Delloye, X. Banse, B. Brichard, PL. Docquier and O. Cornu, Pelvic reconstruction with a structural pelvic allograft after resection of a malignant bone tumor, *J Bone Joint Surg (Am)* **89**, 5795-87 (2007).
7. R.S. Bell, A.M. Davis, J.S. Wunder, T. Buconjic, B. McGoveran and A.E. Gross, Allograft reconstruction of the acetabulum after resection of stage-IIB sarcoma. Intermediate-term results, *J Bone Joint Surg (Am)* **79**, 1663-1674 (1997).
8. N. Dion and F.H. Sim, The use of allografts in orthopaedic surgery. Part I : The use of allografts in musculoskeletal oncology, *J Bone Joint Surg (Am)* **84**, 644-654 (2002).
9. G.E. Friedlaender, D.M. Strong, W.W. Tomford and H.J. Mankin, Long-term follow-up of patients with osteochondral allografts. A correlation between immunologic responses and clinical outcome, *Orthop Clin North Am* **30**, 583-588 (1999).
10. A. Hillman, C. Hoffman, G. Gosheger, R. Rödl, W. Winkelmann and T. Ozaki, Tumor of the pelvis : complications after reconstruction, *Acta Orthop Trauma Surg* **123**, 340-344 (2003).
11. F.J. Hornicek, M.C. Gebhardt, W.W. Tomford, J.I. Sorger, M. Zavatta, J.P. Menzner and H.J. Mankin, Factors affecting nonunion of the allograft-host junction, *Clin Orthop Relat Res* **382**, 87-98 (2001).
12. H.J. Mankin, F.J. Hornicek and K.A. Raskin, Infection in massive bone allografts, *Clin Orthop Relat Res* **432**, 210-216 (2005).
13. E. Ortiz-Cruz, M.C. Gebhardt, L.C. Jennings, D.S. Springfield and H.J. Mankin, The results of transplantation of intercalary allografts after resection of tumors, *J Bone Joint Surg (Am)* **79**, 97-106 (1997).
14. J.L. Sorger, F.J. Hornicek, M. Zavatta, J.P. Menzner, M.C. Gebhardt, W.W. Tomford and H.J. Mankin, Allograft fractures revisited, *Clin Orthop* **382**, 66-74 (2001).
15. R.C. Thompson Jr., A. Garg, D.R. Clohisy and E.Y. Cheng, Fractures in large-segment allografts, *Clin Orthop Relat Res* **370**, 227-235 (2000).

16. D.L. Wheeler and W.F. Enneking, Allograft bone decreases in strength in vivo over time, *Clin Orthop Relat Res* **435**, 36-42 (2005).
17. W.F. Enneking and D.A. Campanacci, Retrieved human allografts. A clinicopathological study, *J Bone Joint Surg (Am)* **83**, 971-986 (2001).
18. W.F. Enneking and W.K. Dunham, Resection and reconstruction for primary neoplasms involving the innominate bone, *J Bone Joint Surg (Am)* **60**, 731-746 (1978).
19. A. Petrie, Stastitics in orthopaedic papers, *J Bone Joint Surg (Br)* **88**, 1121-1136 (2006).
20. J.R. Landis and G.G. Koch, The measurement of observer agreement for categorical data, *Biometrics* **33**, 159-174 (1977).
21. T. Hüfner, M. Kfuri Jr., M. Galanski, L. Bastian, M. Loss, T. Pohlemann and C. Krettek, New indications for computer-assisted surgery: tumor resection in the pelvis, *Clin Orthop Relat Res* **426**, 219-225 (2004).
22. C. Krettek, J. Geerling, L. Bastian, M. Citak, F. Rucker, D. Kendoff and T. Hüfner, Computer aided tumor resection in the pelvis, *Injury* **35**, 79-83 (2004).
23. E. Gautier, R. Bachler, P.F. Heini and L.P. Nolte, Accuracy of computer-guided screw fixation of the sacroiliac joint, *Clin Orthop Relat Res* **393**, 310-317 (2001).
24. F. Langlotz, R. Bächler, U. Berlemann, L.P. Nolte and R. Ganz, Computer assistance for pelvic osteotomies, *Clin Orthop Relat Res* **354**, 92-102 (1998).
25. G. van Hellemondt, M. de Kleuver, A. Kerckhaert, P. Anderson, F. Langlotz, L.P. Nolte and P. Pavlov, Computer-assisted pelvic surgery: an in vitro study of two registration protocols, *Clin Orthop Relat Res* **405**, 287-293 (2002).
26. T. Hüfner, T. Pohlemann, S. Tarte, A. Gänsslen, M. Citak, N. Bazak, U. Culemann, L.P. Nolte and C. Krettek, Computer-assisted fracture reduction: novel method for analysis of accuracy, *Comput Aided Surg* **6**, 153-159 (2001).
27. T. Hüfner, T. Pohlemann, S. Tarte, A. Gänsslen, J. Geerling, N. Bazak, M. Citak, L.P. Nolte and C. Krettek, Computer-assisted fracture reduction of pelvic ring fractures: an in vitro study, *Clin Orthop Relat Res* **399**, 231-239 (2002).
28. PA. Adragna, S. Samper, H. Favrelière, Tolérancement modal - un langage de spécification, 2nde journée européenne du tolérancement, Annecy (June 20, 2006).

PART III

Gripping and Feeding Solutions for Micro-Assembly

Chapter 6

Micro-Gripping Methods and Applications

A CRITICAL REVIEW OF RELEASING STRATEGIES IN MICROPARTS HANDLING

Gualtiero Fantoni, Marcello Porta
Department of Mechanical, Nuclear and Production Engineering, University of Pisa, Italy

Abstract In the last ten years, various grasping principles and releasing techniques suitable for microassembly have been hypothesised and successfully tested. Since in microdomain adhesion forces overcome the gravity one, new grasping principles have been exploited to grasp microparts. Unfortunately, in microassembly the most critical phase is not to grasp a micropart, but to release it. Therefore, the development of releasing strategies plays a fundamental role in the entire assembly cycle. This paper proposes a scheme for classifying many releasing strategies developed in the microassembly field, provides a map of interesting grasping-releasing couples and analyses in detail one of the most reliable grasping principles (i.e. capillary gripper) and the related possible releasing strategies. Finally, a procedure for the selection of grasping and releasing strategies on the basis of the components to be handled and on the boundary conditions is provided.

Key words releasing, design for microassembly, grasping

1 Introduction

In the microassembly process, grippers are very important because they have to pick up microparts and release them with the correct orientation in the right place with high accuracy. Furthermore, they have to not damage or contaminate the microparts they handle. Many strategies have been developed for grasping microparts and the gripper configuration depends in general on the features of the object to be grasped and on the dominant forces at microscale [1]. Some microgrippers have been directly downscaled from the macro world while others, thanks to the very small dimensions and mass of handled microparts, take advantage of surface forces or other physical effects. A few authors have developed microgrippers changing systems [2] to handle many different microcomponents.

With the exception of few contactless grippers (e.g.: Bernoulli's, laser and sonodrote grippers), the main problem of the contact ones (exploiting friction, suction, jaw, surface forces etc.) concerns the releasing task. The gravity force, generally used in traditional assembly to release components, is often less relevant than adhesion forces in microdomain. Hence, components tend to remain stuck to the gripper, causing the need for the development of innovative releasing strategies.

Even if the releasing phase has been deeply discussed in [3, 4], novel strategies and approaches have been developed in the last few years. Therefore an up-to-date survey seems to be necessary. The methodology chosen in this paper for the study and the analysis of the releasing strategies in microassembly consists of the following steps:

- it starts from the collection of literature concerning the releasing phase;
- data are organised to highlight problems, advantages and opportunities of each releasing method;
- grasping strategies are coupled with the considered releasing methods;
- finally, each grasping principle has to be analysed to investigate possible releasing solutions: in this paper the capillary principle is considered as an example.

2 Grasping and Releasing Strategies in Microdomain

The main sources for the analysis of the releasing principles and strategies were papers, websites and technical documentations. This analysis allows the authors to cluster homogeneous papers and to organise them expanding the taxonomy from general principles to detail solutions.

The ontology proposed here splits the releasing strategies into two separate groups: *passive releasing strategies* (Table 1), where suitable gripper features or environment conditions make possible the reduction of adhesive forces between gripper and microparts; and *active releasing strategies* (Table 2), where parts can be released by means of additional actions. *Passive releasing strategies* exploit **microgrippers features** (in terms of shape, surface coatings and material) or **environment conditions** to reduce the force acting at the microscale, such as electrostatic, adhesion, van der Waals ones. On the contrary, the *active releasing strategies* make use of additional actions to allow the grasped object to be detached from the gripper. These additional actions can be (i) **forces** able to overcome the adhesive ones between gripper and object or (ii) **means** to reduce the **contact area**. With regard to additional forces, they can be supplied by external equipment or by suitable substrate (or substrate features) where the objects have to be released.

It is important to highlight that often releasing strategies make use of more than a single approach at the same time: in this case the authors tried to recognise the most significant releasing method. Furthermore obvious adjustments, such as for example a clean environment (dust is recognized to be a factor that modifies adhesion forces and friction [3]), are considered basic requirements for handling and do not appear in Table 1 and 2.

Tables 1 and 2 are organised as follows: each *Releasing Principle* is briefly described and a *Scheme* is provided. The *Releasing Principle* is defined as the physical principle which (i), decreasing the adhesive forces, makes gravity and/or other forces able to detach the part from the gripper or (ii), overcoming the adhesive forces, causes the releasing of the micropart. Table 1 (passive releasing strategies), differs from Table 2 (active releasing strategies) because it has not the columns *Problems* and *Released components*. Actually, while Table 1 contains general design rules applicable in cases of direct contact between the gripper and the micropart, conversely Table 2 also describes the *Problems* derived by the introduction of additional forces and provides a list of the *Released components*.

Type	Principle	Scheme	Description	Force↓
Gripper	Conductive material/coatings -*Grounded gripper*		Conductive materials or coatings (which do not form insulating oxides) reduce static charges. Grounded grippers prevent the charge storage [3, 5]	electrostatic
	Low difference of EV potential		Gripper and object made of materials with a small potential difference reduce "contact interaction" forces [5]	electrostatic
	Hydrophobic coating		Hydrophobic coating reduces surface tension effects: it prevents the adsorption of moisture [6]	surface tension
	Low Hamaker constant Coating		Low Hamaker constant coating reduces van der Waals forces [3]	van der Waals
	Hard materials		Contact pressure causes deformations, increasing the contact area between gripper and object: grippers made of hard material have to be preferred [5]	van der Waals, electrostatic
	Rough surface -*Micro pyramids*		The gripper roughness reduces the contact area and sharp edges induce the self discharge effect [5, 6]	van der Waals, electrostatic
	"Spherical" fingers		Spherical fingers reduce the contact area in comparison with planar ones [5]	van der Waals, surface tension
Environment	Dry atmosphere/ Heating the environment		A dry atmosphere reduces surface tension effects (but increases the risk of triboelectrification and the generation of electrostatic force) [3, 5]	surface tension
	Vacuum		If no moisture affects the contact, there is no liquid bridge and so surface tension is reduced [5] (but risk of triboelectrification)	surface tension
	No O_2 in the environment		If there is no oxygen, native oxide can not arise on the surface of gripper/handled objects [5]	electrostatic
	In fluid releasing		Assembly in fluid eliminates surface tension effects and reduces electrostatic force [5, 7]	electrostatic, surface tension
	Ionized air		Ionized air can neutralize free charges on the surfaces and so it reduces electrostatic force [8]	electrostatic

Table 1. Passive releasing strategies

Type	Principle	Scheme	Description	Problems	Released components
forces	Air pressure 1.Direct 2.Indirect (Adsorption force)	air	1. An air pressure flow [9, 10] overcomes adhesion forces; 2. By heating a suitable end effector the object is released thanks to the adsorption force [8]	Possible lack of precision in the releasing place	1. 50-300µm parts [9]; square silicon chips (4.2*4.2*0.5mm³) of 20.5mg; [10]; 2. Max. adsorption force 0.22µN [8]
	Acceleration Or vibration		An acceleration or a vibration given to the gripper support allows the object to be detached thanks to inertial forces [11, 12, 13]	Possible lack of precision in the releasing place [13]	40µm pollen microspheres [12]; 400µm spherical and 900µm half spherical lenses [13]
	Micro heater 1.Evaporating moisture/liquid 2.Melting ice		1. A micro heater reduces the moisture-liquid (so surface tension-capillary forces) [3]; 2. Melting of ice (ice gripper) [14, 15]	Temperature sensitive parts can be damaged by heat	SMD plastic elements, small copper coils for telecommunication [14]
	Electrostatic force control 1.Shorting the gripper 2.Voltage tuning 3.Inverting polarity		Parts are released by the electrostatic force control: 1. Shorting down the gripper electrodes [3]; 2. Tuning the electrostatic force between gripper and substrate [16]; 3. Inverting the polarity [17, 18]	Problem in releasing conductive components [18]	2. Metallic spheres of d=30µm [16]; 3. Spheres of d=100-800µm and cubic valve (l=180µm) [17]; Spheres, cylinders of 300-1000µm [18]
	Different adhesion force 1.Adhesion on substrate 2.Different adhesion tools 3.Different volume of liquid	Tool A Tool B	Objects pass from a tool A to a tool B exploiting the difference in adhesion force between the tools and the object. The tool B can be: 1. A substrate [17]; 2. A gripper [12]; 3. The force difference can be given by different volume of the same liquid [19]	The object has to be detached from the tool B (if the releasing place on tool B is not the final place)	1. Glass spheres with d=100-800µm and cubic valve flap with edge 180µm [17]; 2. 40µm pollen microspheres [12]
	Engagement by the substrate/ tool 1.Snap 2.Against edge 3.Scraping 4.Rolling 5.Needle	Glue	The object is released by its mechanical engagement on the substrate [10] or another tool. This strategy includes: 1. The use of snaps [20]; 2. Part against an edge [9, 21, 22]; 3. Scraping [12]; 4. Rolling [12]; 5. Use of needle [10]	Often additional features on the substrate are required	2. Metallic/non metallic parts of 50-300µm [9]; 3.-4. 40µm pollen microspheres [12]; 5. Square silicon chips (4.2*4.2*0.5mm³) of 20.5mg; [10]
	Gluing on substrate		Parts are released by gluing them on the deposal place [10]	Not suitable for moving parts	Square silicon chips (4.2*4.2*0.5mm³) of 20.5mg; [10]

reduction of the contact area	3D handling of the gripper *1. Variation the curvature* *2. Tilting* *3. Parallel motion*	*Add. tool*	A decreasing of the contact area through: 1. Varying the gripper curvature from a flat shape to a curved one [23]; 2. Tilting the gripper [24]; 3. Parallel motion of the gripper respect to the substrate [25]	Complex 3D handling of the gripper. Many DOF required.	1. Minimum object weight 98mg [23]; 2. Metallic spheres of d=20-30µm [24]; 3. 40µm pollen microspheres [12]
	Additional tool		An additional tool (with little contact area with the object) allows the object to be first detached from the gripper [24], then released on the substrate by removing the tool	Many devices in a small space	Metallic spheres with diameter of 20-30µm [24]
	Roughness change		The roughness change reduces adhesion forces allowing the part to be released [3]	Difficulties in realization	
	Electrowetting		The modification of the liquid drop by an electrostatic field reduces the contact area [11]	Difficulties in meniscus control	

Table 2. Active releasing strategies

The classification schemes proposed in Tables 1 and 2 have to be considered as necessary steps for the development of rules for the selection of the most suitable couples of grasping-releasing principle. Actually, in order to make grippers able to pick and place microparts in a reliable way, often a grasping approach has to be coupled, from the design phase, with one or more releasing strategies. Table 3 updates the works presented in [26, 27] and has been created to map the grasping principles and the possible releasing strategies adopted and/or suggested in literature. The grasping methods are adapted from the classification of grasping principles carried out by [1], while releasing methods come from the classification reported in Tables 1 and 2. In comparison with [1], the item "Surfaces forces (general)" has been introduced in grasping methods to include authors that generally refer to adhesive forces as grasping forces [24].

			GRASPING PRINCIPLES					
			Friction	Cryogenic	van der Waals	Electrostatic	Capillary force/surface tension	Surface forces (general)	Suction
Passive Releasing	Gripper	Conductive material/coatings	[5]		[3]	[3]		[3]	
		Low difference of EV	[5]						
		Hydrophobic coating					[11, 28]	[4]	
		Hard materials	[5]						
		Rough surface	[5] [6]		[3]			[3]	
		"Spherical" fingers	[5]						
	Environment	Dry atmosphere	[5]					[3]	
		Vacuum						[3]	
		In fluid releasing						[7]	
		Ionized air				[3]			
Active Releasing	Forces	Air pressure					[10]		[9]
		Acceleration or vibration				[13, 18]	[11, 12]	[3]	
		Micro heater		[14, 15]				[3]	[8]
		Electrostatic force control				[17, 18, 29]		[3]	
		Different adhesion force				[17]	[12, 19]		
		Engagement by the substrate				[22]	[10, 12]		[9]
		Gluing on substrate					[10]		[30]
		3D handling of the gripper			[25, 31]	[24]	[11, 12]		
	Contact area	Additional tool				[24]	[11]		
		Roughness contact area	[3]						
		Electrowetting					[11]		

Table 3. Releasing strategies available for grasping strategies

3 Releasing Strategies for Capillary Gripper

Each grasping principle and all the possible releasing methods in Table 3 can be analysed further. Hereinafter the case of capillary gripper has been used as an example. As demonstrated by the numerous papers in literature [19, 21, 23], the capillary gripper is one of the most reliable. The advantages of capillary grippers are [21] the ability to grasp:
- small components (the available capillary force is few mN);
- components with only one upper free surface;
- components with a small available grasping area;
- any kind of component in terms of material and shape;
- fragile components because the meniscus acts as a "bumper".

Moreover there is a favourable downscaling law (the force is proportional to the linear dimension). Finally, the capillary gripper is compliant and exerts a self centering effect due to surface tension.

Most of the releasing strategies in Tables 1 and 2 can be used for releasing objects grasped by capillary grippers. Unfortunately, the releasing phase is difficult even if numerous systems have been tested. Table 4 focuses on releasing approaches developed in literature for capillary grippers.

		Principle	Scheme	Problems/Difficulties	Advantages	Released compo-
RELEASING PRINCIPLES	Passive release / Gripper	Hydrophobic coating		-Manufacturing difficulties -Resistance of coating to few cycles	-Flexibility -Manufacturing by silicon technologies	Prevalently flat parts
	Passive release / Envir.	Dry Atmosphere (heating the environment)		-Energy consumption -Long cycle time -Scarce reliability	-Easy to be realised	Probably all shapes
	Active releasing / Forces	Air pressure		-Precise control of the volume and pressure of air	-Precise if done in contact with the substrate	All parts
		Acceleration or vibration		-Lack of precision in releasing	-Easy to be realised -Reliable releasing	Parts with not too low mass
		Micro heater		-Reach exactly the drop -Risk to damage temperature sensitive parts	-Various types of heating source -Local action	-All parts except temperature sensitive ones
		Different adhesion force		-The precise control of the dispensed volume of liquid is difficult	-Self centering of the object on the releasing drop	All parts

Contact area	Engagement by the substrate/tool		-Often not precise -Releasing from other tool -Structured substrate -Damages on the part	-Reliable releasing	It can depend on substrate features
	Gluing on substrate		-Releasing precision depending on the glue -Glue volume control -Glue curing	-Self centering on the glue drop	All parts
	3D handling of the gripper		-Difficulties in the surface curvature control -Difficult to reach small radii of curvature	-Flexibility -Many actuation strategies are available	Prevalently flat parts
	Additional tool		-Difficult to reach the part without wetting the tool -Damages on the part	-Precision	Probably all, better if flat parts
	Roughness change (*increase gap*)		-Manufacturing difficulties -Need for hydrophobic needles	-Flexibility	Probably all, better if flat parts
	Electrowetting		-Difficulties in meniscus curvature control -Charges induced both in liquid and component	-Direct control of the liquid meniscus (force)	Prevalently spherical parts

Table 4. Releasing strategies related to capillary grasping

4 Selection of Suitable Grasping-Releasing Strategy

As reported in [1], the selection of the suitable microgrip principle depends both on *part characteristics* in terms of physical and functional features (such as material, shape, mass, function, etc.) and on the boundary conditions (i.e. environment and assembly tasks). Once the grip principle has been chosen, it is necessary to verify which are the releasing strategies suitable to be coupled with the selected grasping approach (Table 3). If no releasing strategy shown in Table 1 and 2 makes possible the releasing of the components, a different microgrip principle has to be selected. Hence, only a correct selection of both the grasping and the releasing strategy allows a reliable design of the gripper. Actually, for each grasping method various releasing approaches can be used but, as shown in Table 4 for capillary grippers, every releasing strategy presents some advantages and drawbacks and is suitable to release components with particular features.

The procedure, proposed for the selection of the grasping-releasing strategy, is shown in Figure 1.

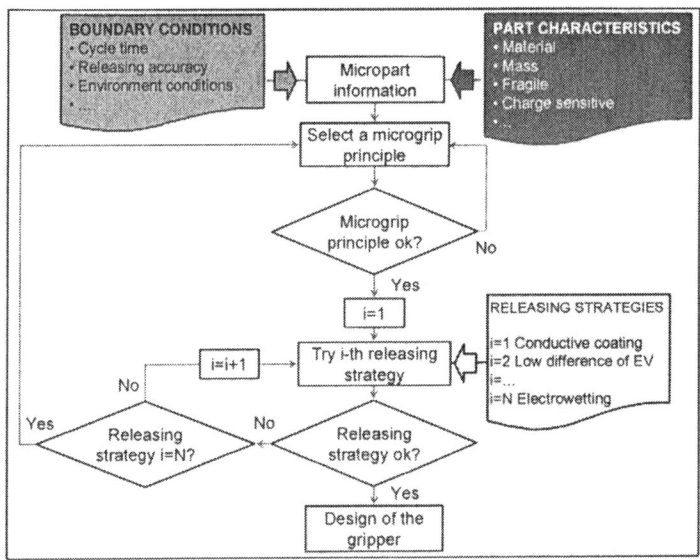

Figure 1. Conceptual procedure for the grasping-releasing selection

With regard to **part characteristics**, *geometrical parameters* are one of the first drivers in the selection. For example (see Table 4), a flat component could be grasped and released with capillary grippers owning hydrophilic and hydrophobic coating but probably the same gripper is not able to handle spheres and cylinders. Moreover, it is necessary to know the *component properties and functions*, i.e. if components are fragile, porous, dielectric, conductive, magnetic, diamagnetic, charge and water sensitive, etc. and/or if their surface is polished or must have optical characteristics. Actually, fragile parts can be damaged by mechanical grippers but also by a high stress releasing such as an engagement on the substrate or the use of an additional tool. For instance, charge sensitive SMD are altered by electrostatic grippers or electrowetting releasing. Optical lenses can be stained by the remaining traces left by liquid grippers. Similar damages can occur when these parts are released with glue or mechanical tool such as needles.

With regard to the **boundary conditions**, information about the *initial status* of the microparts is also necessary to choose the best grasping tool. Therefore different choices can be done if the parts are randomly positioned or correctly fed, if they are separated, regularly spaced (ordered) or they present difficulties in feeding (e.g. they stick, nest, tangle, or they are slippery, flexible, squiggle etc.). For example, an adhesive or an electrostatic gripper is not able to pick only a sphere from a group of some spheres in contact. Conversely if the parts are sufficiently spaced both principles can be successfully exploited. Also the requirements of the *final positioning* (2D, 3D, fine, coarse, etc.) determine the choice of the gripper or conversely prevent the use of some grasping principles. For example, a compliant adhesive gripper cannot be used for a forced peg-in-hole operation, while it could be particularly suitable for pick-and-place operation of delicate micro parts. Finally, the *environment* plays a fundamental role because it can allow or prevent the use of some grasping principles: even in a clean environment the small amount of dust can dramatically reduce the performance of a gripper (such as an electrostatic one) both in the grasping and in the releasing phase.

5 Conclusion and Future Development

A critical survey of the releasing strategies in microassembly has been performed. These strategies are organised in a novel classification scheme to allow a structured selection of the suitable methods for releasing a grasped microcomponent. Then, a matching of the possible releasing approaches related to each grasping principle has been developed and it should help the designer with a simple decision map. Furthermore, a deeper analysis of a particular microgripper (i.e. the capillary one) has been performed: the releasing strategies, allowed in this case, are compared in terms of advantages, drawbacks and handled components. In order to help the designer in the right choice of both grasping and releasing principle a step-by-step procedure has been finally proposed.

Future works concern the development of Design for Microassembly rules by extending this work to other grasping principles and assembly operations as for example microfeeding, microsorting, microinserting, microforcing and microgluing.

Acknowledgements

This research has been partially supported by the Italian Ministry of University and Research (MIUR) within the project 'Development of innovative technologies for the assembly of hybrid microproducts'. The authors thank Prof. M. Santochi for his continuous help and suggestions.

References

[1] Tichem, M., Lang, D., Karpuschewski, B., 2003, A classification scheme for quantitative analysis of micro-grip principles, Proc. of the 1st Int. Precision Assembly Seminar (IPAS'2003), 17-19 March, Bad Hofgastein, Austria.
[2] Clevy, C., Hubert, A., Agnus, J., Chaillet, N., 2005, A micromanipulation cell including a tool changer, Journal of Micromechanical and Microengineering vol.15/10, pp. 292-301.

[3] Arai, F., Andou, D., Fukuda, T., Nonoda, Y., Oota T., 1995, Micro Manipulation Based on Micro Physics -Strategy Based on Attractive Force Reduction and Stress Measurement-, Proc. of IEEE/RSJ Conf. on Robots and Intelligent Systems 2, pp. 236-241.

[4] Van Brussel, H., Peirs, J., Reynaerts, D., Delchambre, A., Reinhart, G., Roth, N., Weck, M., Zussman, E., 2000, Assembly of Microsystems, Annals of the CIRP, vol. 49/2.

[5] Fearing, R.S., 1995, Survey of sticking effects for micro parts handling, IEEE/RSJ International Workshop on Intelligent Robots & Systems (IROS), Pittsburgh.

[6] Arai, F., Andou, D., Nonoda, Y., Fukuda, T., Iwata, H., Itoigawa, K., 1996, Micro end effector with micro pyramids and integrated piezoresistive force sensor, Proc. of the IEEE/RSJ Int. Conf. on Intelligent Robots and Systems.

[7] Gauthier, M., Lopez-Walle, B., Clevy, C., 2005, Comparison between microobjects manipulations in dry and liquid mediums, Proc. of CIRA'05, Espoo, Finland.

[8] Arai F., Fukuda, T., 1997, Adhesion-type Micro Endeffector for Micromanipulation, Proc. of the IEEE Int. Conf. on Robotics and Automation (ICRA'97), pp. 1472-1477.

[9] Zesch, W., Brunner, M., Weber, A., 1997, Vacuum tool for handling microobjects with a NanoRobot, Proc. of the IEEE Int. Conf. on Robotics and Automation (ICRA'97), pp. 761-776.

[10] Bark, C., Binneboese, T., 1998, Gripping with low viscosity fluid, IEEE Int. workshop on MEMS, pp. 301-305.

[11] Lambert, P., 2005, PhD Thesis, Faculté des sciences appliquées, Université Libre de Bruxelles.

[12] Driesen, W., Varidel, T., Régnier, S., Breguet, J.-M., 2004, Micromanipulation by adhesion with two collaborating mobile micro robots, 4th Int. Workshop on Microfactories.

[13] Monkman, G.J., 2003, Electroadhesive Microgrippers, Assembly Automation vol. 24/1, MCB University Press.

[14] Kochan, A., 1997, European project develops "ice" gripper for micro-sized components, Assembly Automation, vol. 17/2, pp. 114-115.

[15] Lang, D., Kurniawan, I., Tichem, M., Karpuschewski, B., 2005, First investigations on force mechanisms in liquid solidification micro-gripping, In C Mascle (Ed.), ISATP 2005; the 6th IEEE Int. symposium on assembly and task planning, pp. 1-6.

[16] Saito, S., Himeno, H., Takahashi, K., Onzawa, T., 2002, Electrostatic detachment of a micro-object from a probe by applied voltage, Int. Conf. on Intelligent Robots and System, vol. 2, pp. 1790-1795.

[17] Hesselbach, J., Büttgenbach, S., Wrege, J., Bütefisch, S., Graf, C., 2001, Centering electrostatic microgripper and magazines for microassembly tasks, Microrobotics and Microassembly 3, Proc. of SPIE, vol. 4568, Newton, USA.

[18] Fantoni, G., Biganzoli, F., 2004, Design of a novel electrostatic gripper, International Journal for Manufacturing Science and Production, 6/4, pp. 163-179.

[19] Obata, K.J., Motokado, T., Saito, S., Takahashi, K., 2004, A scheme for micro-manipulation based on capillary force, Journal of Fluid Mechanics, vol. 498, pp. 113-121.

[20] Prasad, R., Böhringer, K.-F., MacDonald, N. C., 1995, Design, fabrication, and characterization of single crystal silicon latching snap fasteners for micro assembly, in ASME Int. Mech. Eng. Congr. Expo., pp. 917-923.

[21] Lambert, P., 2006, Design of a capillary gripper for a submillimetric application, Proc. of the 3rd Int. Precision Assembly Seminar (IPAS'2006), Bad Hofgastein, Austria, 19-22 February.

[22] Enikov, E.T., Lazarov, K.V., 2001, Optically transparent gripper for microassembly, SPIE, vol. 4568, pp. 40-49.

[23] Pagano, C., Ferraris, E., Malosio, M., Fassi, I., 2003, Micro-handling of parts in presence of adhesive forces, CIRP Seminar on Micro and Nano Technology 2003, Copenhagen, November 13-14, pp. 81-84.

[24] Koyano, K., Sato, T., 1996, Micro object handling system with concentrated visual fields and new handling skills, Proc. of the IEEE Int. Conference on Robotics and Automation, pp. 2541-2548.

[25] Miyazaki, H., Sato, T., 1996, Pick and place shape forming of three dimensional micro structures from fine particles, Proc. of the IEEE Int. Conf. on Robotics and Automation, ICRA'96, pp. 2535-2540.

[26] Santochi, M., Fantoni, G., Fassi, I., 2005, Assembly of microproducts: state of the art and new solutions, Proc.of the AMST'05, Udine June 8-9, pp. 99-115.

[27] Porta, M., 2007, PhD Thesis, Department of Mechanical Nuclear and Production Engineering, University of Pisa.

[28] Fantoni, G., 2005, PhD Thesis, Department of Mechanical Nuclear and Production Engineering, University of Pisa.

[29] Nakao, M., Tsuchiya, K., Matsumoto, K., Hatamura, Y., 2001, Micro Handling with Rotational Needle-type Tools Under Real Time Observation, Annals of the CIRP, vol. 50/1.

[30] Ansel, Y., Schmitz, F., Kunz, S., Gruber, H.P., Popovic, G., 2003, Development of tools for handling and assembling micro components, Journal of Micromechanics and Microengineering, pp. 430-437.

[31] Lambert P., Delchambre A., 2003, Forces Acting on Microparts: towards a numerical approach for gripper design and manipulation strategies in micro assembly, Proc. of the 1st Int. Precision Assembly Seminar (IPAS'2003), 17-19 March, Bad Hofgastein, Austria.

A LOW COST COARSE/FINE PIEZOELECTRICALLY ACTUATED MICROGRIPPER WITH FORCE MEASUREMENT ADAPTED TO EUPASS CONTROL STRUCTURE.

Kanty Rabenorosoa, Yassine Haddab, Philippe Lutz.

Laboratoire d'Automatique de Besançon
LAB UMR CNRS 6596
24, rue Alain Savary
25000 BESANCON
France

Abstract Over the last few years the demand for miniature objects and devices has continuously increased. So the need for MEMS (Micro Electro Mechanical Systems) and MOEMS (Micro Opto Electro Mechanical Systems) in various fields has become more and more important. New micromanipulation technologies must be developed that take into account the specificities of MEMS and MOEMS. For performing manipulation tasks, gripping is one of the fundamental functions. In this paper, the design and the control of a microgripper dedicated to very high precision tasks are presented. The LAB (Laboratoire d'Automatique de Besançon) has developed a microgripper based on the piezoelectric properties [1]. This gripper is equipped with a precise sensor for force measurement. Force measurement is very important in order to avoid damaging or destroying the fragile object. Various approaches were used to model and calibrate the microgripper. The control of the gripper is done using the EUPASS Control Structure [6].

1 Structure of the system

To handle micro-objects in a wide range of sizes, the designed microgripper is made up of two different parts (see Figure 1):
   - A mobile part, shown in Figure 2, composed of a coarse positioning device using a piezomotor from PiezoMotor® and a piezoelectric bending cantilever for fine positioning. This cantilever is the first finger of the gripper.
   - A fixed part, linked to the main support which constitutes the second finger of the gripper.

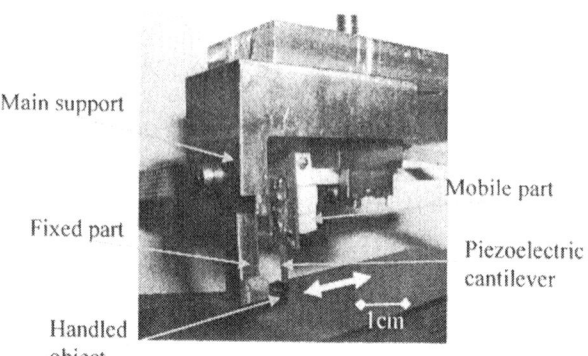

Fig.1. Structure of the microgripper

2 Force measurement

To perform force measurement, a strain gauge (length: 1.27mm, width: 0.38mm) is glued on every side of the piezoelectric cantilever (see Figure 2) which is bonded on a PCB. The force measurement depends on the gauge properties and the physical characteristics and dimensions of the cantilever. Other principles of force measurement are also possible [2] [3] [4].

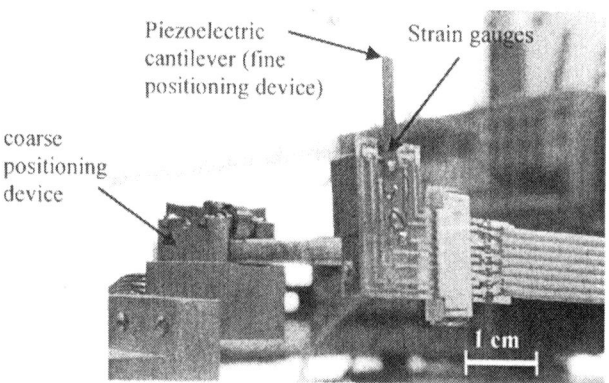

Fig. 2. The mobile part equipped with strain gauges

2.1 Principle of the cantilever piezoelectric beam

Piezoelectricity is the property of some materials (crystals and ceramics) to generate electrical charges in response to a mechanical stress. This is the direct piezoelectric effect. The reverse piezoelectric effect is the production of shape change when an electrical field is applied. The cantilever is composed of two PZT layers (see figure 3) because this structure allows larger deflexion than a cantilever with only one PZT layer. The dimensions of the cantilever are: length: 16mm, width: 2mm and height: 0.5mm. The bending cantilever response is studied and given in [5]. This bending cantilever is considered as two inputs/one output system. The inputs are the force F applied at the tip of cantilever and the voltage V applied across the electrodes. The output is the deflexion δ of the cantilever tip. These signals are linked by the following equation:

$$\delta(s) = \underbrace{\frac{\alpha}{as^2 + bs + 1}}_{dynamic\ behaviour} F(s) + \underbrace{\frac{\beta}{as^2 + bs + 1}}_{dynamic\ behaviour} V(s)$$

where:
s is the Laplace variable and α, β, a, b and c are calculated from the physical properties and dimensions of the cantilever.

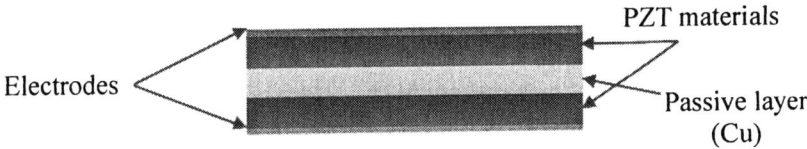

Fig. 3. Bending cantilever structure

2.2 Calibration

For the measurement, the strain gauges are integrated in a Wheatstone bridge. The output signal is amplified (gain 51.3) and filtered with a low pass filter with a cut-off frequency of 247Hz, in order to obtain a suitable range for the feedback signal. Step identification was done using dSPACE® board and a laser sensor from KEYENCE® (resolution: 10nm). The Wheatstone bridge associated with the conditioner (filter + amplifier) provides 50.75mV/mN sensitivity.

The response is:

$$\delta(s) = \frac{1.22}{3.43*10^{-8}s^2 + 9.26*10^{-6}s + 1} V(s) + \frac{1.231}{3.43*10^{-8}s^2 + 9.26*10^{-6}s + 1} F(s)$$

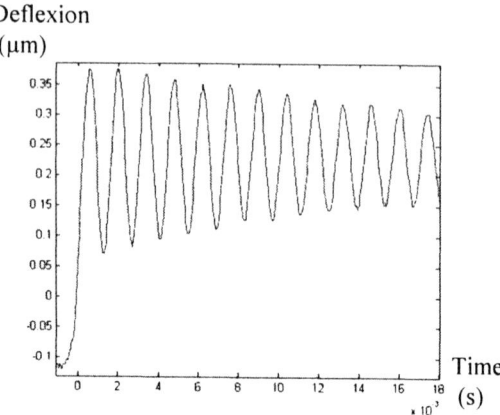

Fig. 4. The cantilever step response when applying the input voltage

The movement of the piezomotor is based on stick-slip principle and has its own controller box PDA 3.1 from PiezoMotor®. The piezomotor provides 10nm resolution in bending mode and 7.3N holding force. The velocity of piezomotor was determined by using laser sensor. The measurements show the non linearity of the velocity according to control voltage and the non symmetrical behaviour (see Figure 5).

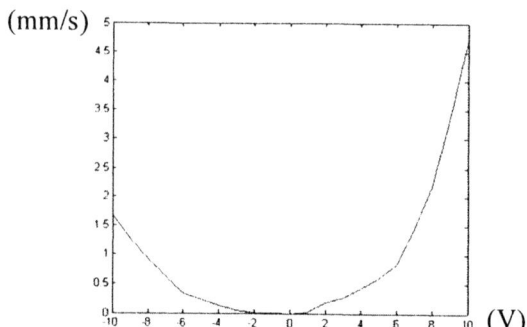

Fig. 5. Velocity of the piezomotor VS control voltage

3 Force controlled gripping tasks

The control of gripping force has many applications. During the grasping task, the force control has to prevent the release of the object without exceeding force limit. The microgripper is able to handle small components in a wide range of sizes (from 500μm to few mm). The applied force on the object is limited to 81mN for staying in the linear range.

The accuracy of force measurement depends on the output signal noise. In steady state, figure 6 shows the measured noise.

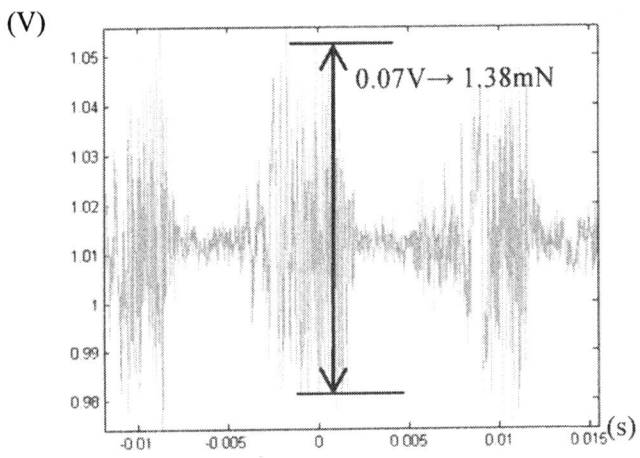

Fig. 6. Noise in steady state

4 Microgripper adapted to EUPASS Control Structure

The system has been adapted to the EUPASS Control Structure. For this, Beckhoff provided to EUPASS members the EUPASS Template Library. EUPASS Template Library is composed of the library of functions and function blocs and, the template program (written in Structured Text – IEC 61131-3) which is used to build up the customer specific modules.

In this application, the bending cantilever is used only for force measurement and the grasping is done by the piezomotor movement. The cantilever associated with strain gauges is used as force sensor. The EUPASS Control uses Beckhoff Embedded PC. This control structure supports OMAC (Open Modular Automation Control) PackML guidelines (State, Mode and Communication). The Beckhoff Embedded PC (CX1020) is used for controlling the microgripper in agreement with the EUPASS hardware specification. In combination with TwinCAT (The Window Control and Automation Technology), it works like a PLC (Programmable Logic Controller).

The CX1020 is associated with an analog input which converts the feedback signal and an analog output which controls the piezomotor. Input/Output work with 16bits resolution.

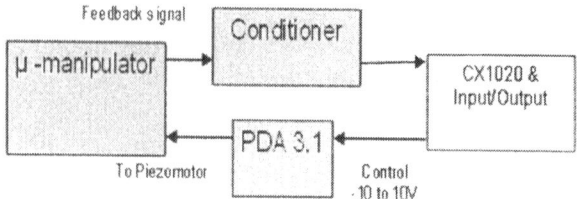

Fig. 7. Block Diagram of the system

Thanks to the structure of the microgripper (see Figure 1), the movement is divided into three steps. In the first step, the microgripper leaves its initial position to go near the object. This displacement is done with a constant speed which is calculated by calibration results. In the second step, the force control is processed by the function bloc PID during a defined time. This function is available in the library ("Tc.utilities.lib"). In the last step, the mobile part goes back to its initial position and grasp work is finished.

This function bloc uses the continuous PID which is discretized with the trapezoidal rule.

$$G(s) = k_p (1 + \frac{1}{T_n s} + \frac{1+T_v s}{T_d s})$$

k_p : Proportionnal gain, T_n : Integral gain,

T_v : Derivative gain, and T_d : Attenuation Derivative gain

Figure 8 shows the force during the grasping task.

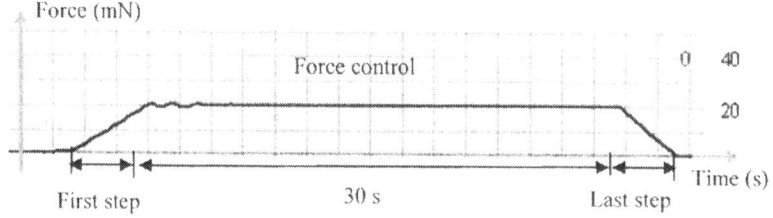

Fig. 8. Stable force control

PID's parameter: k_p :0.045, T_n :0.035, T_v :0 et T_d :1

The different steps of the movement are integrated in "Execute" state according to the EUPASS Template Library. The different steps of the movement correspond to a "Substate". During the test, the microgripper works in Automatic mode and its state switches between "Execute" and "IDLE".

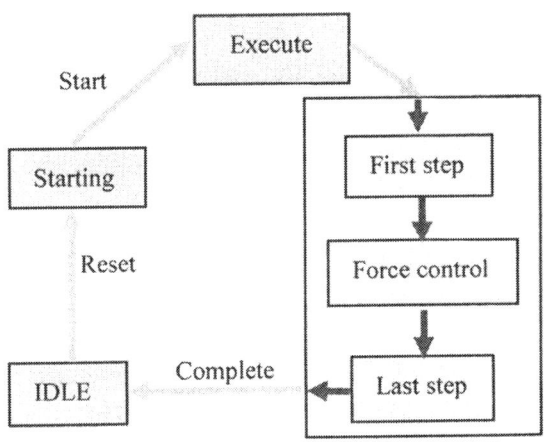

Fig. 9. Bloc diagram of the sequential state and substate

This microgripper can be integrated in an assembly cell or assembly line and co-operate with other EUPASS modules (Pick & Place, Gluing, etc.).The modules are able to communicate using distributed Master-Master Ethernet network and various topologies would be possible. The synchronisation task is done with a Publisher/Subscriber data exchange via real-time network among the modules.

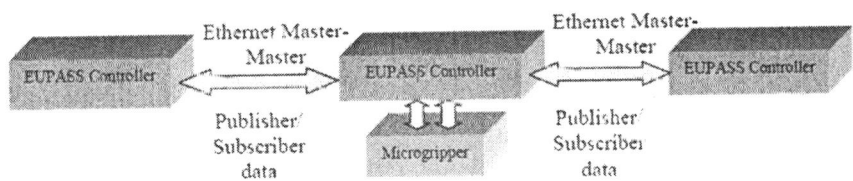

Fig. 10. EUPASS architecture

5 Conclusion

In this paper, we have presented a piezoelectrically actuated microgripper which performs force measurement using bimorph cantilever and strain gauges. The force measurement avoids damaging or destroying the manipulated micro-object (from 500μm to few mm). The maximum measurable force is 81mN and the force measurement error is 1.38mN.

This microgripper is adapted to the EUPASS Control Structure which ensures the standards to integrate an assembly cell or assembly line. This microgripper is able to cooperate with other modules in order to perform micro-assembly tasks.

In a future work, the bending cantilever will be used as an actuator to improve the dynamic characteristics of the gripper and the piezomotor for large displacements.

Acknowledgments

This work has been done in EUPASS project, supported by the public funding of the European Commission.

References

1. Qiao C., Yassine H., Philippe L. (UFC), Gebhard M. (Festo), A Low Cost Coarse/Fine Piezoelectrically Actuated Microgripper with Force Measurement, EUSPEN May (2007)
2. Markus Kemper. Development of a Tactile Low-Cost Microgripper with integrated Force Sensor. International Conference on Control Applications Taipei, Taiwan, September 2-4, 2004.
3. Lu, Z., Chen, Peter C. Y., and Lin, W., Force sensing and control in micromanipulation. IEEE Transactionson Systems, Man, and Cybernetics, Part C. September 2005.
4. Eisinberg A., Menciassi A., Micera S., Campolo D., Carrozza M.C., Dario P.," PI force control of a microgripper for assembling biomedical microdevices", IEE Proc. - Circuits Devices Syst., August 2001.
5. Joël Agnus, Contribution to the micromanipulation: studies, realization and command of a piezoelectric grip, November 2003.

DEVELOPMENT OF A MONOLITHIC SHAPE MEMORY ALLOY MANIPULATOR

Kostyantyn Malukhin[1], Kornel Ehmann[2]

[1]Department of Mechanical Engineering, Northwestern University, Evanston, IL, USA 60208
Fax: (847) 491 3915
e-mail: k-malukhin@northwestern.edu

[2]Department of Mechanical Engineering, Northwestern University, Evanston, IL, USA 60208
(Adjunct appointments at the University of Illinois at Urbana Champaign, Indian Institute of Technology – Kanpur, India and Chung Yuan Christian University, Chung Li, Taiwan)
Tel: (847) 491 3263, fax: (847) 491 3915
e-mail: k-ehmann@northwestern.edu

Abstract The essential steps and issues in the development of a Shape Memory Alloy (SMA) based multi-degrees-of-freedom monolithic manipulator are reflected upon. The use of Direct Metal Deposition (DMD) will be described for the manufacture of the manipulator's structure followed by the exploration of the feasibility of the "self-sensing" functionality of the structure as a basis for the manipulator's control. Finally, a synopsis of a new mathematical model of the kinetics of the temperature-induced phase transformation that was developed and its experimental verification will be given and shown that the analytical predictions closely follow the experimentally measured responses.

1 Introduction

The paper describes an R&D effort in progress that is aimed at the development of a compact monolithic meso-scale manipulator for space-limited applications in which a relatively high work output of the manipulator per unit volume of the application is required. Therefore, it is advantageous to use so-called "active" (functional) materials that can simultaneously function as structural as well as actuating and/or sensing elements in the manipulator's embodiment. Active materials include [1]: shape memory alloys, bimetals, piezoelectric materials, magnetostrictors, elecrtostrictors, electrorheological, magnetorheological, thixotropic and rheopex fluids, chemochromic, electrochromic, hydrochromic, photochromic, thermocromic elements, functional gels and others. The use of active materials may potentially result in fewer discrete parts in the manipulator's structure, thus increasing its robustness, precision, controllability, and work output. In addition, according to [2], there are several ways to improve the work output (power transfer to a load) of actuators that use "active" materials, for example, by matching the impedance of the active material and of the load. SMAs can function as structural, actuating and sensing elements in an active structure. They function as actuators due to the shape memory effect (SME) property of the alloy, resulting in the ability to recover large, mechanically induced strains [3].

In the current work an active material, namely NiTi (SMA), has been chosen for the fabrication of the proposed meso-scale monolithic manipulator. Three aspects of the development will be addressed. These include: (1) the manufacture and material characterisation of the monolithic manipulator's structure, (2) the feasibility of utilizing the SME to build "self-sensing" SMA structures, and (3) the control of the SMA actuation functions.

For the manufacture of the manipulator, mechanically pre-alloyed NiTi powder was processed by DMD [4] to generate its final geometry. The property of the SMA to undergo temperature-induced phase transformations allows the use of SMA elements as sensors as well. This property was evaluated to assess its feasibility to provide position feedback information. To facilitate the control of the manipulator an analytical modeling effort was undertaken to characterise the kinetics of the phase transformation. Theoretical and experimental results will be presented that confirm the suitability of the model to serve as a basis for the manipulator's control. A synopsis of these three developments is given below.

2 Design and manufacture of the manipulator

The conceptual design of the SMA NiTi manipulator is shown in Fig. 1.

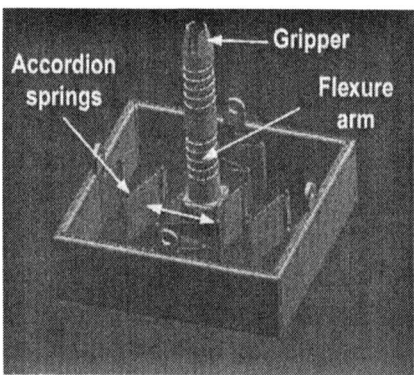

Fig. 1. Conceptual design of the SMA NiTi manipulator: approximate size 40 × 30 × 30 mm

The NiTi manipulator is a monolithic one-piece structure, that has three moving parts – accordion springs (accordion spring actuators), flexure arm and a microgripper. The accordion springs provide the X-Y in-plane motion of the flexure arm. The flexure arm provides the Z-motion of the gripper. All three moving elements posses the shape memory effect (SME), which allows them to generate the necessary motion when heated. The heating of the moving parts can be realized in several ways: (1) direct (resistive) heating, by passing an electrical current through the actuating springs, and (2) indirect heating through external heaters. Due to the relatively high heat capacitance and low thermal conductivity of NiTi alloys (e.g., in comparison to steel), localized heating of the moving elements of the manipulator will not produce a significant heat affected zone.

3 Characterization of the material fabricated by DMD

Preliminary research on the quality of the NiTi alloy manufactured by DMD was conducted. Pre-alloyed NiTi powder (NITINOL) was purchased from the Special Metals Corporation. The DMD apparatus was lent to us by Prof. Dr. R. Kovacevic (Southern Methodist University, RCAM lab). DMD is a "layer-by-layer" laser based additive manufacturing process during which metallic powder is deposited onto a substrate. The laser melts the powder, and the solidification of the melt takes place afterwards. It is possible to create complex 2.5- and 3-dimensional metallic structures by laser-based additive manufacturing processes [5].

Several NiTi samples were produced in the form of round plates (Fig. 2). The samples were deposited either on a Ti substrate (Fig. 2a) or on a NiTi (Fig. 2b) substrate. A smaller piece of the sample shown in Figure 2b was cut and polished for further analysis of the material's quality and properties. The scanning electron microscope (SEM) image of the polished sample shows a very low amount of pores, no voids, and no cracks (Fig. 3).

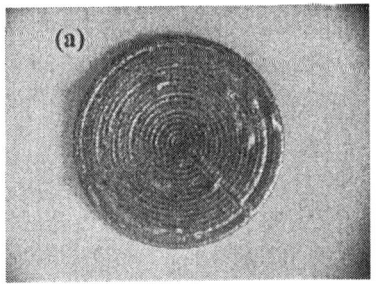
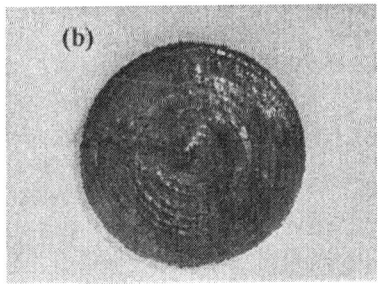

Fig. 2. NiTi samples deposited by DMD: *(a)* on Ti substrate, *(b)* on NiTi substrate

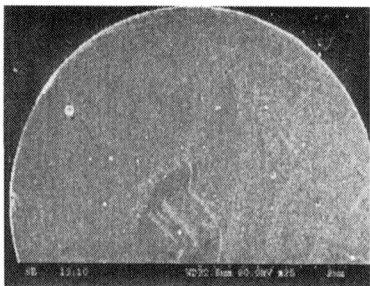

Fig. 3. Scanning electron microscopy (SEM) image of the polished NiTi sample

Secondary ion mass spectroscopy (SIMS) was performed on the sample. The analysis has shown that the amount of impurities in the DMD fabricated NiTi material is lower than in a commercially available NiTi material, manufactured using conventional manufacturing processes.

The DMD fabricated NiTi material was post heat-treated in an oven with a protective Ar atmosphere. Afterwards, the phase transformation properties of the material were measured by differential scanning calorimetry (DSC). The phase transformation temperatures (TTRs) that determine the temperature level at which material transforms from martensite to austenite, and *vice versa* [6] were measured and found to be close to room temperature. Furthermore, the effect of some of the parameters of the DMD process (laser power, powder feeding rate, and heat sink conditions of the DMD apparatus) was also investigated through a factorial design of the deposition experiments [7]. Several thin-wall structures were deposited onto NiTi substrates with varying DMD parameters (Fig. 4). The responses were the geometrical parameters of the walls and their roughness. Statistical analysis of the results has shown that laser power has an about 60% influence on the geometric and surface properties of the fabricated thin-walls. It was also found that the intensification of the heat sink conditions will lead to a better control of the quality and the geometry of the NiTi thin-wall structures.

Several prototypes of high quality accordion springs – actuating parts of the manipulator (Fig. 1) - were fabricated [7]. The springs are shown in Fig. 5.

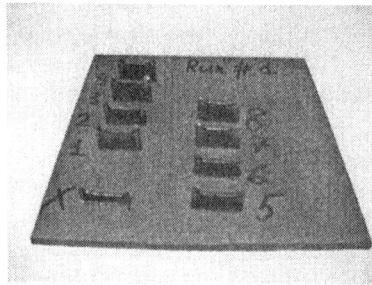
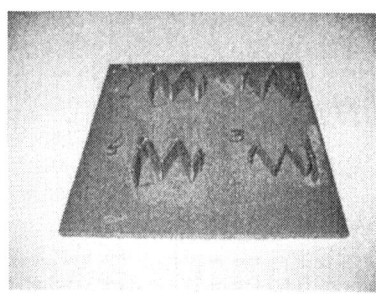

Fig. 4. Thin walls deposited on NiTi substrate   *Fig. 5.* Accordion springs on NiTi substrate

4 "Self-sensing" concept

SMAs can be used as displacement sensors due to the property of the material to change its electrical resistance during heating/cooling. In the present development the possibility of using the structure of the SMA actuator in sensing its motion with a micron level resolution was evaluated. This "self-sensing" property was explored by means of an SMA wire-based actuator [8]. The pre-stretched SMA wire (commercially available Flexinol wire) was subjected to a heat load, allowing the wire to recover its initial non-stretched length due to the shape memory effect. During the recovery process, the change in the electrical resistance of the SMA wire and its contraction displacement were measured. The heat was applied to the wire by passing an electrical current through it. The effect of the heat sink conditions on the response of the SMA wire to the heat load was studied as well. The experimental study has shown that by using the "self-sensing" method it was possible to measure a minimal SMA wire displacement of 17 μm with the resolution of 1.7 μm as seen from Fig. 6.

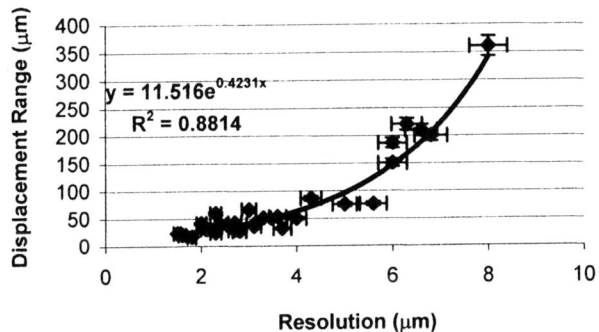

Fig. 6. Resolution of the "self-sensing" method (experimental results)

5 Mathematical modeling of the kinetics of the phase transformation in SMAs

A mathematical model describing the kinetics of the phase transformation in the NiTi material subject to a heat load was also developed. The final outcome of the model represents an analytical function of the austenite fraction, ξ_A, depending on the stress and temperature state in the NiTi material. The purpose of this model was two-fold: (1) to predict the amount of the austenite and, therefore, the magnitude of the recovery displacement of the moving parts of the manipulator during heating [8, 9], and (2) to serve as a basis for the development of the manipulator's controller. The developed model can be expressed as:

$$\xi_A = \sqrt{\frac{2C_p}{r_A}(T - A_{S0})} \qquad (1)$$

where:

C_p - is the specific heat, J/(kg°K),

A_{S0} - is the austenite phase transformation start temperature for the stress-free condition,

r_A - is the latent heat of the phase transformation in NiTi, J/kg.

The phase transformation kinetics model was verified experimentally. NiTi SMA wire actuators were used in the strain recovery experiments conducted. The wires were pre-stretched and heated by resistive heating – electrical current was passed through the wires. The voltage drop across the wires during the heating process was measured and correlated to the amount of the contraction displacement of the SMA wire, generated during the heat induced recovery process. The experimental values of the austenite fraction were calculated as the ratio of the recovered length of the wire to the initial pre-stretched length of the wire. Afterwards, they were compared to the values of the austenite fraction, computed from Eq. 1. The comparison between the experimentally and theoretically obtained values shows a very good agreement (Fig. 7).

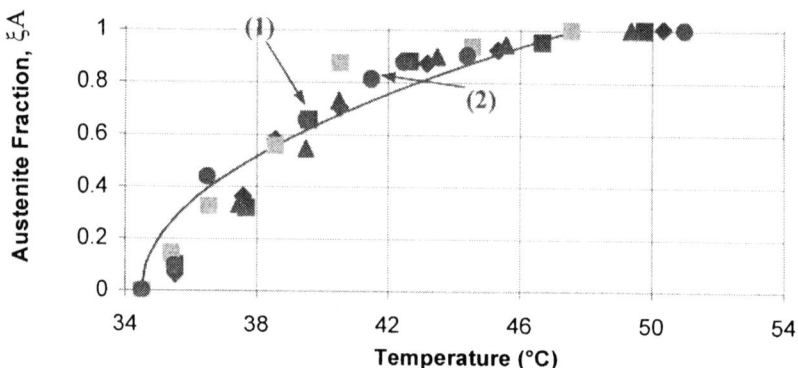

Fig. 7. Experimental (1) and analytical (2) values of the austenite fraction evolution during temperature-induced phase transformation

6 Mathematical modeling of the linear motion of the manipulator

The suggested phase transformation model serves as a basis for the development of the control law of the motion of a NiTi SMA manipulator. Firstly, the motion model of the linear stage of the manipulator was developed based on the equations of motion written for an "agonistic-antagonistic" layout of the accordion spring actuator according to [8, 9, 10]. Step response experiments were carried out using an SMA NiTi wire. The numerical simulations of the step response of the model of the motion of the "agonist-antagonist" actuator were undertaken using the "Simulink" toolbox in "MatLab" environment. The comparison of the simulation and experimental results is shown in Fig. 8.

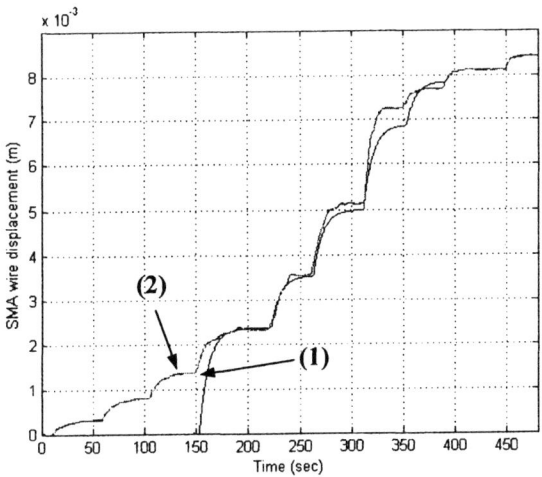

Fig. 8. Comparison of the numerical simulation (1) and experimental results (2) of the step response of the "agonist-antagonist" actuator model

Figure 8 shows an excellent agreement between the modeled and the experimentally measured displacements of the SMA wire-based actuator. Currently, work is in progress on applying this model to the accordion spring-based actuator.

7 Conclusions

The DMD process was used to process NiTi powder into high quality (non-porous, non-cracked, low contamination level) structures. A factorial design of experiments was used to identify optimal parameters for the deposition of thin walls by DMD from NiTi powder. It was possible to DMD fabricate several a high quality thin-wall NiTi structures – accordion springs. A mathematical model of the kinetics of temperature-induced phase transformation was developed and experimentally verified. A model of motion of an SMA actuator, based on the kinetics model, was developed and experimentally verified. The model will be used in the development of a control model of motion of arbitrary SMA actuators.

Acknowledgments

We are thankful to Professor Dr. R. Kovacevic and his colleagues at Southern Methodist University/RCAM for giving us the opportunity to use the DMD equipment and for their valuable help in all the DMD experiments. We also acknowledge the financial support of the National Science Foundation (USA) under grant #DMI-0400316.

References

1. H. Janocha, Actuators: basics and applications (Springer Berlin Heidelberg, New York 2004).

2. G.A. Lesieutre, J. Loverich, G.H. Koopmann, and E.M. Mockenstrum, Increasing the Mechanical Work Output of an Active Material Using a Nonlinear Motion Transmission Mechanism, Journal of Intelligent Material Systems and Structures, 15, pp. 49-58 (2004).

3. L.C Brinson, One-Dimensional Constitutive Behavior of Shape Memory Alloys: Thermo-mechanical Derivation With Non-Constant Material Functions and Redefined Martensite Internal Variable, Journal of Intelligent Materials Systems and Structures 4 (2), pp. 229-242 (1993).

4. R. Kovacevic R (2007); http://engr.smu.edu/rcam/rcamweb/index.htm.

5. J.K. Wessel. Handbook of advanced materials: enabling new designs (J. Wiley, Hoboken, N.J. 2004).

6. K. Malukhin and K. Ehmann, K., Material Characterization of NiTi Based Shape Memory Alloys Fabricated by the Laser Direct Metal Deposition Process, Journal of Manufacturing Science and Engineering, 128, pp. 691-696 (2006).

7. K. Malukhin, K. Ehmann, Identification of Direct Metal Deposition (DMD) Process Parameters for Manufacturing Thin Wall Structures from Shape Memory Alloy (NiTi) Powder, Transactions of the North American Manufacturing Research Institution of SME, 35 (2007).

8. K. Malukhin, K. Ehmann, The Possibility of the Development of a "Self-Sensing" Shape Memory Alloy Based Actuator, Proceedings of the 1st International Conference on Micromanufacturing ICOMM 2006, pp. 231-236 (2006).

9. K. Malukhin and K. Ehmann, Model of Motion of an Actuator Based on a NiTi Shape Memory Alloy. Proceedings of the 2nd International Conference on Micromanufacturing ICOMM 2007, (2007).

10. C.H. Wu, J.C. Houk, K.-Y. Young, L.E. Miller in: Nonlinear Damping of Limb Motion/Multiple Muscle Systems: Biomechanics and Movement Organization, edited by J.M.. Winters and S.L-Y Woo, (Springer-Verlag, New York 1990).

STATICALLY DETERMINED GRIPPER CONSTRUCTION

Ronald Plak, Roger Görtzen, Erik Puik
TNO Science & Industry, Eindhoven, the Netherlands

Abstract The current approach to reducing impact forces during component placement in micro-assembly is to couple the gripper with 5 Degrees Of Freedom (DOF) to the drive unit of the placement device, wherein the mass of the gripper has been reduced maximally. At the end of the placement motion the drive unit will move relatively to the gripper, as linear as possible given the limitations of the construction. During this motion generally stress builds up in the gripper and it becomes over-constrained due to the extra constraints added by the contact between gripper/component and the substrate. Due to this tension the gripper will start vibrating when the gripper rebounces or when the component is released, resulting in significant placement inaccuracies. A solution has been found to prevent the gripper from becoming over-constrained by adding an extra tilting member to the gripper, leading to a reduction of the placement inaccuracies.

Keywords assembly, collision, gripper, impact, micro system, membrane, placement accuracy

1 Introduction

In micro-assembly placement devices are used for assembling components or placing components on a substrate, e.g. a printed circuit board. These placement robots manipulate usually 4 DOF of the component to be placed [1-10]. For the placement devices it is desired to perform a high number of pick and place actions per minute, limit impact forces and achieve high placement accuracies.

The current approach to reduce the impact forces is to couple the gripper to the drive unit of the placement device while leaving 1 DOF unconstrained using a flexible part, wherein the gripper that contacts the component, has a relatively low mass [11-19]. In this setup the drive unit will move relatively to the gripper at the end of the component placement. During this relative motion generally also small lateral displacements occur (see Figure 1).

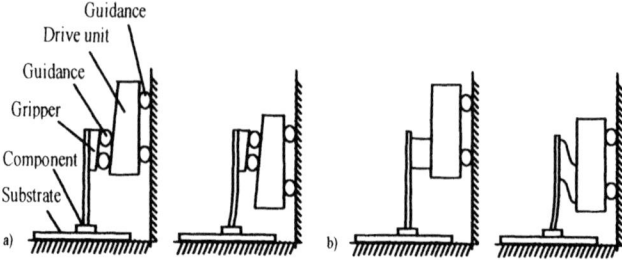

Fig. 1. Movement of drive unit out of line relative to main axis gripper, tension build-up in gripper needle due to the relative lateral displacements: a) roller bearing or sleeve bearing guidance, b) material hinge or leave spring guidance.

These lateral displacements during the component placement can lead to a tension build-up in the gripper. The gripper generally becomes over-constrained due to the extra constraints caused by the high contact forces between substrate, component and gripper and the coefficient of friction between these parts. Due to the tension the gripper can start vibrating when the gripper rebounces or when the component is released, causing significant placement inaccuracies [20-23].

In this article a gripper construction will be presented that does not become over-constrained during the placement collision, leading to a better placement accuracy.

2 Problem analysis

During the collision, the base structure adds constraints to the component when the friction forces at the contact between component and substrate are bigger then the lateral forces. The friction force is proportional to the contact force, which is relatively big (more then 100 times the gravity force) during the collision. The lateral forces depend largely on the angle between the z-axis of the gripper and the axis perpendicular to the top surface of the substrate, which is minimised in most assembly processes. The substrate will therefore generally add three main constrains to the component and gripper: x,y,z (see Figure 2). The rotations R_x and R_y are usually not constrained by the substrate due to the relative small width of the contact area compared to the length of the gripper. The rotation R_z is not likely to be constrained by the substrate due to the small area of the contact point. Therefore during the collision three of the six constrains from the guidance between gripper and drive unit must disappear to keep the gripper statically determined [19, 21, 24].

One of the three constraints that must disappear during the collision is the constraint of the degree of freedom in z-axis. This is because the drive unit will keep moving in the z-direction during and after the collision, and during this motion the gripper is not allowed to make a significant rotation. During the collision the drive unit must also allow freedom (in the order of 10 micrometer) in the x and y direction without changing the x and y position of the component. Therefore the gripper needle must be able to tilt a little around the x and y axis. This can be achieved by

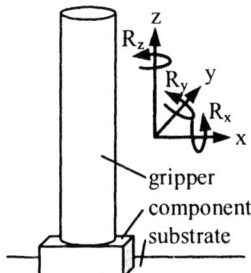

Fig. 2. Axis nomenclature for degrees of freedom gripper

either removing the x and y constrains or the R_x and R_y constrains from the guidance between gripper and drive unit.

3 Concept

Before the component is in contact with the substrate, all six degrees of freedom of the gripper must be constrained by the guidance between gripper and drive unit. When the component is in contact with the substrate, the remaining three constraints must disappear from the guidance. Releasing the constraint in z-direction can be realised relative easily by using a one-directional constrain between gripper and drive unit. This can be done by pushing part of the gripper on a support structure of the drive unit. When the gripper/component comes into contact with the substrate the drive unit moves relative to the gripper causing the contact between gripper and support structure to be broken and the constrain in z-direction will disappear. The constraints R_x and R_y of the guidance can be controlled in a similar way by implementing a tilting member which is pressed in a reference orientation when the gripper is in its reference position.

With degrees of freedom in z, R_x and R_y direction for the guidance between gripper and drive unit when the gripper is pushed out of its reference position, the gripper will not become over-constrained when the gripper/component comes in contact with the substrate. A practical embodiment for such guidance has been found to be a membrane placed transverse to the main motion axis (z-direction) of the gripper. The membrane will constrain the x, y and R_z degrees of freedom of the guidance between gripper and drive unit. A membrane theoretically results in an over-constrained structure in x and y direction but in practice the limited stiffness of the membrane will add an extra internal degree of freedom to the structure, which prevents the membrane of becoming over-constrained when it is around its neutral position.

The z, R_x and R_y degrees of freedom can be constrained when the gripper is in its reference position by implementing a stop piece to the top of the gripper which can be pressed at three points on a support structure of the drive unit (see Figure 3). In the gripper a force is needed to press the gripper to its reference position and orientation. In the prototype this has been achieved by placing a pressurised air chamber

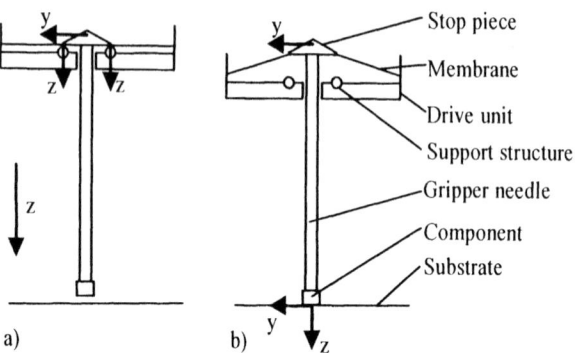

Fig. 3. Schematic gripper concept with membrane as guidance between gripper needle and drive unit. Stop piece controls the gripper in its reference position and orientation when component is not yet in contact with substrate. Arrows indicate constrains in x-z plane. a) gripper approaching substrate, b) component placed on substrate.

on top of the membrane. This force can also be used to prevent the gripper of re-bouncing at the end of the placement collision. Components can be gripped for example by implementing a magnetic force on the gripper needle or by adding a vacuum supply to the hollow gripper needle. To allow the component to align with the top surface of the substrate an additional tilting member between gripper needle and component is recommended e.g. a rubber gripper tip.

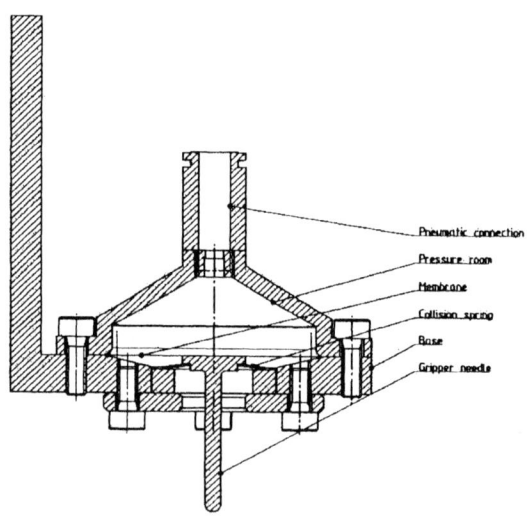

Fig. 4. Prototype of developed gripper for micro assembly: a) drawing gripper prototype, b) realized gripper prototype.

4 Results

A prototype (see Figure 4) of the gripper has been built and tested. The experiments showed that with this concept the drive is allowed to make lateral displacements when the component is already in contact with the substrate without significant tension build-up in the gripper or displacement of the component. An additional advantage of this gripper design is the low weight achieved by using a single membrane as guidance. With an improved design of the gripper prototype a mass of less then 1 g of the components involved in the placement collision was achieved for a gripper equipped with a vacuum supply for gripping the components. A patent is pending for this gripper design.

5 Conclusion

The placement inaccuracies during component placement in micro-assembly can be significantly reduced by preventing the gripper of becoming over-constrained. This has been achieved by implementing a membrane as tilting member in the gripper construction, which adds extra degrees of freedom between gripper and drive unit when the gripper/component contacts the substrate.

Acknowledgments

Manuscript received July 1, 2007. This work was supported in part by the MicroNED program, project number 3-B-1: Concept design of miniaturized unit for micro part joining and assembly.

R. Plak was with TNO Science and Industry, department of Micro Device Technology, Eindhoven, The Netherlands.

R. Görtzen is with TNO Science and Industry, department of High-end equipment, Eindhoven, The Netherlands (phone: +31-40-265-0928; fax: +31-40-265-0305; e-mail: Roger.Gortzen@tno.nl).

E. Puik is with TNO Science and Industry, department of Electronic System Integration, Eindhoven, The Netherlands. He is professor, chair of Micro-Systems-Technology at university of applied sciences Utrecht, The Netherlands (e-mail: Erik.Puik@tno.nl).

References

[1] R. Escorpizo, A. Moore, Applied Ergonomics 38 (2007) 609-615.
[2] I. Giouroudi, H. Hotzendorfer, J. Kosel, D. Andrijasevic, W. Brenner, Precision Engineering In Press, Corrected Proof 615.

[3] T. Huang, P. F. Wang, J. P. Mei, X. M. Zhao, D. G. Chetwynd, CIRP Annals - Manufacturing Technology 56 (2007) 365-368.
[4] M. T. Das, L. Canan Dulger, Simulation Modelling Practice and Theory 13 (2005) 257-271.
[5] O. V. Makarova, D. C. Mancini, N. Moldovan, R. Divan, C.-M. Tang, D. G. Ryding, R. H. Lee, Sensors and Actuators A: Physical 103 (2003) 182-186.
[6] E. K. J. Chadwick, A. C. Nicol, Journal of Biomechanics 33 (2000) 591-600.
[7] I. Mizuuchi, M. Inaba, H. Inoue, Robotics and Autonomous Systems 28 (1999) 99-113.
[8] G. i. Yasuda, International Journal of Production Economics 60-61 (1999) 241-250.
[9] J. Angeles, Z. Liu, R. Akhras, Mechanism and Machine Theory 28 (1993) 261-269.
[10] M. Rygol, S. Pollard, C. Brown, Image and Vision Computing 9 (1991) 33-38.
[11] S. K. Nah, Z. W. Zhong, Sensors and Actuators A: Physical 133 (2007) 218-224.
[12] H. Choi, M. Koc, International Journal of Machine Tools and Manufacture 46 (2006) 1350-1361.
[13] Z. W. Zhong, C. K. Yeong, Sensors and Actuators A: Physical 126 (2006) 375-381.
[14] Z. A. Soomro, Materials & Design 27 (2006) 591-594.
[15] O. Millet, P. Bernardoni, S. Regnier, P. Bidaud, E. Tsitsiris, D. Collard, L. Buchaillot, Sensors and Actuators A: Physical 114 (2004) 371-378.
[16] X. Yin, X. Wang, International Journal of Engineering Science 40 (2002) 231-238.
[17] B.-Z. Sandler, Manipulators, in: Robotics (Second Edition), Academic Press, San Diego, 1999, pp. 314-384.
[18] W. Backe, Robotics 2 (1986) 45-56.
[19] P. R. Ouyang, Q. Li, W. J. Zhang, L. S. Guo, Mechatronics 14 (2004) 1197-1217.
[20] J. H. Oh, D. G. Lee, H. S. Kim, Composite Structures 47 (1999) 497-506.
[21] J. De Schutter, D. Torfs, H. Bruyninckx, Robotics and Autonomous Systems 19 (1996) 205-214.
[22] Y. G. Kim, K. S. Jeong, D. G. Lee, J. W. Lee, Composite Structures 35 (1996) 331-342.
[23] C. W. Jen, D. A. Johnson, R. Gorez, Mathematics and Computers in Simulation 41 (1996) 539-558.
[24] C. Melchiorri, Robotics and Autonomous Systems 20 (1997) 15-38.

PRECISION POSITIONING DOWN TO SINGLE NANOMETRES BASED ON MICRO HARMONIC DRIVE SYSTEMS

Andreas Staiger and Reinhard Degen

Micromotion GmbH, An der Fahrt 13, 55124 Mainz, Germany
info@micromotion-gmbh.de

Abstract At present the Micro Harmonic Drive is the world's smallest backlash-free micro gear. Its favourable properties such as high repeatability, high torque capacity, low mass of inertia and high degree of efficiency are proved applied in different applications for semiconductor manufacturing, measuring machines, aerospace and other industries. However for high precise assembly applications often linear adjustments of a few nanometres are required. The common solution for positioning in the nanometres range is the use of piezoelectric effect actuators. The drawbacks are the need for a closed loop control system, overshooting during positioning, local wear, short travel ranges and loss of position without power supply. To overcome these obstacles the advantages of the Micro Harmonic Drive are exploited together with a stepper motor, an eccentric drive and monolithic flexure hinges are combined in a compact unit, called Nanostage. With this setup a travel range of 40 microns and a resolution < 3 nm is achieved. The single sided and double sided repeatability was measured in an open loop control system with < 5 nm and < 10 nm respectively.

1 Requirements for micro actuators in positioning systems

For products consisting of micro parts the assembly can add up to 80 % of the production costs [1]. For the automation of the assembly processes specially designed machines are necessary to handle micro parts. Typically movements with several degrees of freedom realised by motors, gears and lead screws are necessary. As a result of using conventional drive technology the assembly machines need a multiple of space against the handled parts. To increase the accuracy and reduce the cycle times the size of the precision assembly machines had to be reduced to the range of the micro parts. Therefore fast, precise and reliable micro actuators for rotational as well as for linear movements are needed. Moreover further integration of functionalities like vacuum feed-throughs or sensors are needed [2]. The development of micro motors [3, 4] and backlash-free micro gears [5] allow densely packed micro-electro-mechanical systems for innovative positioning applications [6]. The advantages of such micro actuators are their low mass of inertia, small footprints and often long tools that amplify inaccuracies are dispensable. Due to the low mass of inertia which accounts for high acceleration the power dissipation is reduced and therefore effects of changing temperatures ranges on the accuracy are almost negligible. If the assembly has to be done under special environmental conditions, smaller footprints of the assembly machines led to lower costs.

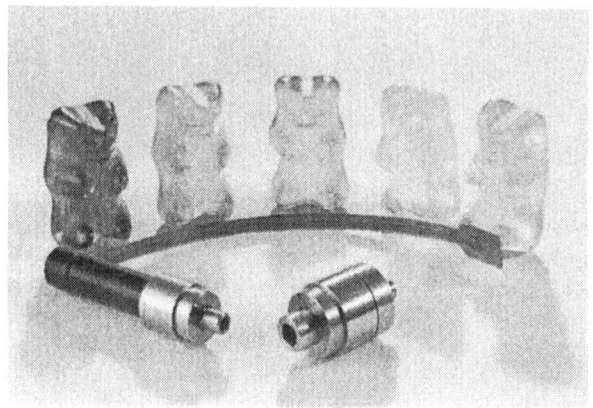

Fig. 1 Micro Harmonic Drive and servo actuator

The key elements of such micro systems are compact sized gears like the backlash-free Micro Harmonic Drive (Fig. 1). This gear principle provides a unique combination of precision, torque capacity, power density, high degree of efficiency, low mass of inertia, hollow shaft, robustness and a wide range of reduction ratios within the same dimensions and the same number of gears. They have proved their ability for assembly processes for years in semiconductor and other assembly machines.

For high precision adjustment applications the advantages of the micro gears are transferred over in linear movements. For travel ranges between 0.5 mm and 2 mm an eccentric drive consisting of a backlash-free eccentric, a micro harmonic drive and a stepper motor are applied. With such a system a resolution of 0.15 µm can be achieved (Table 1).

Table 1. Technical data of eccentric drive

	Unit	Value
Cross section	mm	10
Length	mm	43.5
Travel range	mm	2
Linear speed	mm/s	2
Force	N	12
Resolution (half step)	µm	0.15
Repeatability	µm	+/- 1
Mass	g	10

2 The Nanostage

For some adjustments, such as movements of optical lenses, accuracy in the single nanometre range is requested. To realise positioning units for the single nanometre range with conventional drive technology, new adaptation mechanisms are necessary to cope with requests regarding stiffness, guidance and controllability [7-9].

2.1 Boundary conditions for the Nanostage

Current systems with a resolution in the nanometre range are mostly based on piezoelectric effect actuators. The drawbacks of such systems are:
- Loss of position if power supply is interrupted
- Local wear, especially by inch-worm drives
- Need for a closed loop control system
- Need for position measuring system
- Overshooting during positioning
- Short travel ranges compared to the actuator size

The aim for the development of a new positioning mechanism was the use of common control systems within the resolution of single nanometres. Therefore a backlash-free eccentric drive (Table 1) is combined with monolithic flexure hinges (Fig. 2). Because of the stepper motor, only a common open loop stepper control unit is necessary. Moreover by the use of a high reduction ratio gear set a high resolution is achieved.

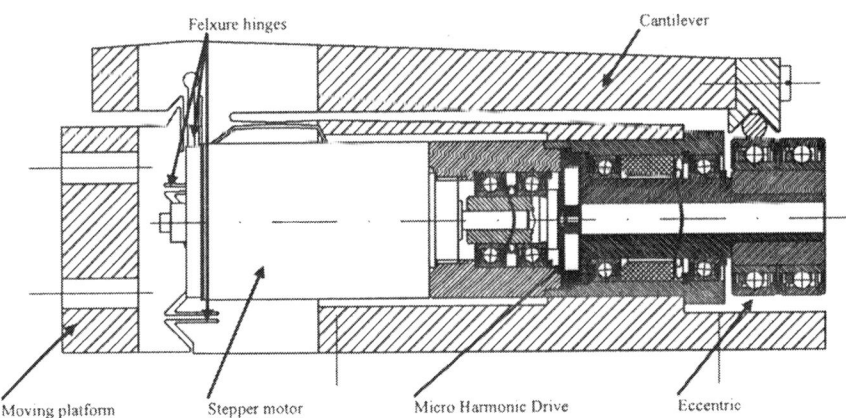

Fig. 2. The Nanostage. Conventional drive technology for high precise positioning

The flexure hinges provide on the one hand the guiding system and on the other hand an additional reduction ratio. Through the monolithic manufacturing, influences on the accuracy caused by assembling such as asymmetric stress or unbalanced tightened screws are avoided. Another important influence on the accuracy is the quality of the output shaft bearings. True running deviations caused by manufacturing, assembly and the bearings itself are visible in the transmission accuracy of the system. Therefore the true running requirements of these parts are in the single micrometre range. As a result of this requirement the bearing carriers have to be manufactured in one chucking and the ball bearings are preloaded axial so that the balls are running on the high surface finished inner and outer rings of the bearings.

2.2 Structure synthesis

The kinematic train of the Nanostage consists of the following components:

- Stepper motor with 20 full steps per revolution
- Micro Harmonic Drive with ratio 1000 : 1
- 1 mm eccentric
- Flexure hinges with ratio 50 : 1

The aim of the design was a couple hinge guiding to avoid the more complex manufacturing of a four-point spring guidance. It was designed as follows:

- Material AL7022
- Length of spring: 5.37 mm
- Width of spring: 4 mm
- Minimum thickness: 0.2 mm
- End thickness: 0.266 mm
- Allowed linear deformation at 2.2 N: 17 µm
- Allowed angular deformation at 2.8 Nmm/rad: 16.2 mrad

This led to a length contraction of just 0.04 µm with allowed linear deformation. Due to better manufacturing capabilities and the sufficient guidance accuracy resulting from the spring contraction, the concept of a four-point spring guidance was chosen. It is regarded as the best solution both technically and economically. Two more points had to be regarded in the design of the flexure hinge. First, at z-accelerations the cantilever could lift-off. This is prevented by a flexure pre-load of the eccentric bearings. Second, at acceleration across the moving direction, the cantilever could oscillate so that the eccentric bearings could be affected. An additional leaf spring between cantilever and body for prevention is included.
This led to the following structure of the flexure hinge:

- 2 x 2 guidance springs induce the intended short linear movements.
- 2 cantilever springs to induce an additional ratio of 50 : 1. They are deformed angular (+/- 16 mrad). For this hinge no requirements regarding the length constancy or linear deformation exist.
- 1 leaf spring to guide the cantilever. It is 70 µm linear and +/- 16 mrad angular deformed.

2.3 Manufacturing of the flexure hinges

The body is manufactured of Al7022 (Fig. 3). Short and therefore thin springs with high tension load are preferred to reduce cutting time in the eroding process. The cutting width for the flexure hinges could be to a large extent optimised to process parameters. The accuracy of the outline of the five hinges that are remaining as bars in the body is critical. Therefore the tolerance and the roughness have only to be minimized in these areas. For the other parts the cutting speed could be higher.

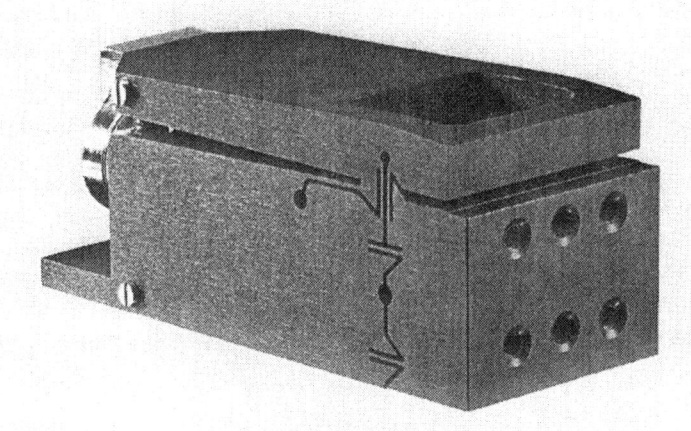

Fig. 3. The Nanostage. The moving platform contains four M2 threads and two 2H7 fits. Size: 20 x 20 x 50 mm³

3 Measurements

To measure the resolution of the Nanostage a capacity sensor with a resolution of 5 nm was utilized. Because of the A/D converter used for the analysis the resolution was reduced to 10 nm. The measured resolution is 10 nm, that is equivalent to one digit. As the resolution is limited by the used measuring system it is assumed that the resolution of the Nanostage is below 10 nm. This has to be verified with a laser interferometer.

Fig. 4 shows the Fast Fourier Transformation of the transfer function. The fundamental wave shows the maximum travel range. The first harmonic wave shows the deviation from an ideal sinus caused by nonlinearities in the guiding and the flexure springs. The higher harmonics are caused by concentricity faults of the ball bearings. The highest influence on the position accuracy is the slightly nonlinear ratio of the cantilever, followed by the nonlinearities of the flexure hinges and the concentricity faults of the ball bearings caused by the quality of the surface finishing and assembly.

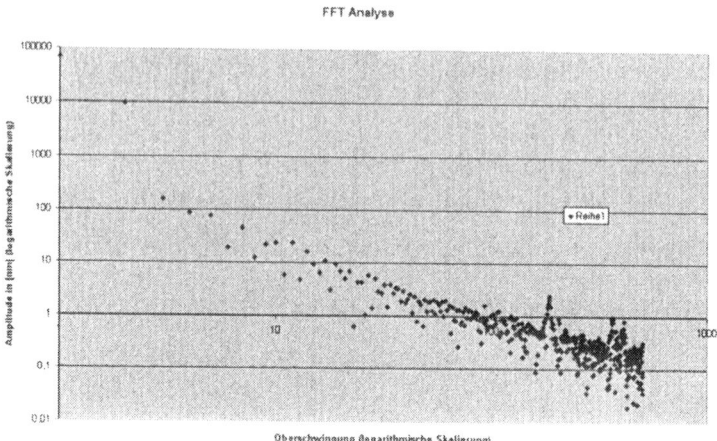

Fig. 4. FFT-analysis of one cycle to identify failure influences. Ordinate: amplitude in nm, logarithmic. Abscissa: Overshooting, logarithmic.

4 Conclusion

The Nanostage provides a compact unit for positioning in the nanometre range by using conventional micro drive technology. Consequently the advantages of an open-loop stepper motor control unit and hold of the position without power supply can now be utilized in nanometre positioning and assembly systems (Table 2).

Table 2. Technical data of the Nanostage

	Unit	Value
Dimensions	mm	20x20x50
Travel range	μm	40
Resolution (half step)	nm	<10
Linear speed	μm/s	2
Force	N	12
Mass	g	0.15

The theoretical resolution of < 3 nm and the repeatability have to be verified with a laser interferometer.

References

1. J. Hesselbach and A. Raatz, mikroPRO – Untersuchung zum international Stand der Mikroproduktionstechnik (Vulkan Verlag, Essen, 2002).
2. R. Degen and R. Slatter, High Speed And Low Weight Micro Actuators For High Precision Assembly Applications, Proceedings of IPAS, Bad Gastein (2006).
3. C. Thürigen, W. Ehrfeld, B. Hagemann, H. Lehr and F. Michel, Development, fabrication and testing of a multiple-stage micro gear system, (Proceedings of Tribology issues and opportunities in MEMS, Columbus (OH), (Kluwer Academic Publishers, Dordrecht, 1998) pp. 397-402.
4. S. Kleen, W. Ehrfeld, F. Michel, M. Niehaus and H.-D. Stölting, Ultraflache Motoren im Pfennigformat, F&M, Vol. 108, No. 4, (Carl Hanser Verlag, München, 2002).
5. R. Degen and R. Slatter, Hollow shaft micro servo actuators realized with the Micro Harmonic Drive, Proceedings of Actuator, Bremen (2002).
6. R. Slatter, R. Degen and A. Burisch, Micro Mechatronic Actuators for Desktop Factory Applications, Proceedings of the Joint Conf. on Robotics, München (2006).
7. S.T. Smith and D.G. Chetwynd, Foundations of Ultraprecision Mechanism Design (Taylor & Francis Books Ltd, London, 1998).
8. A.H. Slocum, Precision Machine Design (Society of Manufacturing Engineers, Dearborn (MI), 1992).

ASSEMBLY OF A MICRO BALL-BEARING USING A CAPILLARY GRIPPER AND A MICROCOMPONENT FEEDER

C. Lenders, J.-B. Valsamis, M. Desaedeleer, A. Delchambre, P. Lambert

Bio Electro and Mechanical Systems (BEAMS) Department
Université Libre de Bruxelles (ULB)
Avenue F.D. Roosevelt, 50 CP 165/14
B-1050 Bruxelles, Belgium
http://beams.ulb.ac.be

1 Introduction

Assembly of microsystems is still a challenge today. The numerous parasitic forces often make the manipulation behaviour unpredictable, hence difficult to automate. Therefore special designs of grippers and adequate strategies have to be implemented. Recent works have shown that capillary forces are strong enough to be used to manipulate components in microassembly technology [1]. However, many investigations need to be performed concerning the manufacturing of a microgripper for an industrial use. If the feasibility has been theoretically shown, there is still work to be achieved over the practical implementation of such a gripper.

The purpose of this paper is to present an assembly station for a watch bearing including a component feeder: the balls delivered in bulk are separated and arranged in specific holes. Then a gripper picks five balls successively and places them in the housing of the bearing.

The station we present here uses the capillary gripper designed by Lambert [2]. This gripper is integrated on a precision *xyz* positioning table. Since electrostatic repulsion and capillary adhesion are still an issue with this kind of gripper during the gripping step, a microcomponent feeding device has been implemented to supply the gripper.

This paper falls into 5 parts: Section 2 describes the problem while Section 3 focuses on the design of the solution, concerning the feeder and the gripper. We present the experimental setup in Section 4, and the results in Section 5. Finally, we give conclusions in Section 6.

Please use the following format when citing this chapter:

Lenders, C., Valsamis, J.-B., Desaedeleer, M., Delchambre, A., Lambert, P., in IFIP International Federation for Information Processing, Volume 260, *Micro-Assembly Technologies and Applications*, eds. Ratchev, S., Koelemeijer, S., (Boston: Springer), pp. 265-274.

2 Problem definition

The problem can be divided into several independent sub-problems. A first set of problems is linked to the balls feeding. A second one is linked to the micro-manipulation of the balls (picking, moving and placing).

2.1 Microcomponent feeding system

Because of the scaling effects, many parasitic forces are becoming predominant at microscale, which may disrupt the gripping process. The two major actors are the electrostatic forces and the capillary forces generated by liquid bridges between solids.

The electrostatic forces arise mainly from the unpredictable triboelectrification and may either repulse the component from the gripper, or attract it. In the case of repulsion, the gripper may never reach the component. This problem is illustrated in Fig. 1(a).

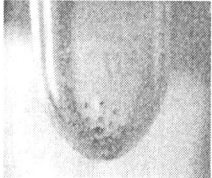

(a) Illustration of electrostatic repulsion: the ruby balls (⌀0.5mm) repel each other due to triboelectrification

(b) The capillary forces due to liquid bridges may be large enough to stick the components (alumina ball, (⌀0.5mm) together

Fig. 1. Example of repulsive and attractive parasitic forces

The capillary forces are indeed present in this kind of problem either by putting liquid for the purpose of surface tension gripping [1] or because of the relative humidity [3]. In some cases, the force may be large enough to stick different components together, as illustrated in Fig. 1(b). It is then difficult to grip them one by one.

The solution proposed to minimise these parasitic effects is to work in an immersed environment, as already proposed by the *Automation and Micromechatronics Department, FEMTO-ST Institute, Besançon, France*. The design shall comply with the following specifications (illustrated in Fig. 2):

- The components are supplied in bulk
- The components are supposed to be spherical, with a diameter of 500μm
- The final location of the components must be known
- The capillary gripper should be able to catch one component
- When the capillary gripper catches a component, it must grip only one

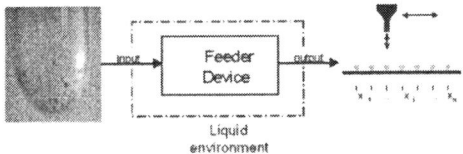

Fig. 2. Illustration of the specifications for the microcomponent feeding system

2.2 Microassembly system

Once the balls are arranged, the bearing is assembled with a capillary gripper. The gripper is dived in a liquid bath to form a liquid drop on its tip. Then the gripper picks one ball at a time, places it in an empty hole on the ball bearing and releases it. Problems appearing during the manipulation are described below.

- *The picking phase* consists in taking the ball out of its hole. The main problems are the checking of the presence of a ball in a feeder, the strength of the capillary forces which should be large enough to extract the ball.
- *The placing phase* is the positioning of the ball into a determined hole of the ball bearing. The precision of placing should involve a control of the relative position of the ball with regard to the gripper and a good precision of the *xy*-axes.
- *The releasing phase* consists in breaking the ball away. Several strategies exist: deforming the liquid bridge (changing the curvature [4], by using external hydrophobic materials), using external forces (pulling forces such as sticking or soldering the component, and shearing forces).

The next step is now to repeat the sequence five times in order to fill each housing of the bearing.

3 Solutions design

3.1 Feeding device

As announced in the previous section, the cell of the feeding device will be filled with liquid to avoid parasitic electrostatic and capillary forces. Since there is no free surface, capillary forces are totally cancelled in immersed environment, while the permittivity of water, which is 80 times larger than that of dry air, reduces the electrostatic force drastically. Moreover, if the liquid is a conductive media, the electrostatic charges accumulated on the components could disappear. Of course, because we use a capillary gripper to catch the components, we have to extract the liquid from the cell before the picking step.

In order to fulfil the requirements, we propose to create a housing for each component. This housing will separate the balls and maintain them on a specific location. It will also allow the gripper to catch the component. Hence the housing shall be large enough to allow the gripper to pass. The conformity with the ball must also be as low as possible, meaning that the residues of liquid should not develop a capillary force larger that the gripping force (see Fig. 3).

The geometry chosen (Fig. 4(a)) is a thin metal plate (1), the mesh, with in-line holes (3). To prevent the ball from passing through and yet to allow the liquid to flow in order to empty the cell, we have placed a vertical plate (2) beneath the holes. This also gives us a low conformity.

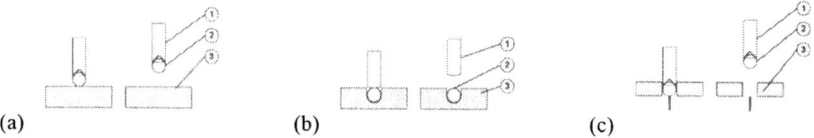

(a) (b) (c)

Fig. 3. Illustration of the conformity issue. If the shape of the component matches better the gripper than the housing (a), the capillary force generated by the liquid bridge between the component and the gripper is larger than that of the liquid bridge between the component and the housing. Otherwise, the gripper shall not be able to extract the component (b). With our design (c), the conformity is kept low thanks to the thin vertical plate beneath the housing.

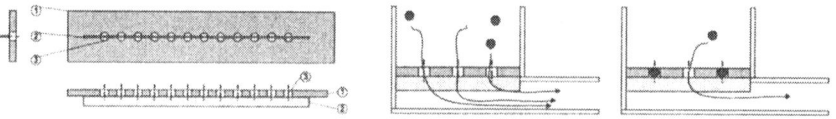

(a) Schematic view of the ball feeder

(b) Schematic view of the balls flow. When a housing is filled with a component, the streamline is cut and the remaining components are guided to the free locations

Fig. 4. Design and working of the feeder

3.1.1 Perspective

So far, the cell is tilted manually in order to move the components toward the housings. We intend to improve the device by using the hydrodynamic forces in this liquid. By flowing the fluid to the housings, we will guide the components to these locations. To achieve this, we have placed a channel below the mesh to allow the fluid to flow. It is also used to empty the cell when the components have reached their location.

The advantage of this design is that when a location is filled with a component, no more streamline passes through this hole, hence guiding the remaining components to the other free locations (see Fig. 4(b)).

To generate this streamline, we shall use a pumping device ((IV) on Fig. 6) to create a velocity and henceforth a depression in the lower channel, and inject the fluid back in the cell (III) through a buffer (I). Finally, a valve (II) will be used to control the flow between the buffer and the cell. A complete sequence is described schematically below (Fig. 6).

3.2 Modelling

This principle has been checked by simulation on Comsol Multiphysics software (see Fig. 5).

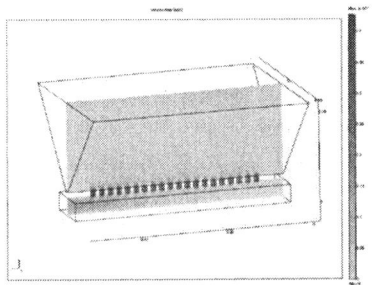

(a) 3D simulation of the flow in the feeder. With the particle tracing option, we have an information about the behaviour of the ball in the cell.

(b) Evolution of the velocity of the fluid above the different housing of the mesh (outlet is under the right-hand side)

Fig. 5. Flow simulation using Comsol Multiphysics. With this simulation, we can see what are the paths of the streamlines. This should help us to estimate the best position to inject the components

We have approached the geometry of the feeder cell, and computed Navier-Stokes equation to find the pressure and velocity field. In this simulation, we have supposed that the liquid feedback was injected uniformly on the free surface.

The particle tracing option of Comsol Multiphysics makes an estimation of the movement of particles in the fluid (the flow is assumed not to be modified by the particle movement). This will help us to optimise the parameters and the geometry of the device.

Since we are pumping the fluid on one side of the cell, the flow will not be uniform in all the housings. This simulation allows us to check if the flow is not too low on the opposite side of the mesh.

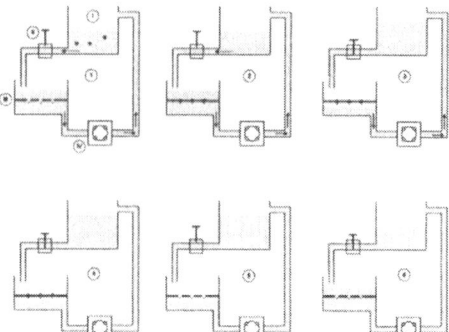

Fig. 6. General diagram of the whole component feeder. The components are unloaded in the buffer (1). They are directed to the housing in the cell (2). Then the valve is closed to eliminate the liquid surrounding the components (3). The gripper picks up the components (4), and then the valve is opened (5) to refill the cell (6).

3.3 Microassembly system

The gripper (design taken from [1]) is shown in Fig. 7. It is made of a stainless steel cylindrical tip with a conical concavity, in order to improve the conformity between the gripper and the ball. The cylinder diameter agrees with the ball diameter which is 0.5mm.

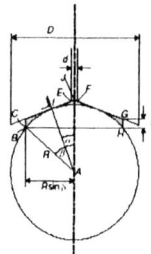

Fig. 7. Gripper design, from [1]. Main parameters are the diameter D, the angle spread $\pi-2\alpha$

Capillary forces depend in a first approximation on the surface contact. In our design, cohesive forces of the ball are higher with the gripper than the feeder. Indeed, if the ball is centred on the gripper, the contact line is $\pi D \cos(\alpha)$ while the contact between the feeder and the ball is 2 points as shown in Fig. 8(a). Fig. 8(b) shows the evolution of the capillary forces on successive picking of a ball [5] with only one feeding in liquid.

(a) Contact zone were capillary forces will appear

(b) Capillary forces generated by the gripper

Fig. 8. Capillary forces generated with our gripper during almost 10000 picking

During the placing phase, the main advantage of our gripper design is the autocentring of the ball, avoiding having to control the ball position with regard to the gripper (Fig. 9(a)). Furthermore the ball can roll in the destination hole (Fig. 9(b)). Theoretically the accuracy should be below the size of the radius that is widely achieved (see Sect. 4). As a result, we will study the link between the initial moving off centre (Δx) on the residual moving off centre (δx) and the maximum tolerance on the placement accuracy (Δu).

Finally the last stage consists on releasing the ball. In this paper, we will use the cage as a stop that will produce a shearing on the liquid bridge (Fig. 9(c)).

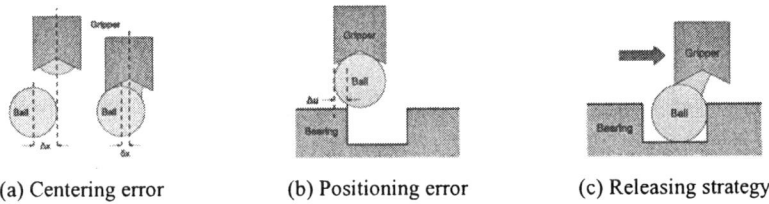

(a) Centering error

(b) Positioning error

(c) Releasing strategy

Fig. 9. Concept in placing and releasing phases

4 Experimental setup

The manipulation is mounted on a station with three degrees of freedom (one per translation axis). The station is made of a gripper, a component feeder and a ball bearing placed on a support.

Horizontal axes are decoupled from the vertical one according to the split-axes concept [6]: the gripper has the vertical motion (z-axis). The feeder and the bearing are mounted on a table equipped with the two horizontal motions (xy-axes).

The vertical motion is controlled by a linear motor (a SMAC linear actuator LAL95-050-71 and a LAC-1 controller, accuracy 1μm). The motion of xy-table is controlled by two brushless DC servo-motor (SMART SM2315D, from Animatics, accuracy 20μm). Rotative motion is converted into a linear motion thanks to linear guides with a worm screw (THK, KR2001A). Two optical devices (a Keyence laser displacement sensor LC-2440, accuracy 0.2μm and a controller unit LC-2400W) allow us to get the position of the table according to the gripper.

5 Results

5.1 Feeding device

So far, the feeder has been tested manually. The components were injected in still water, on one end of the mesh, and the device was tilted until the components reached the other side. A small pool was placed next to the cell to catch the balls that did not fall into a housing. Fig. 10 shows the feeder used manually. With the manual feeder, we have measured the filling ratio (number of filled housing on total number of housing). The results are shown in Table 1.

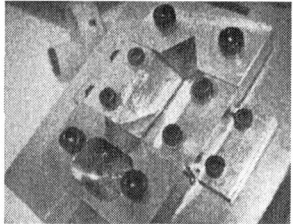

Fig. 10. Picture of the feeder with the recovery pool. The components are injected on the other end. The device is tilted and the components slide towards the recovery pool

Repetition	Housing n° (injection on the right, recovering pool on the left)																				Ratio
---	1	2	3	4	5	6	7	8	9	10	11	12	13	14	15	16	17	18	19	20	---
1	1		1		1	1	1		1	1	1		1	1		1			1	1	1	75%
2	1	1	1	1	1	1	1	1	1	1	1	1	1	1								70%
3						1	1		1	1	1	1	1	1	1	1	1	1	1	1	1	70%
4				1	1	1	1	1			1	1		1	1	1			1	1	1	75%
5	1	1	1	1		1	1	1	1	1	1	1		1	1					1		75%
6		1	1		1	1	1	1				1	1	1	1	1	1			1	1	75%
7	1				1	1	1		1	1	1	1	1	1	1	1		1	1	1	1	80%
8	1	1	1	1		1					1	1	1	1	1	1	1	1	1			75%
9	1	1	1	1			1	1		1	1	1	1	1	1	1	1		1	1		80%
10	1	1								1			1	1	1		1	1				65%
11	1	1	1	1	1	1	1	1		1	1	1		1	1	1		1				75%
12	1	1		1	1	1	1	1	1	1		1	1				1					65%
13	1			1		1		1		1	1		1		1	1	1		1	1	1	80%
14	1		1			1	1	1	1	1	1	1		1	1		1		1	1	1	80%
15				1	1	1	1	1	1	1	1	1	1	1	1	1	1	1	1	1	1	85%
Average	73%	53%	67%	67%	53%	80%	93%	87%	100%	87%	73%	87%	100%	93%	67%	80%	60%	60%	73%	53%		75%

Table 1. Filling ratio measurement on the manual device. The average filling ratio is 75%

We have also shown that it was possible to extract the balls from the mesh (Fig. 11). It is however necessary to wait several minutes for the liquid remaining in the housing to evaporate. If the liquid is water, the time before picking the balls is larger than 10 minutes. To reduce it, we have used a solution of ethanol and ether.

In this case, the time to wait before picking the components is reduced down to 5 to 10 minutes. To reduce this delay, we propose to blow heated air above the mesh, or to heat the mesh itself by flowing an electrical current. Otherwise, the influence of the delay on the entire assembly station could be reduced by parallelizing several feeders.

Fig. 11. Extraction of a ball out of the mesh with the capillary gripper

With the automation proposed in Section 3.1, we still have to optimise the system. The main parameters are the viscosity and density of the fluid, the velocity of the fluid and the geometry of the device. We shall estimate the influence of these parameters on the filling ratio (i.e. how many housing are filled with a component) and the cycle time, which is mainly due to the time to wait for the residues of liquid to evaporate. This will allow us to find the best combination to enhance the performances of the device.

Concerning the tolerance of the positioning for the picking phase, we have observed that the auto-centring is only working if Δx is below 31µm (measured in the worst case, i.e. with minimum amount of liquid). This value allows us to estimate the positioning accuracy needed during the picking phase.

The maximal tolerance on the placement accuracy for the release task is mainly due to the geometry of the gripper and the final location (see Fig. 12). The calculation gives a result of $\Delta u = 164$µm, while we experimentally encounter problems to insert the ball in the bearing if $\Delta u \geq 125$µm.

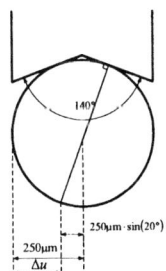

Fig. 12. Positioning tolerance in releasing task is defined by the geometry of the system

We also measured the cycle time of picking, placing and releasing steps on our experimental setup. Typical values are 2.56s for picking a component, 0.14s to release it, and a total of 6.15s to wet the gripper tip in a water tank and to move the gripper to the different locations (component feeder, assembly area and water tank).

The picking phase is slow in order not to create a shock on the system, but the placing phase is constrained mainly by the actuators speed.

6 Conclusion and future work

In this paper, we have designed a manipulation station from the balls in bulk to their assembly in a ball bearing. We try to characterise the feeder, the gripper and the station by showing the success rate of the manipulation, the cycle time and the tolerance on the accuracy.

As future work, we could continue the automation of the station. The feeder could work automatically with different pumps and a system that drain the excess of balls. A vision system could be attached to watch if holes are provided with balls (since the feeding ratio is below 100%) and the number of balls leaving in order to restart the feeder. Vision system could equally check the presence of a ball on the gripper and presence in the bearing.

Another investigation is to study the scale effect by using a smaller ball with a diameter of 300µm.

Acknowledgements

J.-B. Valsamis is a researcher funded by a grant of the F.R.I.A. - Fonds pour la Formation à la Recherche dans l'Industrie et l'Agriculture.

References

1. P. Lambert, F. Seigneur, S. Koelemeijer, and J. Jacot, "A case study of surface tension gripping: the watch bearing," J. Micromech. Microeng., vol. 16, no. 7, pp. 1267–1276, 2006.
2. P. Lambert, "A contribution to microassembly: a study of capillary forces as a gripping principle," Ph.D. dissertation, Université libre de Bruxelles, Belgium, 2004.
3. A. Chau, S. Régnier, A. Delchambre, and P. Lambert, "Three dimensional model for capillary nanobridges an capillary forces," Model. Simul. Mater. Sci. Eng., vol. 15, pp. 305–317, 2007.
4. F. Biganzoli, I. Fassi, and C. Pagano, "Development of a gripping system based on capillary force," in Proceedings of ISATP05, Montreal, Canada, 19-21 July 2005, pp. 36–40.
5. J.-B. Valsamis, A. Delchambre, and P. Lambert, "An experimental study of prehension parameters during manipulation task," in Proc. of the 5th International Workshop on Microfactories, 25-27 October 2006.
6. S. van Gastel, M. Nikeschina, and R. Petit, Fundamentals of SMD Assembly. Assembléon Netherlands B.V., 2004.

Chapter 7

High Precision Positioning and Feeding Techniques

PNEUMATIC CONTACTLESS MICROFEEDER, DESIGN OPTIMISATION AND EXPERIMENTAL VALIDATION

Michele Turitto, Carsten Tietje, Svetan Ratchev

Precision Manufacturing Group, The University of Nottingham,
University Park, Nottingham, NG7 2RD, UK
epxmt2@nottingham.ac.uk, epxct1@nottingham.ac.uk, svetan.ratchev@nottingham.ac.uk

Abstract A novel contactless pneumatic microfeeder based on distributed manipulation is proposed. The part to be conveyed floats over an air cushion and is moved to the desired location with the desired orientation by means of the coordinated action of dynamically programmable microactuators. The design was optimised by CFD (Computer Fluid Dynamics) simulations. A proof of concept prototype was manufactured to obtain experimental indications that confirmed the soundness of the solution presented

1 Introduction

Highly integrated products such as those based on telecommunication technology or precision biomedical devices require the production and assembly of microparts. Though the need for such production has increased rapidly, the scale of manufacturing systems has not changed accordingly, thus becoming inadequate. In order to overcome this mismatch the concept of microfactory has been developed. The reduction in size of the manufacturing equipment leads to an improved space utilisation, lower energy requirements, and an overall cost reduction. Moreover microfactories are characterised by easier machinery replacement and dynamic reconfigurability that grants a prompter response to changing customer requirements [1].For further progress of the microfactory concept, the introduction of a compact mode and mechanism to facilitate transfer of work between its components is necessary [2]. The presented microfeeder aims at filling this gap: it is designed not taking a specific object as reference. The only requirement is that the object is big enough to cover a few nozzles (each microactuator has a top surface of about 300 µm²). Such design grants the device with the much needed flexibility in part feeding which allows the introduction of new parts into the assembly system with minimal reconfiguration. Ultimate flexibility in feeding would require a device capable of accepting new parts without any or, at least, with a very short pause in the production [3].

The proposed microfeeder relies on distributed manipulation to carry out the conveying task: it is based on an array of microactuators each of which is made up of four nozzles. The nozzles are closed and opened by electrostatic forces giving the possibility to move objects in four different directions. Air is provided from the lower surface of the feeder so that the parts float over an air cushion and can be conveyed without any contact. A prototype, whose design is the outcome of an optimisation process by CFD simulations, is used to prove the concept on which the microfeeder is based.

2 Microfeeding – contactless manipulation

Feeders have the function of presenting parts that were previously randomly oriented to an assembly station at the same position, with the correct orientation and the correct speed. Distributed manipulation (Fig.1) is quite a common approach for conveying microparts.

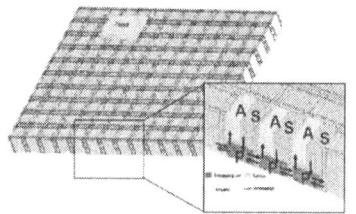

Fig. 1. Pneumatic contactless microfeeder

It is based on arrays of tiny actuators, each able to provide a simple motion. Even though the motion imparted by a single element is within a small range, it is possible to move objects over relatively long distances through the cooperation of a large number of microactuators.

Contactless manipulation is a feasible solution for microassembly because of the small size and light weight of the objects to be moved.

It is advantageous as [4]:
- Surface forces can be completely neglected
- It is suitable for handling fragile, freshly painted, sensitive micron-sized structured surfaces
- It allows the handling of non-rigid microparts
- There is no contamination of and from the end effector

3 Four directions microactuator

The microfeeder consists of an array of micronozzles. Air is used for keeping the parts suspended and moving them through the control of the micronozzles. A single microactuator is made up of four nozzles formed by a central electrode and four walls around it (Fig.2). The nozzles are opened or closed by electrostatic actuation. In neutral position the four nozzles are all open: the airflow, coming from the bottom of the microactuator, is equally divided among the four nozzles because of the symmetry of the structure (Fig.3). The outcoming airflows result in a force field that causes the micropart to hover above the microactuator (Fig.4).

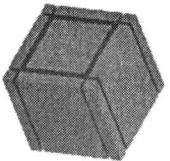
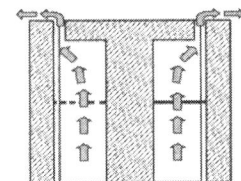

Fig.2. Electrostatic microactuator

Fig.3. Microactuator - cross section

For moving the object, the central cursor is attracted towards one of the walls and the corresponding nozzle is closed. In Fig. 5 the upwards and downwards jets compensate each other hence there is a net force that pushes the micropart leftwards. A similar working principle was presented in [5]. The proposed design is advantageous because the single microactuator is more compact as it keeps the dimensions of the airflow channel constant. Moreover, movement in four orthogonal directions is achieved with a single microactuator whereas in [5] the same result is obtained combining four different microactuators capable of conveying objects in two directions only. This feature is of paramount importance as distributed manipulation becomes more effective if two conditions are satisfied: the microactuators have to be as small as possible, as their size directly affects the minimum size of the parts that can be moved, and the density of microactuators has to be high because this directly influences the position resolution that can be achieved. Hence, having smaller individual microactuators improves the performance of the microfeeder. Further details about the initial design and a sequence based on IC-compatible fabrication can be found in [6].

The performance of the microfeeder in directing the airflows when in neutral and active position was assessed by means of CFD simulations. The effect of geometry modifications on the issuing airflows was determined and the design consequently modified.

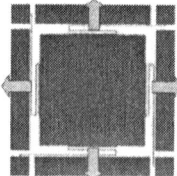 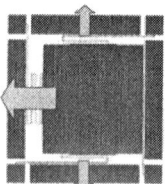

Fig. 4. Top view of the microactuator in neutral position

Fig. 5. Top view of the microactuator in active position

Further details can be found in [7]. The final outcome of this refinement process can be seen in Fig. 7 and Fig. 8.

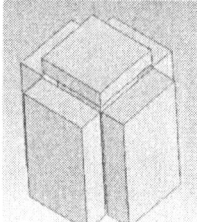 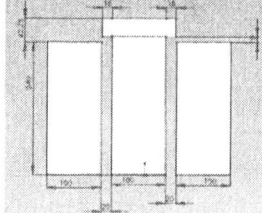

Fig. 7. Modified microfeeder design

Fig. 8. Modified microfeeder - cross section

4 Experimental validation

A prototype based on the optimised design was built. The prototype is 50 times bigger than the original design and it was used to validate the theoretical approach and assess the influence the relevance of the weight and size of the part on the conveyance process.

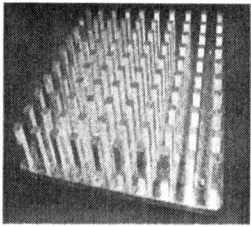 

Fig.9. Prototype - central elements

Fig.10. Prototype - side walls

As can be seen in Fig. 9 the prototype has a simplified structure in which all central elements of the actuators are connected together. Hence they move all at the same time whereas the original design requires all the microactuators to be individually addressable. The side walls which, together with the central elements, constitute the nozzles have been changed accordingly (Fig.10). In the prototype, the nozzles are 300 μm wide. The central elements are moved in two orthogonal directions by two PI M-110.2 closed loop micro-translation stages that have 0.0085 μm resolution and a maximum speed of 1.5 mm/s.

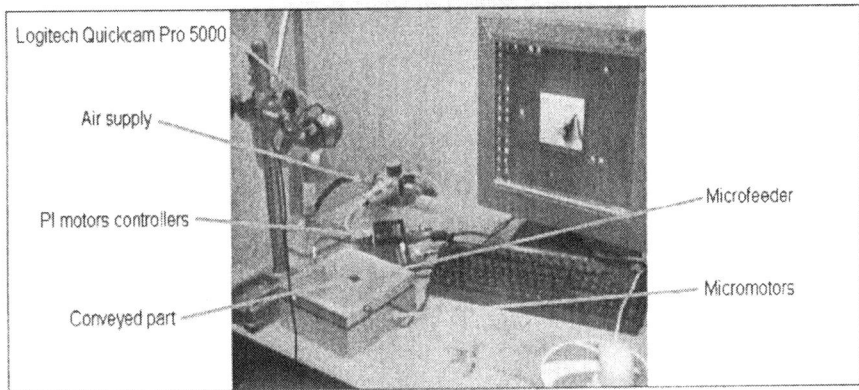

Fig. 11. Experimental testbed

Fig. 11 shows the experimental testbed. The sidewalls enclose the central elements and the motors thus creating a confined space that let the incoming air (which is provided from the sides) escape the device through the nozzles on top only. A Logitech Quickam Pro 5000 completes the system and allows a performance evaluation of the device. For this reason, an algorithm was developed in a Matlab environment that processes the images acquired by the camera, identifies the center of mass of the part and follows its movement during the conveyance.

Because all the information about the part movement is obtained by image processing, it is expressed in numbers of pixel. For this reason it is necessary to determine the value of each pixel in the acquired image. The part conveyed is a flat dark (so that it stands out against the light background of the microfeeder surface) square component with an area of 0.001289 m^2 (measured with a Mitutoyo Absolute Digimatic vernier caliper). The same area expressed in pixels after the image processing is 2292 pixels^2. Hence the following relation stands:

1 pixel = 750 μm

A certain level of approximation has to be taken into account considering that the camera is not perfectly perpendicular to the microfeeder surface and that there is a certain distortion in the image acquired. However these effects are negligible and

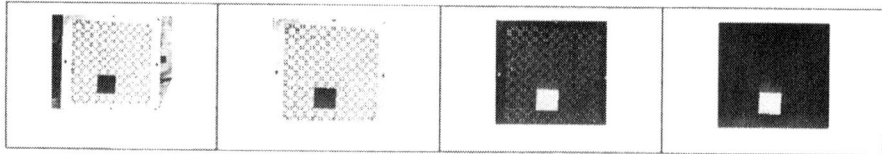

Fig.12. Image processing

the correct value is not too far from the one calculated. Fig. 12 shows the different stages in the image processing. The acquired image is cropped in order to remove the portions that do not belong to the microfeeder and therefore reduce the computational time.

This, together with other programming improvements, has reduced the computational time per loop to about 0.1 seconds. The part is then identified by creating a negative of the acquired image and by a thresholding operation that cuts out the connected areas whose value is below a certain figure which was experimentally determined.

To further assess the accuracy of the vision system a series of measurements were carried out leaving the part on the microfeeder surface without any air being supplied. Ideally the values of the coordinates of the centre of mass should remain constant throughout the whole measurement. In fact the values obtained from image processes are quite constant. The standard deviations s_x and s_y respectively for the x and y coordinate of the centre of mass were calculated as follows:

$s_x \approx 60$ μm $\qquad s_y \approx 70$ μm

These values are low enough to consider the error coming from the vision system negligible.

Measurements were carried out with the central elements in neutral position (Fig. 4). The part behaves as expected: under the action of an equal number of forces pushing in opposite directions, it maintains its position. Fig. 13 shows the variation of the position of the centre of mass of the part in the case of a test in which the image processing loop was repeated 30 times. The origin of the axes is in the top left corner in order to match with the frame of reference associated with the acquired image. The variation along x is around 10 pixels corresponding to 7.5 mm whereas the variation along y is around 8 pixels corresponding to about 6 mm. These values are fully acceptable considering that at this stage the microfeeder is operated in open loop and the inevitable instability due to the transfer of shear stress from the impinging airflow to the part.

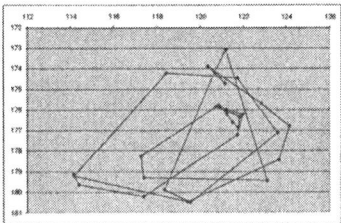

Fig. 13. Mapping of the variation of the position (expressed in number of pixels) of the part's centre of mass with central elements in neutral position

The following step was assessing the capability of the microfeeder transporting the part along a linear trajectory. Fig. 14 shows the results in a case similar to that shown in Fig. 5. The central elements are moved towards the right hand-side so that there is a resulting force that pushes the part towards the left. As expected the part moves towards the left along a linear trajectory.

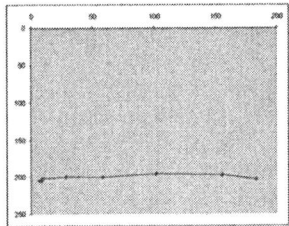

Fig. 14. Tracking of the position (expressed in number of pixels) of the part's centre of mass with central elements in active position

5 Conclusions

A new pneumatic microfeeder based on contactless distributed manipulation was presented. The initial design was refined by CFD simulations. A proof of concept prototype was built and used as a testbed to collect experimental results that confirmed the validity of the proposed solution. Future work will entail the use of closed loop control based on image processing to compensate for the instability that characterises the motion of the part due to the transfer of shear stress from the impinging airflows. The outcome of these further investigations will be reported in due course.

Acknowledgments

The presented work is part of the ongoing UK EPSRC "Grand Challenge" research project "3D-Mintegration" which aims to provide radically new ways of thinking for end-to-end design, processing, assembly, packaging, integration and testing of complete 3D miniaturised/integrated '3D Mintegrated' products.

References

[1] Okazachi Y., Mishima N. and Ashida K, Microfactory - concept, history and developments, Journal of Manufacturing Science and Engineering, Vol. 126, November 2004, pages 837-843
[2] Ashida K., Mishima N., Maekawa H., Tanikawa T., Kaneko K. and Tanaka M., Development of desktop machining Microfactory, Proceedings of Japan-USA Symposium on Flexible Automation, 2000, pages 175-178
[3] Viinikainen H., Uusitalo J. and Tuokko R. New flexible minifeeder for miniature parts, Proceedings of the International Precision Assembly Seminar IPAS'2004, Bad Hofgastein, Austria, Feb 11-13 pages 87-94
[4] Vandaele V., Lambert P., and Delchambre A., Non-contact handling in micro-assembly: Acoustic levitation, Precision Engineering Vol 29 (2005) pages 491-505
[5] Fukuta Y., Yanada M., Ino A., Mita Y., Chapuis Y.A., Konishi S. and Fujita H., Conveyor for pneumatic two-dimensional manipulation realized by arrayed MEMS and its control, Journal of Robotics and Mechatronics, 2004, Volume 16, No. 2, pp 163-170
[6] Turitto M., Chapuis Y-A and Ratchev S., "Pneumatic Contactless Feeder for Microassembly", Proceedings of IPAS'2006 - The Third International Precision Assembly Seminar, Bad Hofgastein, Austria, 19-21 February 2006, published by Springer as "Precision Assembly Technologies for Mini and Micro Products", pages 53-62
[7] Turitto M., Ratchev S., Xue X., Hughes M. and Bailey C., Pneumatic contactless microfeeder design refinement through CFD simulation, Proceedings of 4M 2007, The third international conference on multi-material micro manufacture, Borovets, Bulgaria, 3-5 October 2007, pages 65-68

PNEUMATIC POSITIONING SYSTEM FOR PRECISION ASSEMBLY

Martin Freundt, Christian Brecher, Christian Wenzel, Nicolas Pyschny

Fraunhofer IPT, Steinbachstraße 17, 52074 Aachen, Germany
martin.freundt@ipt.fraunhofer.de

Abstract Micro assembly is typically characterised by positioning tolerances below a few micrometers. In the case of a hybrid micro system assembly, such as optical glass fibres, micro ball lenses or micro probes for measurement tasks, even positioning accuracies in the sub-micrometer range have to be achieved. Due to the need for highly accurate assembly systems and extensive alignment procedures the assembly of hybrid microsystems is characterised by customised solutions. In this context the Fraunhofer IPT develops a concept on how to realise a highly flexible, fast and cost-efficient hybrid assembly system, consisting of a conventional assembly device and an active assembly head.

The active assembly head will be pre-positioned by imprecise but dynamic conventional handling devices like an industrial robot. By means of its integrated 6-axes fine positioning system and its sensor system it will be able to first detect position and orientation deviations and second compensate the deviation and execute the final positioning and alignment of the micro part. In this context, a matchbox-sized air bearing stage with an integrated non contact interface for a transfer of pneumatic energy between stage and slides was conceived, allowing ultra precise and frictionless guidance for travel ranges up to 3 mm. In order to apply this distinct design to the whole system, even actuators and sensors must be wireless or pneumatically driven in order to ensure a friction free stage movement.

In this context the requirements caused by a conventional pre-positioning device will be discussed. It will be shown that travel ranges of about 1 mm must be achieved in order to allow a reliable and safe use of the hybrid handling device. Based on that analysis, a design of a friction free, damped pneumatic actuator design as well as a concept of a pneumatic sensor will be presented. Both components will target a travel range of 1 mm combined with a resolution in the sub-micron range.

1 Introduction – Initial Context

Technological miniaturisation is a major trend and thus a key characteristic of future product innovations [1]. Highly sophisticated machining processes such as ultra precision diamond cutting, micro-erosive forming and lithographic processes enable the manufacturing of microscopic components as small as 100 µm³. These highly precise components are in most cases mounted onto larger macroscopic components, to be incorporated into innovative medical, telecommunication or sensor technology products.

Key factors of micro systems production are technologies for the automated handling, alignment and assembly of micro parts. Although available systems for these purposes are sufficiently precise, they are, compared to macroscopic assembly equipment, highly specialised and inflexible, i.e. can hardly be applied or adapted to different assembly tasks. For the flexible automation of macro assembly processes a wide range of standardised handling systems like industrial robots is currently available. But as yet, even the most precise industrial robots cannot be applied to micro assembly processes with accuracy requirements in positioning and adjustment of less than one micrometer.

The conditions that apply to micro component assembly are generally very different to those of macro assembly, e.g. thermal fluctuations, vibrations and the fundamental tolerances of the tools and equipment have a strong influence on assembly results and product quality.

The lack of versatile, modular standard components for micro component handling increases the complexity of an automated micro assembly in terms of engineering, and mechanical design efforts as well as the controls needed to customize independent and oftentimes incompatible automation solutions. As a result, the field of micro assembly keeps being dominated by manual assembly operations, especially for the production of small and medium lot sizes, where highly specialised automation components cannot be operated cost-effectively.

It therefore becomes very difficult to integrate micro assembly operations into production processes related to macroscopic products, and it remains indispensible to synchronise separated microscopic and macroscopic assembly processes, which is both costly and technically complex.

2 State of Technology and Technical Concept

For micro assembly purposes many different positioning and alignment systems have been developed. In order to be able to deal with the entire handling process, an assembly system must incorporate the following process steps in one device:

- Picking up the micro part from a component holder,
- Moving the micro part to the assembly location,
- Positioning and aligning the micro part.

Highly precise positioning systems for the assembly of micro systems, individual components such as positioning axes as well as highly precise actuator and sensor components used to build specialised equipment are well established and readily available. Examples include the systems from Klocke Nanotechnik, MicRohCell (Rohwedder AG), AutoPlace 400 (Sysmelec), RP series (Mitsubishi), and micropositioning stages (Physik Instrumente, Mechonics).

Further systems for micro assembly can be found in the areas of surface mount devices (SMD) and die bonders. SMD assembly is characterised by very high dynamics, but only achieves a precision of 40-50 µm, and offers no more than 3 degrees of freedom (DOF). Die bonders for industrial chip assemblies are Cartesian pick-&-place systems with integrated sensors (mostly cameras) to achieve exact positioning of microelectronic components. However, although die bonders offer accuracies down to ±1 µm these systems are very capital intensive and restricted in their available degrees of freedom due to the Cartesian setup [1].

The typical serial Cartesian design of gantry assembly systems, based on individual, high precision but relatively massive axis components, inevitably leads to large solid cantilever structures. Such very massive systems usually include granite elements with a closed-off structure. This design reduces the usability of the systems to a few specific applications. Depending on the number of assembly partners involved, the type of packaging and storing as well as the size of micro parts and component holders, motion paths for all three process steps can easily exceed 400mm.

Generally, conventional assembly systems have enough structural rigidity to work in spaces of some 100 mm³ with a positioning accuracy in the sub-micrometer range, but have a very small working space compared to their large base and frame size. Large structures are associated with problems in terms of thermal stability. Systems that are sensitive to thermal influences are hardly suitable for applications where the thermal conditions are largely undefined as is the case in macroscopic production processes. With mostly less than six degrees of freedom and a common system structure that limits the construction space, and thus prevents a flexible integration into the flow of material, the assembly process must be redesigned and modified accordingly to fit the assembly device.

Indeed, more flexible, macroscopic robot systems with six DOF only achieve a repeating accuracy down to 20 µm. They are therefore unsuitable for the assembly of micro components which are about 100 to 200 µm in size. For such products – such as sensor tips for micro probes or surgical suture material for complex surgical procedures (e.g. eye operations) - a positioning accuracy in the sub-micrometer range is necessary. These robots do not fulfil the requirements of alignment procedures for larger components, such as optical components of a diode laser.

Available stationary positioning stages providing the required accuracy for the alignment in more than three or four DOF restrict the flexibility of the assembly system (as positioning and alignment processes have to be located at the station), and cause difficulties in integrating the separated module controls into a common system control. Therefore, different approaches of mounting portable alignment modules to conventional robot systems have been pursued to achieve maximum flexibility regarding assembly and positioning tasks as well as different products and components [2].

In this hybrid concept an active alignment module is mounted to the imprecise but dynamic handling robot serving as a portable multi-axis alignment device. The head will be pre-positioned within the large work space of the robot, reference its position and orientation at the assembly location and compensate position and alignment errors with micrometer-accuracy. Thereby the necessary precision for the alignment of optical components can be realised at the tip of the assembly head compensating for positioning errors and guaranteeing full system functionality (Figure 1).

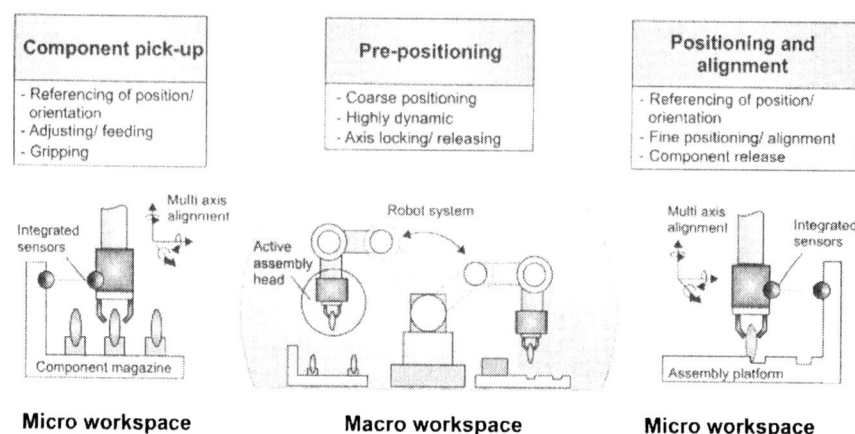

Fig. 1. Hybrid assembly system concept with active assembly head

This hybrid system concept combines the advantages of standardised, conventional robot systems for macroscopic handling processes, i.e. the large work space, high availability and dynamics with the required accuracy for micro assembly tasks. By a flexible changeover between macroscopic and microscopic operating environments a fast integration of high-accuracy assembly and alignment processes into conventional assembly lines will be enabled. This approach has already been adopted in research projects to design fine adjustment units for a very limited range of specialized assembly applications. As examples for these specialised units investigations were conducted into the active adjustment of single mode glass fibres and the camera-aided positioning of electronic components. [3, 4]

To gain broader acceptance of the hybrid concept, a universally applicable active assembly head has to be developed that is qualified for mobile mounting on standard pre-positioning units (e.g. industrial robots) allowing the flexible application of different end-effectors' geometries for multiple highly-precise assembly tasks. Resulting standardised micro assembly systems that are able to cope with a wide range of micro assembly operations would help to shorten cycle times, increase the flexibility of the positioning and adjustment systems and improve the set-up process. These improvements could significantly increase the efficiency of automated micro assembly processes, opening up a new range of micro system applications.

3 Deduction of Design Requirements

In order to improve the precision of an industrial robot system by means of an active assembly head mounted in series, system vibrations caused by active robot drives have to be suppressed or compensated. Measurements show that in conventional systems the stroke of these drive vibrations can range to several micrometers (Figure 2). Vibrations can significantly be reduced when the robot is operating with activated brakes and deactivated drives. The brake activation prevents any agitation of the robot's structure, but results in a load alternation from the drive train to the brakes, associated with non-reproducible movements in the robot's joints causing the robot arm to sag.

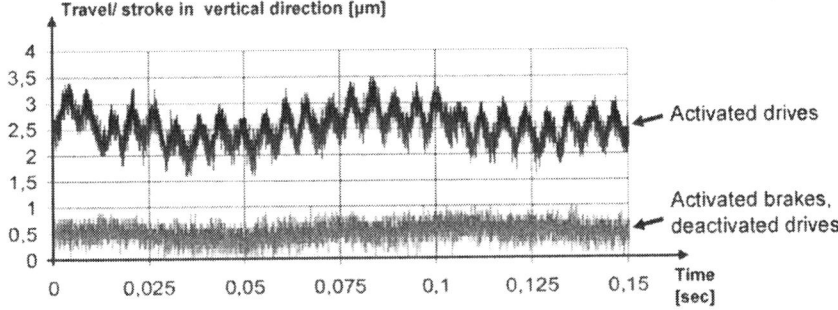

Fig. 2. Vibrations in a 6-axis-robot (arm acting against gravity)

Practical investigation carried out on a 6-axis-robot with an arm length of approx. 600 mm have shown that the robot hand loses approx. 300 µm in height and moves 0.04° from position when brakes are activated. Variations range ±100 µm and ±0.02° due to changing frictional and bracing characteristics of the robot arm.

Thus, a hybrid system of a conventional robot and an active assembly head can only perform stable and reliable positioning operations if the fine positioning axles are designed to cope with travel ranges of at least 1000 µm. In this case, the assembly head can compensate for sagging of between 200 and 500 µm, still providing enough travel (several 100 µm) to carry out positioning and alignment operations.

4 Design Approach

As part of the SFB 440 project "Assembly of hybrid micro systems", current research activities at Fraunhofer IPT aim at the development of an assembly head that meets the aforementioned criteria, and includes an integrated, highly precise, compact and robust fine positioning unit with 6 degrees of freedom and integrated sensors for referencing at the assembly location. The work covers the development of system components in line with the identified requirements, the design of the system structure as well as the development of the controls needed to regulate the interaction between the assembly head and the conventional pre-positioning system. Main tasks in this context are the appropriate miniaturization of guideways as well as the design of compact actuator and sensor components.

State-of-the-art design of portable fine positioning systems incorporates solid hinges that so far only achieve a maximum travel range of about 200 µm preventing the formation of flexible positioning systems for various micro assembly applications. The hybrid concept requires a mechanical design that enables travel ranges about 1mm with angles up to 1°, and possesses an integrated sensor system to reference and control the assembly head.

This in a first step necessitates the development of a guideway technology that can perform compensational and positioning movements with sub-micrometer accuracy over the required travel range. As solid hinges do not fulfil the deducted requirements the design approach focuses on the miniaturisation of air bearing technology and its qualification for the set-up of a 6-axis positioning module.

5 Guideway with Air Bearings and Pneumatic Energy Transmission

Fundamental research activities pursue the development of a friction-free, miniaturised guiding unit with air bearings and up to 3 mm travel range.

The downsizing of an axis with air bearings for the application in highly precise motion guides requires several design changes towards miniaturisation. The miniaturisation of vacuum pre-loaded air bearings to a pad diameter of 14 mm for example demands a reduction of the bearing gap from already very narrow 2 - 4 µm to under one micrometer in order to achieve similar rigidity and damping properties.

Due to the technically highly challenging reduction in bearing clearance, it becomes clear that – due to restrictions in manufacturing accuracy – it will be difficult and expensive to miniaturize the air bearings without compromising on bearing rigidity and damping.

A prototype in operation at the Fraunhofer IPT has a relatively large bearing gap of approx. 2 µm. With a bearing pad diameter of 14 mm this leads to an operating point with relatively low rigidity and damping. Nevertheless technical feasibility of the miniaturised air bearing technology with vacuum pre-loading meeting the requirements in terms of accuracy and travel range of at least 1 mm could be demonstrated (Figure 3). Despite the large bearing clearance, the bearing stability could be proved: according to the calculations from W.J. Barz [5], the operating point lies below the region of bearing instability for average load ratings, while rigidity and damping are however significantly lower than for conventional, non-miniaturised bearing stages where the load capacity has been optimised for greater stability.

In order to qualify the air bearings for a highly precise multi-axial positioning system, bearing loads have to be reduced to an extent where the negative effects of low rigidity and damping on the system behavior can be tolerated.

In the context of the presented design approach, bearing loads arise from three factors: the acceleration of the carrying robot systems, the interfering forces created by the actuators and sensors (parasitic shearing forces and moments), and the interfering forces caused by the rigid component wiring.

This demands all components – the guides, the actuators, the sensors and the system supply lines – to be designed with regard to light weight, i.e. low inertia, and resistance-free signal, energy and fluid transmission. If the entire system can be build light enough that varying loads do not affect operability of the fine positioning system, it becomes possible to implement the miniaturised air bearing technology without reducing critical design parameters, like bearing clearance.

Fig. 3. Bearing pads in a matchbox-sized prototype air bearing with vacuum pre-loading

In this context, the Fraunhofer IPT developed a concept for non-contact compressed air transmission between guideway and slide. The transmission of compressed air via special ducts in the bearing surface avoids friction, force and hysteresis. The transmission ducts are found in both the slide (entrance) and the guideway (exit) aligned opposite to one other. The outlet in the guideway - an elongated slot - allows for axle movement without disconnecting the compressed air or vacuum supply lines between guide and slide. The transition between entrance and exit outlet is sealed and throttled by a non-contact annular gap. The low sealing gap, corresponding to the bearing gap, reduces leaks to a minimum and ensures contact-free operation, without friction, force or hysteresis.

The prototype shown in Figure 3 demonstrates that five serially stacked guideways with air bearings can be supplied with, and run on compressed air without any rigid supply lines between stage and slide. Hence, each guideway can be operated without affecting the others, thereby reducing the requirements on actuator force and rigidity, which in turn enables the miniaturisation of other components.

6 Design of a Pneumatic Actuator

Primarily objective of the pneumatic actuator design is highest possible accuracy. In accordance to the aforementioned requirements, the actuator system has to be small and lightweight, as well as frictionless to prevent negative side effects, like stick-slip.

Conventional designs prove unusable due to contacting sealing elements. Possible contactless and hence frictionless solutions are bellow- and membrane structures, both working without contacting parts. Membrane structures have the advantage of exhibiting high stiffness in one specific direction. But the resulting excellent guideway abilities unfortunately incorporate very small travel distances in the range of a few hundred micrometers. Bellows on the other hand do not have guiding abilities, but even in very small dimensions they allow travel distances in the range of millimetres, suitable to meet the requirements discussed before.

For both designs - bellow or membrane - once the actuator works friction free, no damping occurs, and the mechanical structure resembles a spring. Additionally the compressed air in a pneumatic actuator has a similar behaviour with very low damping and almost no mass.

A completely friction free pneumatic actuator system is not suited for highly precise positioning, as with very low damping existing disturbing forces would cause vibrations and oscillation of the device. Therefore, a damping element must be integrated into the pneumatic actuator which works without causing stick-slip effects, but still damping the oscillations caused by internal and external forces.

An approach, conducted at the Fraunhofer IPT is the integration of an oil based damping device. Because of the required accuracy in the air bearing stages, an oil based device has to be concealed in order to avoid contamination of ultra-precisely manufactured components. Therefore, in contrast to a squeeze film damper arranged in direction of the stage movement, the developed solution is based on an oil filled bellow - the same technology used for the pneumatic actuator. The developed sealable, contamination free damping technology is expected to have significant advantages in long term stability of the system behaviour.

To avoid any parasitic forces the damping and actuating forces of the pneumatic actuator have to have exactly the same contact point and direction. This can be realized by a setup, where the actuator bellow is integrated inside the damping bellow (Figure 4).

This design allows optimised exploitation of construction space and the reduction of parasitic forces. In order to maintain a sealed system, a third bellow is needed as fluid reservoir, whereby the pipe between damping bellow and reservoir bellow works as a resistor causing the damping effect. Applying the bellow technology to a pneumatic actuator and a fluidic damper, the design of a stick-slip free, but damped actuator system is made possible.

Every pneumatic actuator, based on a piston, membrane or bellow design is in fact a pressure-travel converter. Thus, once high precision in travel is required, high precision in the adjustment of the pressure is essential. The specific precision depends on the targeted travel range and the stiffness of the bellow system described before.

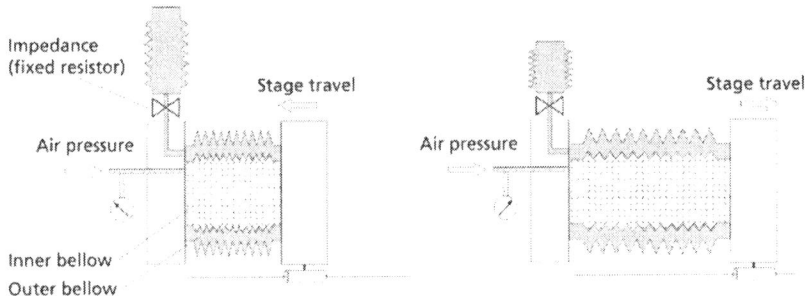

Fig. 4. Stick-slip free, oil-damped pneumatic driven actuator

For estimating calculations of the bellow system's stiffness, both bellows and compressed air can be represented by a conventional spring component. Therefore following simplification will be presumed: Each bellow is represented by the stiffness coefficients c_A, c_D, c_R. The inner volume including a part of the pipe system is represented by the stiffness coefficient c_{air}. The over all stiffness can be calculated with following equation:

$$\frac{1}{c_{ges}} = \frac{1}{c_{air} + c_A + c_D + c_R}$$

The achievable travel range X can be calculated as:

$$X = \frac{F}{c_{ges}} = \frac{(P_{Actuatorbellow} - P_0) \cdot (A_{Actuatorbellow} - A_{Damperbellow})}{c_{ges}}$$

For a required travel range of about 1 mm and a step width of 0,1 μm following sample calculation - based on the assumption, that the common air supply delivers a pressure of 6 bar absolute - shows the requirements on a pressure adjustment device.

$$\frac{s}{\Delta s} = \frac{P_{aktuator} - P_0}{\Delta p} \Rightarrow \frac{1mm}{0,0001mm} = \frac{6bar - 1bar}{\Delta p} \Leftrightarrow \Delta p = 0,0005 bar$$

This example shows that running the actuator based its pressure-travel converter principle requires a pressure resolution of 0,5 mbar. As the systems underlies operating conditions with pressure variations in air supply and environment, which will have significant impact on step width and achievable accuracy of the actuator, a closed-loop positioning control is necessary.

This demands respective sensor solutions for position recognition supporting the overall lightweight design approach.

7 Pneumatic Sensor Concept

In correspondence with the approach of reduced system size, mass and actuator forces, further investigations are conducted into the research of force-free and frictionless pneumatic based sensor systems which can be interpreted by measuring air flow or pressure to unify fluidic and energetic supplies for sensor, air bearing and actuator components.

The common measuring principle nozzle and deflector as it is used in pneumatic valves (Figure 5 left) can not be used in this case. Preliminary tests have shown the nozzle-deflector principle works best in a region of a few micrometers. For the use as absolute measuring device, precision and range are interdependent, mostly determined by the available digitalisation resolution, interferences in the signal, and the limitations of the physical principle. The requirement for travel ranges over 100 µm does hence not allow the use of this technology. Therefore, a relative measuring pneumatic sensor principle is needed. Inspired by the common principle of a linear measuring scale used in wide range of applications, the following concept was developed (Figure 5 right).

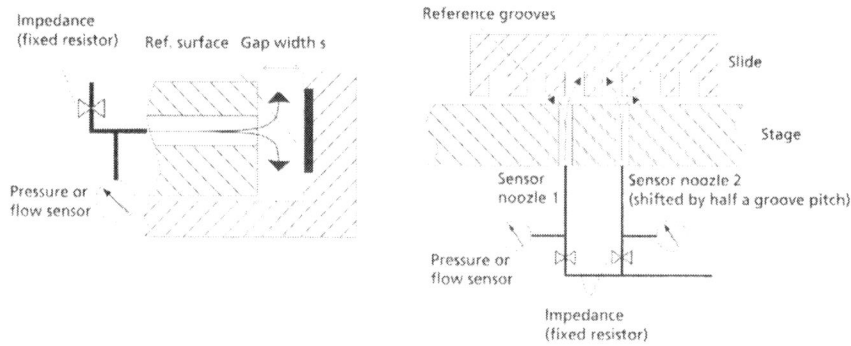

Fig. 5. Pneumatic sensor concept

Therefore it is used to scan a surface of the slide in which small grooves are inserted. Once the slide moves, the impendence in front of the nozzle is influenced by the groove structure. With a groove in front of the nozzle the resistance is reduced and a high air flow is established. If the nozzle is almost closed by the bar, the flow is reduced. The system thereby produces a pressure or air flow signal, similar to that which is generated in conventional optical linear measurement scale. Using two nozzles, placed with an offset compared to the groove structure, it is possible to read the direction of the movement.

8 Outlook

The technology of pneumatic based actuator and sensor systems presented here is currently being investigated in terms of its ability to be integrated into a miniaturised stage system. Therefore the Fraunhofer IPT is already investigating to what precision such system can be utilised. The next stage in this investigation is to set up a test stand in order to prove the technological feasibility and characterise the achievable precision. The main research issue will be how to handle the capacity effects caused by the compressibility of compressed air, which influence both actuator and sensor components.

Acknowledgements

The authors would like to thank the »Deutschen Forschungsgemeinschaft (DFG)« for supporting the »Greifer und Montagemaschinen« research project as part of the special research topic SFB 440 »Assembly of hybrid micro systems«.

References

[1] M. Höhn: Sensorgeführte Montage hybrider Mikrosysteme, Dissertation, München, Technische Universität, 2001
[2] A. Schubert, H.-J. Koriath: Precision-Tilt-Gripper with rigid 3d-structure. In: Proceedings of the 4th euspen international conference, Glasgow, 2004
[3] C. Peschke: Mehr-Achs-Mikrogreifer zur Handhabung von Drähten und Fasern, Dissertation, RWTH Aachen 2007
[4] J. Hesselbach: mikroPRO – Untersuchung zum internationalen Stand der Mikroproduktionstechnik. Essen: Vulkan-Verlag, 2002
[5] W.J. Bartz: Luftlagerungen – Grundlagen und Anwendungen. Esslingen: expert-verlag, 1993, ISBN 3-8169-0992-2

MANUFACTURING OF DEVICES FOR THE PARALLEL PRECISION ALIGNMENT OF MULTIPLE MICRO COMPONENTS

Christian Brecher, Martin Weinzierl

Fraunhofer Institute for Production Technology IPT,
Steinbachstraße 17, 52074 Aachen
martin.weinzierl@ipt.fraunhofer.de

Abstract Micro components are available in a variety of shapes typically sizing from 1 mm down to 0.01 mm. Their mass production is quite common using state of the art production technology. Micro spheres for example, are on the one hand available in lot sizes up to some thousands in a constant quality within micron accuracy [1]. On the other hand they are quite commonly provided as bulk material with diameter variations of up to 10 % in each lot size [2]. In both cases, the bulk micro components are usually arranged in incoherent batches which are packed in plastic bags or small jars for handling and shipping. The decollating of single components for follow-up micro assembly processes is complicated by the well-known effects in micro handling such as dominating adhesion and friction forces [3]. These effects limit the post processing of bulk micro components to manual work in order to sort and align the single micro components prior to their exposure to an automated assembly line. To enable a sophisticated and automated handling of the single micro components, automated sorting and alignment mechanisms are necessary to arrange the bulk micro components in a well defined pattern structure which is essential to achieving an efficient automated micro assembly.

1 Development of passive alignment structures

In preceding researches of the Collaborate Research Centre SFB 440, passive alignment structures have been designed for elementary geometrical structures such as spheres, cubes and cylinders [4]. For each of those geometrical elements, statically determinant structures have been designed and optimised regarding the positioning accuracy of the micro components. As those researches have shown, passive alignment allows for positioning accuracies down to the single digit micrometer range. However, in these investigations each micro component has been handled and aligned manually. The manual alignment of the single micro components prior to their exposure to an automated assembly processes is state of the art in micro production today but reveals a gap in the fully automated process chain ranging from the bulk production of components to the finished micro products. For an efficient production of micro systems, multiple components have to be handled and aligned with a reproducible precision. In the current research work which is conducted at the Fraunhofer IPT, mechanisms for the simultaneous alignment of multiple micro components are developed to bridge the gap between the automated bulk production of micro components and the automated assembly of micro systems. The bases for these mechanisms are arrays composed out of passive alignment structures and the use of active energies, e.g. airflow, which support the alignment of the micro components in these arrays. Such arrays have been realised for the parallel alignment of micro spheres as well as cubical and cylindrical micro components. Due to the widely spread application of micro spheres in micro assembly, the development of array structures for passively aligning an unsorted bulk of micro spheres is introduced in the following.

1.1 Array structures for the parallel passive alignment of micro spheres

The design of the passive alignment array structures is based on the fundamental results from the investigations on passively aligning single spheres where flat, round and pointed elements have been deployed and analysed regarding their alignment accuracies. Here, three-sided pyramid cavities not only provide a simple solution to realise a statically determined three point contact but also have been proved to be least susceptible to misalignment due to form tolerances of the structure and micro sphere. Another essential element which influences the design of the array structures is the ability for the rapid manufacturing by ultraprecision machining and replication through hot embossing [5]. The combination of ultraprecision machining and replication processes like hot embossing not only has a great potential for low cost mass production but also enables the manufacturing of passive alignment concepts which cannot be machined directly by conventional cutting processes. A good example for this is the realisation of the array concepts which is introduced in this paper. These passive alignment arrays are designed in such a way that they are composed of linear elements like intersecting bridges (Figure 1). The tools which are used for the hot embossing of these arrays feature column structures which are formed by intersecting grooves which can be manufactured very efficiently within a few hours of time by ultraprecision machining with accuracies down to the submicron range.

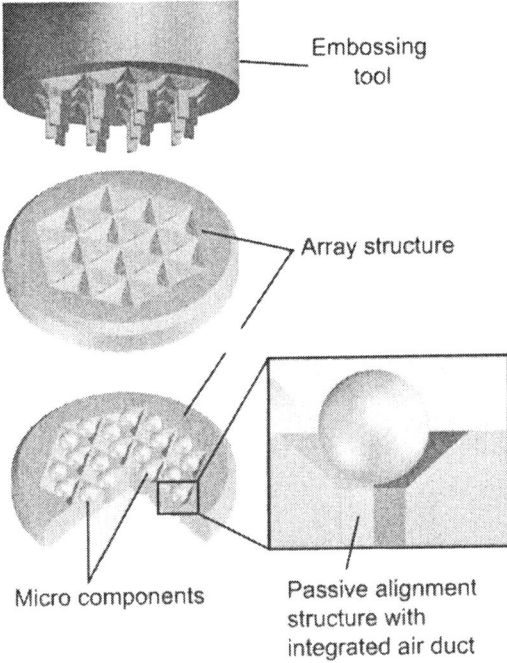

Fig. 1. Concept for ultraprecision machined embossing tools for the replication of passive alignment array structures

2 Ultraprecision machining of passive alignment structures

In ultraprecision machining, conventional metal cutting processes such as turning and milling, are used to manufacture micro structures with form accuracies down to 0.1 mm / 100 mm and an optical surface finish with an average roughness < 10 nm Ra. The classical application of ultraprecision machining is the manufacturing of complex master tools for optical applications such as reflectors or display components (Figure 2). Due to the high form accuracy and surface finish which can be realised by ultraprecision machining, this manufacturing process is advanced at Fraunhofer IPT to manufacture high precision micro assembly devices such as vacuum grippers and passive alignment structures.

The extremely high form accuracies and the optical surface finish is mainly achieved by using ultraprecision machine tools and single crystal diamond cutting tools. Ultraprecision machine tools mainly differ from conventional machine tools in the machine bases, drives, guides and spindles which are used. As a material for the machine base, solid granite is used to achieve high thermal stability as well as excellent damping properties. Minimum step sizes below 50 nm are realised with the use of highly dynamic direct drives in combination with high precision linear glass scales which allow for an interpolated resolution in the single digit nanometer range. The guidances of the feed and positioning axes as well as the spindle rotors are designed with hydro- or aerostatic supports. These allow for a contact less support of the moving parts and therefore eliminate the so-called "stick slip" effect which reduces the positioning accuracy of conventional ball bearing supported machine tools to a couple of microns. Due to this, the single axes of ultraprecision machine tools can be moved extremely smoothly within nanometer steps. This results in sub-micron form accuracies and optical surfaces without requiring any further finishing.

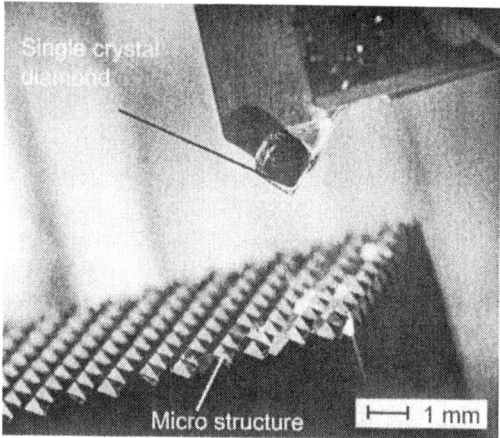

Fig. 2. Ultraprecision machining using single crystal diamond tools

The single crystal diamond tools which are mainly responsible for realising the optical surface finish of ultraprecision machined work pieces are available in a great variety of shapes such as v-groove-, radius- or facets. Their most significant feature is an extremely sharp cutting edge with an edge radius < 50 nm. These sharp cutting edges as well as the generally low friction coefficient of diamond enable a sub-micron chip removal and very low cutting forces which minimizes tool chattering and tool deflection during the machining process. In many application of ultraprecision machining the geometry of the micro structure is a projection of the diamond tool geometry. This requires an absolutely notch-free cutting edge since the smallest irregularities in the surface flanks can disturb the function of the work piece. In optical applications, score marks on the work piece surface cause diffraction patterns. In the case of passive alignment structures, such score marks cause a misalignment of the micro parts.

2.1 Geometric variety of ultraprecision machined micro structures

By combining different ultraprecision machining processes such as turning, fly-cutting or planing and using varying shapes of diamond tools, a great variety of micro structures can be cut into non iron metals or nickel coated steel bodies. In ultraprecision machining, linear groove structures are usually manufactured by fly-cutting or planing. Fly-cutting is a milling process using one single diamond tool which is revolving on an aerostatic supported rotor of a high precision spindle. This rotating tool is slowly moved across the work piece surface with a maximum infeed up to 2 mm. In planing processes, the tool is mounted in a fixed position on a highly dynamic feed axis and is moved across the work piece surface with up to 60.000 mm/min at maximum in feeds below 10 µm. In both cases a groove is cut into the work piece surface which corresponds exactly to the cutting edge of the diamond tool. By machining intersecting grooves in different directions and applying different tool geometries, a great variety of complex micro structures can be manufactured. The decision which ultraprecision machining process is applied usually depends on the properties of the micro structure and the resultant machining time. Furthermore, fly-cutting has the advantage of far lower tool wear compared to planing. The process-related cinematic roughness in fly-cutting can be reduced down to the sub-nanometer range by optimally adjusting the feed rate to the spindle revolution and can thus be considered as negligible. Therefore, fly-cutting is well suitable for the machining of high aspect ratio micro structures with sub micron form accuracies as they are required for the embossing tools for the passive alignment arrays.

3 Hot embossing of passive alignment structures

Within the Collaborate Research Centre SFB 440, hot embossing is used to manufacture passive alignment structures by replication. During hot embossing, a thermoplastic substrate is softened plastically by heating in an evacuated chamber. Subsequently, the micro structured embossing tool is slowly pressed into the molten substrate. After cooling down, the substrate keeps the imprint of the embossing tool which is withdrawn when the substrate has solidified.

Fig. 3. Embossing chamber of a Jenoptik HEX02 hot embossing machine

In contrast to other replication processes, e.g. injection molding, the substrate yields very little which leaves almost no inner tension in the replica. Therefore hot embossing is especially suitable for the highly accurate replication of micro structures. The quality of the replicated micro structures is directly dependent on the quality of the embossing tool. Therefore, the form tolerances and surface finish which are achieved in the manufacturing of the embossing tool are decisive for the replicated micro structures.

4 Manufacturing and practical testing of the passive alignment arrays

The machining of the appropriate embossing tool requires a considerable effort for set up and referencing the tool and the work piece. Since the three-sided cavities and channels are the negative imprint of triangular column structures, the ultraprecision machining process has to be set up in such a way that the grooves which are machined by fly-cutting from three different directions (included angle:60°) cross in exactly one point at each intersection. Otherwise, artifacts would remain between the cavities which would decrease the functionality of the array. Therefore, three grooves which intersect by 60° have been machined in sample cuts which have been analysed by optical microscopy prior to the machining of the actual structure. By measuring the offset between two intersecting grooves and the remaining groove, correction values have been determined to position the tool in such a way that all tree direction intersect within one point with an accuracy < 0.5 μm. This set-up in combination with an appropriately shaped diamond tool enables the manufacturing of regular three-sided pyramids with straight columns on top (Figure 5). The imprint of these columns form funnels which merge into squared channels which serve as air ducts to suck in the micro spheres.

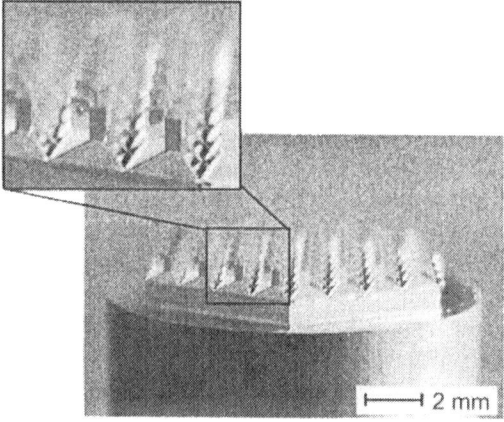

Fig. 4. Embossing tool with three sided column structures

Each cavity can hold exactly one micro part in a defined position. When the array is connected to a suction gripper and positioned over an incoherent batch of micro components the air flow sucks the micro components into the cavities which are then plugged and the air flow is cut off. When the array is filled up, excessive micro components fall off or are not even grasped in the first place. In this way, a defined quantity of micro components can be picked up and is aligned with a defined and constant pitch simultaneously (Figure 6). The symmetry of those micro components supports their self-alignment in the single cavities of the array structures.

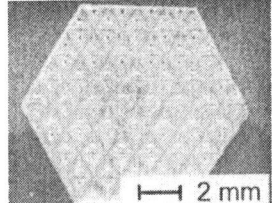 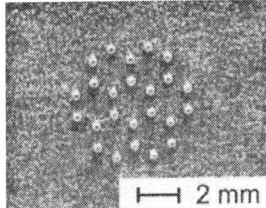

Fig. 5. Micro sphere-array (left) and aligned micro spheres (right)

5 Outlook

The passive alignment arrays are suitable for the secure handling and shipping of the sorted micro components as well. For this purpose, a cap with clip elements has been designed, in which the array can be locked in place. A soft inlay at the bottom of the cap adapts to the shape of the micro components and holds them in position when the array is released from the suction gripper. An adequate pick-up and alignment mechanism can then be realized as follows (Figure 7):

1. Gripping the array with a vacuum gripper
2. Sucking the micro spheres into the cavities of the array
3. Engaging the array into the cap
4. Cutting off the air flow and removing the suction gripper

This mechanism enables the alignment and handling of a fixed quantity of micro spheres in a defined position. Optionally, the arrays can be used in the same way to assort a loose batch of micro spheres right before an automated micro assembly line. Due to their regular three sided passive alignment features, the arrays can cope with varying diameters and still keep up a lateral positioning of the micro spheres at constant distances. The upcoming research activities will focus on the realization of the pick-up and alignment mechanism as described above. With the aid of active energies such as air flow and vibration, the fully automated alignment and temporary fixation of a loose batch of micro spheres will be realised.

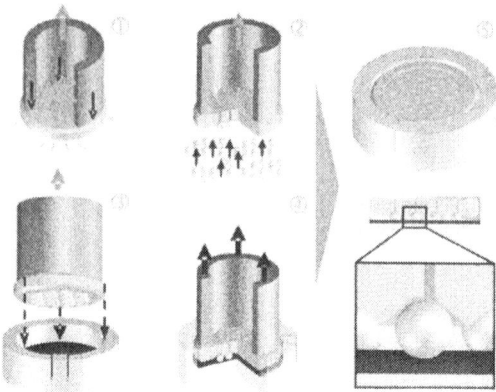

Fig. 6. Mechanism fort the parallel alignment and fixation of micro spheres

Furthermore, the machining of three dimensionally arranged passive alignment structures as well as the use of passive alignment structures to support high precision joining processes will be investigated within the Collaborate Research Centre SFB 440. Therefore the advanced design of passive alignment micro structures will aim for providing a fixed reference in the joining layer to prevent the shape distortion of micro components during gluing or brazing.

Acknowledgements

The authors would like to thank the German Research Foundation (DFG) for their support within the Collaborate Research Centre SFB 440. Through their funding of the sub-project A7 »Assembly Supporting Mechanical Structures for the Passive Alignment of Micro Components« the DFG has enabled the research activities described above.

References

1. ISO 3290 / DIN 5401
2. Brandau, T., Mikroverkapselte Wirkstoffe und Mikrokugeln in pharmazeutischen und kosmetischen Anwendungen, Chemie Ingenieur Technik 75, No. 11, Weinheim, 2003, pp. 1741 - 1745
3. Petersen, B., Flexible Handhabungstechnik für die automatisierte Mikromontage, Dissertation RWTH-Aachen, Aachen, 2003, pp. 12 - 14
4. Brecher, C.; Weinzierl, M.; Lange, S., Development of Passive Alignment Techniques for the Assembly of Hybrid Micro systems, Precision Assembly Technologies for Mini and Micro Products, Proceedings of the IFIP 12, Springer Verlag New York, USA 2006, pp. 611 - 615, ISBN 0-387-31276-5
5. Brecher, C.; Weinzierl, M.; Lange, S.; Peschke, C., Active and Passive Tool Alignment in Ultraprecision Machining for the Manufacturing of Highly Precise Structures, Production Engineering – Research and Development, Annals of the German Academic Society for Production Engineering WGP 8 (2006), XIII/1, pp. 193 - 196

Chapter 8

Micro-Metrology

TOWARDS A TRACEABLE INFRASTRUCTURE FOR LOW FORCE MEASUREMENTS

Richard K Leach, Christopher W Jones

Industry & Innovation Division, National Physical Laboratory
Hampton Road, Teddington, Middlesex TW11 0LW
richard.leach@npl.co.uk

Abstract Over the past ten years or so the need for the measurement of low forces ranging from newtons down to attonewtons has become increasingly important. As we begin to manufacture and manipulate structures on the micrometre to nanometre scale, the forces that are exerted in such processes must be controlled. To control such forces requires some form of measurement, either a direct measurement of the force, or a measurement of the effect the force has on the structure it is applied to. This paper is primarily concerned with the development of a traceability infrastructure for forces in the range from 1 nN to 10 µN. The lower end of this force range does not cover chemical or most biological forces (usually in the femto- to piconewton range) despite the increasing importance of accurately measuring such forces. Further work is still required to push the limits of force traceability to these levels. At the upper end of the force range considered here, more traditional methods for measuring forces can be used that are traceable to the unit of mass, i.e. the force is realised as a mass in a gravitational field. The force range discussed in this paper applies to many nano- and micrometre scale manipulation and assembly applications, including micro-grippers, handlers and force feedback devices. Further applications that fall into the force range discussed here include the force exerted on a surface by atomic force microscopes and other scanning probe instruments, forces in the area of materials property measurement using indentation technology, the forces found in micro-electromechanical systems (MEMS) and the forces exerted by artificial biological tissues, for example muscle fibres. The two main force generation mechanisms that are found in nature and engineering are the weight of the mass of an object in a gravitational field and the deflection of an element with a finite spring constant. On the micro- to nanometre scale the spring force is more usually used to produce or react to a force, for example an AFM cantilever.

1 Introduction

Traceability for force measurement is usually carried out by comparing to a calibrated mass in a known gravitational field. However, as the forces (and hence masses) being measured decrease below around 10 µN (approximately equivalent to 1 mg), the uncertainty in the mass measurement becomes too large and the masses become difficult to handle. For this reason it is more common to have a force balance that gains its traceability through electrical and length measurements.

The current force traceability route is at least a two-stage process. The first stage is to develop a primary force standard instrument deriving traceability directly from the base unit definitions realised at the world's National Measurement Institutes (NMIs). These primary instruments will typically sacrifice practicalities in order to obtain the best possible metrological performance. Various groups have developed such instruments, with the current best performance held by examples at the National Institute of Science and Technology (NIST) in the USA and the National Physical Laboratory (NPL) in the UK.

The second stage in the traceability route is to design a transfer artefact, or sequence of artefacts, to transfer the force calibration to target instruments in the field. These artefacts may sacrifice uncertainties, resolution or range of force measurement, in exchange for cost reductions, portability or compliance with other physical constraints, such as size or environmental tolerance.

2 Primary Force Balances

The leading examples of force measurement instruments are based on the electrostatic force balance principle. The force to be measured is exerted on a flexure system, which deflects. This deflection is interferometrically determined. The deflection of the flexure also changes the capacitance of a set of parallel capacitor plates in the instrument. This is usually achieved either by changing the plate overlap, or by changing the position of a dielectric, with flexure deflection. In this way the capacitance changes linearly with deflection. The interferometer signal is used in a closed-loop controller to generate a potential difference across the capacitor generating an electrostatic force that servos the flexure back to zero deflection. Measurement of the force exerted is derived from traceable measurements of length, capacitance and potential difference. The exerted force is calculated using equation 1, in which z is the flexure displacement, and C and V the capacitance of and voltage across the parallel plates respectively. The capacitance gradient, dC/dz, must be determined prior to use and is a source of statistical uncertainty that must be quantified. Uncertainties are associated with external system interaction, as well as with the measurement of z, V and C, and with contributions from flexure misalignment, corner loading and hysteresis.

$$F = -\tfrac{1}{2} V^2 \frac{dC}{dz} \tag{1}$$

2.1 The NIST Electrostatic Force Balance

NIST's Electrostatic Force Balance (EFB) [1] was the first major example of a fully traceable primary low force balance. The NIST EFB, which is currently operated in a vacuum, has a working range of 10 nN to 1 mN with accuracy of parts in 10^4 and a sub-nanonewton resolution [2]. The instrument, schematically shown in Figure 1, is currently being used to characterise prototype transfer artefacts. The capacitor system is created from two coaxial cylinders, the insertion of one inside the other varied with balance deflection leading to the desired linear capacitance gradient.

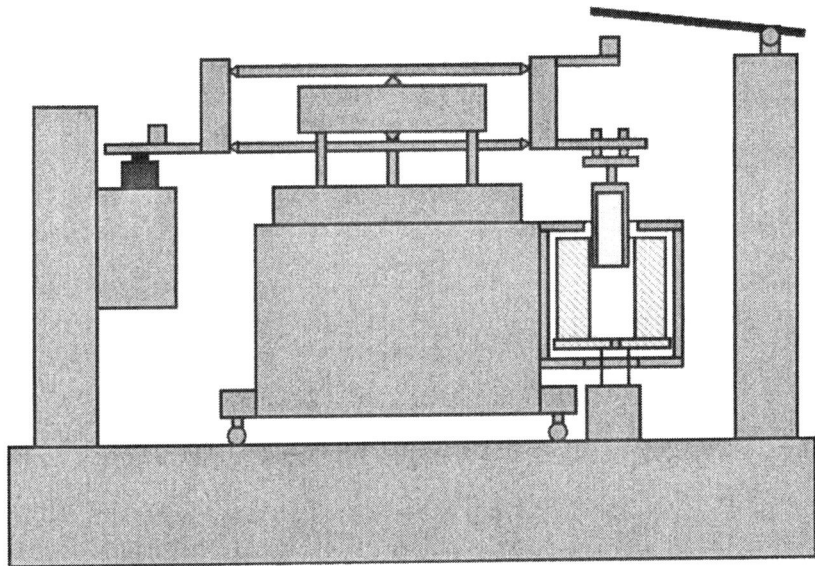

Fig. 1. Schematic diagram of the NIST EFB, from [2]. The hatched item on the right is the outer capacitor cylinder, with the inner cylinder suspended from the flexure system across the top. The red electromagnetic actuator moves the balance for dC/dz determination; the black line shown in the top right corner is a mass loader for wire test masses for the higher force range.

2.2 The KRISS Electrostatic Force Balance

The Korea Research Institute of Standards and Science (KRISS) also has plans in place to construct a low force balance. Much of their effort so far has been concerned with modelling of the motion of the basic low force balance parallelogram flexure arrangement, focussing on the dependency of good results on parallelism and mechanisms to fine-tune the geometry of the flexure [3].

2.3 The NPL Low Force Balance

The unique, monolithic construction of the flexure arrangement in the NPL Low Force Balance (NPL LFB) [4, 5] combined with its smaller size promises better low-end force measurement performance than its current international equivalents. The instrument's range is 1 nN to 10 µN with a design resolution of 50 pN. The

NPL LFB, schematically shown in figure 2, was designed and constructed under contract by the Technical University Eindhoven [6]. Displacement of a dielectric blade in the capacitor, connected to the flexure, is measured using a plane mirror differential interferometer. The feedback voltage generates a force on the blade that returns the flexure displacement to zero. The NPL LFB is currently undergoing characterisation work and preliminary results will be presented shortly.

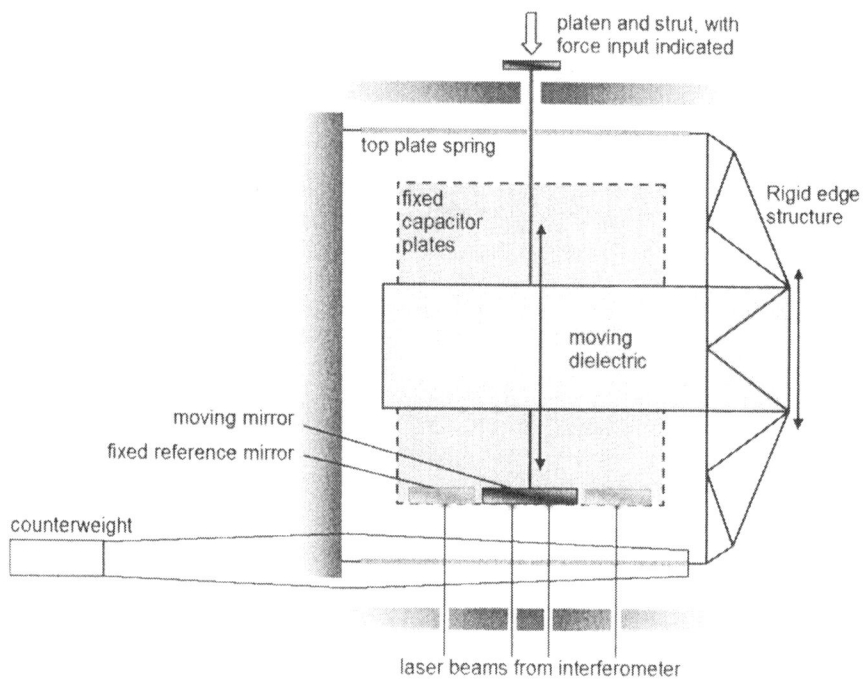

Fig. 2. Schematic diagram of the NPL LFB working mechanism. Force exerted on the platen deflects the dielectric vertically against the plate springs, with motion constrained by the rigid edge structure. The flexure, including the edge structure and solid supporting frame, were monolithically manufactured. The counterweight can accept additional loads to offset test masses. The spring constant of the balance is around 1 N m^{-1}.

2.4 High-end performance: deadweight measurement

An important partial verification of the performance of a low force balance occurs at the overlap of the operating ranges of the deadweight calibration system and the balance. This has been carried out successfully [7] on the larger NIST EFB, which has a relatively high upper force range limit allowing larger and more practical test masses. The significantly lower upper force limit of the NPL LFB adds to the complexity of mass intercomparison with the balance's electrostatic force measurements. Efforts are underway at NPL to produce a suite of low mass artefacts using novel materials that may be calibrated and handled such that relative uncertainties are significantly reduced. Calibrated on the NPL primary mass balance, these will be weighed on both the NPL Low Force Balance and NIST EFB.

2.5 Low-end performance: radiation pressure experiments

Verification of the lower end of the low force balance's force range may be obtained by comparison with radiation pressure, adapting similar examples (such as [7]). A medium power laser would be directed onto the end of the balance platen, to which a very high reflectivity (> 99.99 %) mirror is attached, of suitable size to catch all of the laser light. Assuming the ideal case with no losses, a nanonewton of force would be generated by 150 milliwatts of input optical power. Care would need to be taken to eliminate spurious radiometric effects, which would need to be traded off with lower heat dissipation when deciding on the use of a vacuum for operation. Some noise will originate from the kinematic effect of particles excited off the mirror. The design of the NPL LFB combined with operation in position-nulled mode is such that thermal expansion in the vicinity of the platen would have a negligible effect on the measured force. Furthermore, with a high-reflectivity mirror in place the transmitted power would be of the order of a milliwatt or less, acceptable provided exposure times are kept short.

3 Low force transfer artefacts

A number of groups worldwide have worked to develop technologies that may be implemented as low force transfer artefacts. These projects were motivated by different industrial needs and, therefore, vary in force range, tolerance to environment and target instrument. It is the task of the NMIs in collaboration with industry to develop working transfer artefacts.

3.1 Springs

After gravitational forces from calibrated masses the most intuitive and common technology used for calibrated force production is an elastic element with a known spring constant. The element, such as a cantilever or helical spring, is deflected by a test force. The deflection is measured, either by an external system such as an interferometer, or by an on-board MEMS device such as a piezoelectric element. With the spring constant previously determined by a traceable instrument such as an electrostatic force balance, the size of the test force can be calculated. Recent examples of artefacts use modified AFM cantilevers with piezoresistive deflection measurement, as shown in Figure 3. These examples also have fiducial markings along their length to reduce a major source of uncertainty for this type of elastic element device, the position of the artefact-target interaction. Effective spring constants are known for several positions along the length of the cantilevers, allowing calibration transfer in a number of ranges. An alternative approach, also developed at NPL, is a MEMS device known as MARS (Microfabricated Array of Reference Springs). The device consists of a number of reflective discs supported on helical springs of different nominal spring constants [9]. Lower sensitivity to test probe interaction position is a distinct advantage of this type of spring device but the added complexity of the elastic system makes it harder to model and evaluate.

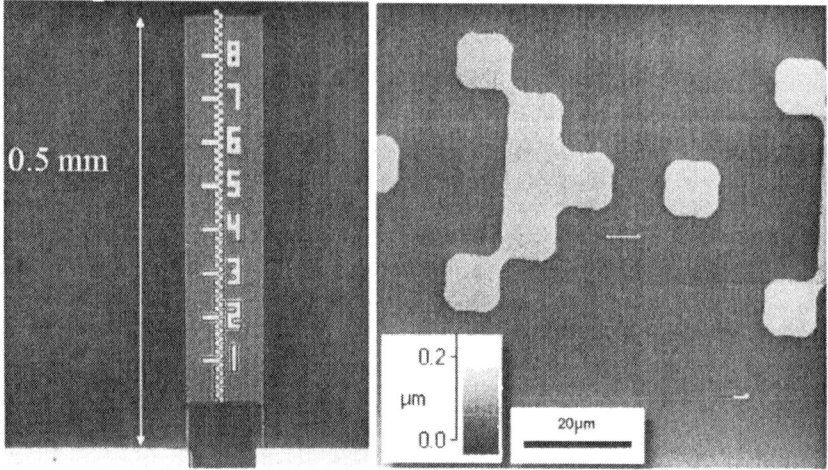

Fig. 3. (left) the NIST (from Pratt 2007); (right) detail of the fiducial markings on the NPL C-MARS device; the 10 μm squares form a binary numbering system along the axis of symmetry (from Cumpson *et al* 2004 [10]).

3.2 Resonant structures

Alternative propositions take advantage of the noise advantages of frequency modulation. Changes in the tension of a stretched string can be detected *via* related changes in its resonant frequency. If a force is exerted on one of the string anchor points along the string axis, the tension in the string will decrease. For a well-characterised string the force exerted can be calculated from an accurate determination of the frequency shift. In this way a low-force measurement device is created. With a careful choice of working frequency through device design, the effect of noise on the measurement signal can be dramatically reduced in comparison to other methods. Stalder and Dürig's 'nanoguitar' [12, shown in Figure 4, is one example of such a device. Similar devices resembling double-ended tuning forks replace the string with narrow plate oscillators [12]. The challenge for resonance methods is in making the frequency measurement system, usually optical interrogation by laser, robust and compact enough for use in a transfer artefact.

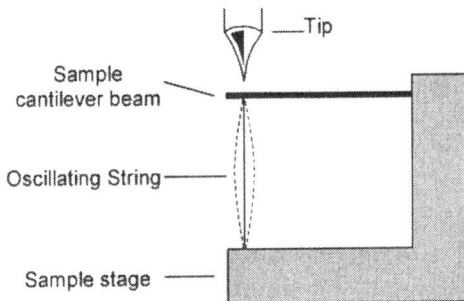

Fig. 4. Schematic of a resonant string sensor (after [10])

4 The future – intrinsic force standards

Completion of the traceability link at the micro- and nanonewton scale will see development efforts shifting towards even smaller forces. Traceability at the levels fundamental to biochemistry, such as covalent bonds (100 pN) and thermal motion of DNA molecules (100 fN) is now required, for example to verify theoretical models that may be inaccurate.

Further radiation pressure experiments may present a possibility, as may the electrostatic force between a probe and a surface of well-defined geometry.

It is hoped that at this scale it will be possible to develop intrinsic force standards based, for example, on reversible conformation changes in proteins. To date results have been mixed. Well-defined force steps in the 100 pN to 300 pN range, as protein structures are unfolded in sequence, have been reported [13]. However extreme hysteresis has been observed when reversing such extensions in similar experiments [14].

As collaborative efforts are developed between biochemical institutions and traditional engineering metrologists significant advances of force traceability into the piconewton range should be made.

References

1. J. R. Pratt, D. T. Smith, D. B. Newell, J. A. Kramar and E. Whitenton, Progress toward Système International d'Unités traceable force metrology for nanomechanics, J. Mat. Res. **19**, 366-379 (2004)
2. J. R. Pratt, private correspondence (2007)
3. I.-M. Choi, M.-S. Kim, S.-Y. Woo and S. H. Kim, Parallelism error analysis and compensation for micro-force measurement, Meas. Sci. Technol. **15**, 237–243 (2004)
4. R. K. Leach, S. Oldfield, S. Awan, J. Blackburn and J. M. Williams, Design of a bi-directional electrostatic actuator for realising nanonewton to micronewton forces, NPL Report **DEPC-EM-001**, 1–32 (2004)

5. R. K. Leach, S. Oldfield, D. Georgakopulos, Traceable nanonewton force measurement at the National Physical Laboratory, UK, Proc. 6th euspen Int. Conf., Baden, Austria, May, 414-417 (2006)
6. R. Henselmans, N. Rosielle, L. Cacace, P. Kappelhof, F. Klinkhamer and H. Spierdijk, Low force measurement facility: mechanical design report (Technical University of Eindhoven) (2004)
7. D. B. Newell, J. A. Kramar, J. R. Pratt, D. T. Smith and E. R. Williams, The NIST Microforce Realization and Measurement Project, IEEE Transactions on instrumentation and measurement **52**, 508-511 (2003)
8. M. Feat, C. Zhao, L. Ju and D. G. Blair, Demonstration of low power radiation pressure actuation for control of test masses, Rev. Sci. Instr. **76**, 036107 (2005)
9. P. J. Cumpson, J. Hedley and P. Zhdan, Accurate force measurement in the atomic force microscope: a microfabricated array of reference springs for easy cantilever calibration, Nanotechnology **14**, 918-924 (2003)
10. P. J. Cumpson, C. A. Clifford and J. Hedley, Quantitative analytical atomic force microscopy: a cantilever reference device for easy and accurate AFM spring-constant calibration, Meas. Sci. Technol. **15**, 1337-1346 (2004)
11. A. Stalder and U. Dürig, Nanoguitar: oscillating string as force sensor, Rev. Sci. Instrum. **66**, 3576-3579 (1995)
12. K. Fukuzawa, T. Ando, M. Shibamoto, Y. Mitsuya and H. Zhang, Monolithically fabricated double-ended tuning-fork-based force sensor, J. Appl. Phys. **99**, 094901 (2006)
13. M. Rief, M. Gautel, F. Oesterhelt, J. M. Fernandez, H. E. Gaub, Reversible unfolding of individual titin immunoglobulin domains by AFM, Science, **276**, 1109-1112 (1997)
14. A. Oberhauser, P. Hansma, M. Carrion-Vazquez and J. M. Fernandez, Stepwise unfolding of titin under force-clamp atomic force microscopy, PNAS **98**, 468-472 (2001)

WHEN MANUFACTURING CAPABILITY EXCEEDS CONTROL CAPABILITY: THE PARADOX OF HIGH PRECISION PRODUCTS, OR IS IT POSSIBLE TO ASSEMBLE FUNCTIONAL PRODUCTS OUT OF COMPONENTS WE ARE UNABLE TO MEASURE?

S. Koelemeijer Chollet, M. Braun, F. Bourgeois, J. Jacot

Laboratoire de Production Microtechnique, Institut de Production et Robotique,
Ecole Polytechnique Fédérale de Lausanne, SWITZERLAND
Laboratoire de Production Microtechnique
EPFL
Station 17
1015 Lausanne
Switzerland
Tél : ++ 41 21 693 5997
Fax : ++ 41 21 693 3891
sandra.koelemeijer.epfl.ch

F. Chautems

MPS AG, Bonfol, Switzerland

Abstract Sorting and mating is a very common practice in high precision manufacturing, as well in the watch industry, biomedical or automotive industry. This strategy helps to increase the assembly yield, but is costly and time-consuming. The question is, when is it necessary to apply sorting and mating, and when can it by avoided without loss in product quality? We furthermore show in this paper that the measurement precision is often lower than that of manufacturing. New and more precise manufacturing equipment and assembly devices allow for narrower distributions, and this raises many questions: is it possible to stop sorting? how can we be sure of the quality of the components and of the products? A case study at the company MPS AG, manufacturer of miniature ball bearings, illustrates this trend. We propose another manufacturing and assembly strategies, and show what the conditions are that are necessary to permit this approach.

1 Introduction

Manufacturing of high precision products such as watches is a secular activity. Despite the low quality of the old manufacturing processes, manufacturers have found pragmatic solutions to reach high precision. Over the years, empiricism and intensive human labor have made this possible. Nowadays, cost becomes an important issue, and a more effective approach is necessary.

To guarantee the functionality of high precision products, particular care has to be taken with the functional dimensions of the components. Several steps in the design phase and in manufacturing have a decisive importance: tolerancing, choice of the manufacturing process, process control, assembly, and finally the functional control of the finished product.

The manufacturers of microproducts are often faced with the following problem: manufacturing processes aren't as precise as the required tolerances. In this case, the process capability is lower than 1 [1,2]. The process capability is defined as follows (Eq.1), where USL and LSL are the upper and lower limits of the tolerance interval.

$$C_{process} = \frac{USL - LSL}{6 \cdot \sigma_{process}} \qquad (1)$$

The reason is that the main kinds of production machines are used for macro and microproducts, milling machines, automatic lathes, etc. Their capability is largely sufficient for macroparts, but is no longer when the components size drops. The process is capable if the capability value is of 1 and more, but it isn't unusual to have values of 1.5 to 3 in micromachining. Sorting and mating, or selective assembly, is the pragmatic solution to this problem.

2 State of the art

Selective assembly has been widely used in the precision industry over the years. This consists in classifying the batches of the two parts to assemble into classes of same dimension, and to select the convenient classes to obtain a functional assembly. On the shop-floor, the width of classes is fixed empirically, often corresponding to the resolution of the available measurement equipment, and is often of 2 μm in the watch making industry (Fig. 1). Bourgeois [3] has made a thorough review on the topic of selective assembly. Academic work proposes how to optimise classes, for example with equal areas, but those methods are difficult to apply on the shop-floor [4, 5, 6, 7, 8, 9].

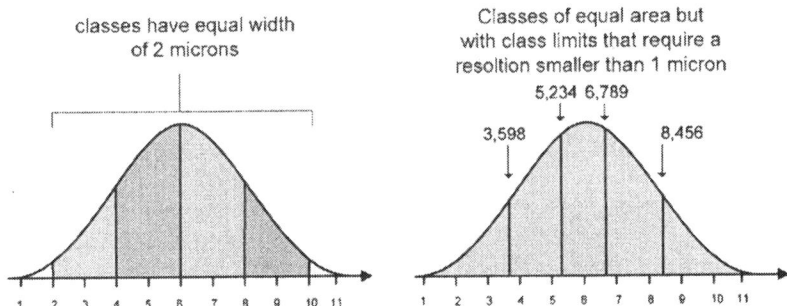

Fig. 1. Classes with equal width are commonly used in the industry, whereas classes with equal area may lead to a higher assembly yield

3 The paradox

The main parameters of the assembly yield – the percentage of good assemblies – are the standard deviation and the off-centering of the batches to be assembled and the class width. Their effects are shown in the case study in the next paragraph. Recent developments in machine design, especially in automatic lathes, allow for narrower distributions, and thus may completely change what was considered best practice until now.

Furthermore, shop-floor people as well as academics make the same assumption: they assume that the measurement device is precise compared to the distribution of the components. Practical studies showed that a significant part of the distribution of the components is in fact due to the dispersion of the measurement (Eq. 2).

$$\sigma^2_{component} = \sigma^2_{process} + \sigma^2_{measurement} \qquad (2)$$

We are now confronted with the following paradox: the dispersion of the machining may be tighter than that of the measurement. It is vital to redefine new strategies taking in account this paradox. How do we define classes while we are not sure we can measure the components correctly? What is the risk we take if the components aren't put in the right classes because of errors made during the measurement process?

Is it still necessary to make those classes? What is the risk we take if we don't make classes?

To answer those questions, we have to evaluate the costs related to all those cases. There is a direct cost related to measurement and sorting, as well as to production. Those costs have to be compared to the yield of the assembly, that is the number of functional parts that are manufactured. We also have to take into consideration the cost related to the risk of a non functional product arriving at the customers, and the possibility to realise a functional control at the end of the assembly process.

4 Case study: assembly of miniature ball bearings at MPS AG.

MPS is one of the world leader in the manufacturing of miniature ball bearings that are mainly used in mechanical watches. The inner cage is realized in two parts (cone and noyau) which are press fitted together, at the last step of the assembly process (fig. 2). Their diameters are critical parameters of the precision of the ball bearing. If the interference is too small, the force to press fit will be low, and the ball bearing won't hold together when axial forces are applied. A too high interference will lead to the deformation of the components and the bearing won't turn smoothly.

 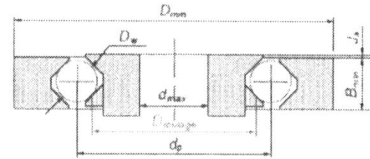

Fig. 2. Image and section off the ball-bearing – outer diameter is from 2 to 5 mm according variant, and balls are 300 µm diameter

MPS has practised sorting and mating for years. Parts are sorted every 2 µm, and classes are selected to obtain a nominal interference of about 6 to 16 µm. Parts are press fitted with a manual press, and the functional check is done by the operator, by "feeling" the smoothness of the gliding of the balls. The sorting and mating process is expensive while very time consuming. The question about the possibility to stop this process has often been raised, but technically it has not been possible up to now. Two main technical improvements have recently been introduced: more precise automatic lathes on one hand, and a numerical press that measures the press fitting force and the z position on the other hand. The question now arises again: is it possible to stop sorting [10]?

Several simulations have been run [11], in order to calculate the assembly yield with classes and without classes. Each batch is characterized by the standard deviation and the off-centering of its distribution (Fig. 3). The standard deviation is due to the capability of the manufacturing equipment. The introduction of the new lathes will lead to smaller values of the standard deviation. On the other hand, the off-centering of the batch is given by the operator when tuning the machine to manufacture a new batch. Important parameters are thus the standard deviation and the off-centering of the batch, as well as the width of the classes.

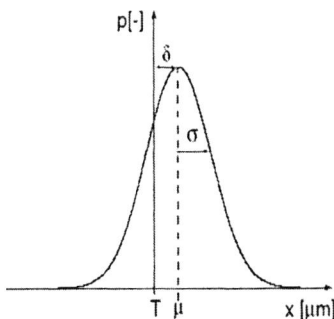

Fig. 3. Part distribution with target T, mean μ, standard deviation σ, off-centering δ=μ-T

Typical parameters for the parts to be assembled are standard deviation of 2.5μm and 1.5μm with the old lathes, and of 1μm and 0.8μm with the new equipment. The assembly yield for different parameters is shown in the Table 1. One can notice that selective assembly is the most interesting with large standard deviations and rather small values of off-centering, whereas with small standard deviations, the yield isn't improved. It is thus very important to monitor the standard deviations for the different components manufactured by MPS, and the evolution due to the new equipment in order to be able to decide if sorting can be skipped.

Assembly yield (%)					
Off-centering (μm)	0	1	2	3	4
	Std 2.5 μm and Std 1.5 μm				
Random assembly	82,1	81,3	81,3	74,4	62,3
Selective assembly (cw 2μm)	93,0	92.1	88,7	75,1	60,0
	Std 1 μm and Std 0.8 μm				
Random assembly	99,1	96,2	95,8	79,9	75,4
Selective assembly (cw 2μm)	99,4	96,1	93.9	63,0	51,1

Table 1. Selective assembly is particular interesting for large standard deviations and small off-centerings

In a second step, variance of the measurement device has been taken into account. The measurement device has been carefully characterized. The same batch of 10 parts has been measured 3 times by the same operator, and by 3 different operators. In the end, the measurement capability is computed and compared to the acceptable tolerance interval. This has been done for the outer diameter of the "cone" and the internal diameter of the "noyau". The variability of the total measurement system is defined as (Eq. 3):

$$\sigma^2_{measurement\ Error} = \sigma^2_{Gauge} = \sigma^2_{Re\ peatability} + \sigma^2_{Re\ producibility} \qquad (3)$$

A system is considered acceptable with a P/T ratio of P/T <0,1, marginally acceptable with 0,1<P/T <0,3, and unacceptable with P/T>0,3. A high precision industry like the watch industry is especially prone to measurement problems. The P/T ratio is often in the range of P/T=0.5, when measuring holes it is even mostly around P/T=1 (Eq. 4).

$$\frac{P}{T} = \frac{6 \cdot \sigma_{Gauge}}{USL - LSL} \qquad (4)$$

The values measured at MPS, with a tolerance interval USL-LSL of 8 µm for both components, give P/T ratio values of 0,19 and 0,39. Those values are at the limit of what is acceptable. But if we take a tolerance interval of 2 µm which is the class width, we see that the precision of the measurement is really out of scope.

The variance of the measurement process leads to the risk of misclassification of a part, either a false accept risk (risk of accepting a bad part) or a false reject risk (risk of rejecting a good part). Additional misclassifications occur with selective assembly, when putting a part in the wrong class (Fig 4).

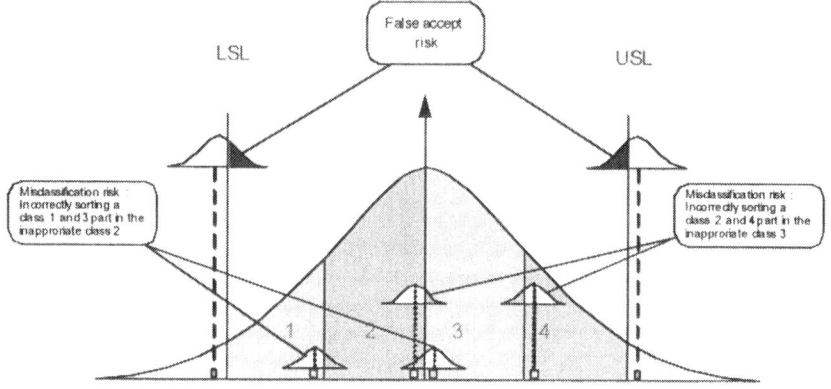

Fig. 4. The measurement error leads to general false accept risks, as well as false accept risks for each class

The values of the false accepting risk are almost negligible. The values of false rejected risk are of 3% and 5% for the internal diameter cone and the external diameter of the core with random assembly With selective assembly and classes of 2 µm, they are of 15% and 17%.

If we now again calculate the assembly yield, the values for the selective assembly drop significantly, and are around 70% (Table 2). These results are very interesting for the shop-floor managers. They do corroborate their intuitive approach, as well as their first practical results. On the other hand, the number of parts rejected during the last assembly stage that controls the assembly force doesn't reach 30%. This can have several meanings. Either the interference isn't as critical as generally admitted, at least not for all models of ball-bearings. Either

other parameters such as the friction coefficient significantly affect the press fit force [12, 13].
The following manufacturing and assembly strategy can be suggested:
- Respect of a very narrow off-centering on both components
- Statistical dimensional control of the whole batch
- Random assembly
- Force control while press fitting

The idea is to replace the expensive 100% dimensional control on each component by a statistical dimensional control of the batch as a whole, and a 100% functional control of the assembly force during the assembly process [14].

Assembly costs are thus reduced while there is no more measuring, sorting and mating. Risks of a defect ball-bearing reaching the customers are excluded by the functional force control off the last assembly operation. This kind of manufacturing strategy is valid only if the manufacturing equipment is under control, and if the risk of a bad assembly going further into production or reaching the customer can be excluded.

Assembly yield (%)					
Off-centering (µm)	0	1	2	3	4
Std 2.5 µm and Std 1.5 µm					
Random assembly	75,5	74,8	74,8	68,8	57,3
Selective assembly (cw 2µm)	67,1	66,9	66,7	58,1	47,9
Std 1 µm and Std 0.8 µm					
Random assembly	91,2	88,5	88,1	73,4	69,3
Selective assembly (cw 2µm)	74,5	72,6	70,5	51,0	45,0

Table 2. Values of assembly yield for different class widths and standard distributions, in the case of random and selective assembly, taking in account the measurement error

5 Further work

Results presented in this paper are simulation results based on measurements on only two different components of one model of ball-bearing. Further measurements have to be made on a larger amount of different components, and the assembly yield predicted by the simulation has to be compared to the real assembly yield given by the assembly force. A general model for the assembly yield with the following parameters: width of classes, standard deviation, off-centering and variance of measurement device will than be proposed.

Furthermore, the functional interference leading to an acceptable range of assembly forces has to be investigated in more detail. The commonly admitted interval range may be enlarged for several ball-bearings, and the effects as well on the yield as on production control are to be analysed [15, 16, 17, 18].

References

1. F. Bourgeois, Y. L. de Meneses, J. Jacot, "Routes & Déroutes - Sur les traces d'un jeune ingénieur qui se lance dans la microtechnique", Revue Polytechnique, Novembre 2005

2. W. L. Pearn and P. C. Lin, "Measuring process yield based on the capability index Cpm," *International Journal of Advanced Manufacturing Technology*, No. 24, pp: 503–508, 2004

3. F. Bourgeois, Vers la maîtrise de la qualité des assemblages de précision, Thèse, EPFL, Ecole Polytechnique Fédérale de Lausanne, Lausanne, 2007

4. F. Bourgeois, Y. L. de Meneses, S. Koelemeijer Chollet, J. Jacot, "How much can sorting improve the capability in assembly tasks?", IEEE International Symposium on Assembly and Task Planning (ISATP03), Besançon, July 2003

5. F. Bourgeois, Y. L. de Meneses, S. Koelemeijer Chollet, J. Jacot, "De l'utilisation du tri pour augmenter le rendement d'un procédé d'assemblage", Journée d'étude de la Société Suisse de Chronométrie (SSC03), Bienne, Septembre 2003

6. Sm. Kannan, and V. Jayabalan, "A new grouping method to minimize surplus parts in selective assembly for complex assemblies," *International Journal of Production Research*, vol. 39, no. 9, 1851 – 1863, 2001

7. Sm. Kannan, A. Asha, A., and V. Jayabalan, "A New Method in Selective Assembly to Minimize Clearance Variation for a Radial Assembly Using Genetic Algorithm," *Quality Engineering*, Vol. 17, pp: 595 – 607, 2005

8. H. M. Kwon, K. J. Kim, and M. J. Chandra, "An Economic Selective Assembly Procedure for Two Mating Components with Equal Variance," *Naval Research Logistics*, 46: 809-821, 1999

9. G. A. Pugh, "Selective Assembly with Components of Dissimilar Variance," *Computers and Industrial Engineering*, Vol. 23, No. 1-4, pp: 487 – 491, 1992

10. Y. Fu, « Tolerance Analysis for Ball Bearing Assembly », Internal report, EPFL, Ecole Polytechnique Fédérale de Lausanne, Lausanne, 2007

11. M. Braun, "Optimization of the production process for high-precision ball-bearings", Internal report, EPFL, Ecole Polytechnique Fédérale de Lausanne, Lausanne, 2007

12. R. S Srinivasan, K. L. Wood, and D. A. McAdams, "Functional Tolerancing: A Design for Manufacturing Methodology," *Research in Engineering Design*, Vol. 2, pp. 99-115, 1996

13. F. Bourgeois, J. Jacot, "Comprendre le chassage à l'échelle horlogère", Congrès International de Chronométrie (CIC04), Montreux, Septembre 2004

14. F. Bourgeois, L. Charvier, J. Jacot, G. Genolet, H. Lorenz, "La maîtrise du procédé de chassage dans le domaine submillimétrique", Congrès International de Chronométrie (CIC07), Colombier, Septembre 2007

15. J. Gfeller, F. Bourgeois, S. Koelemeijer Chollet, J. Jacot, "La mesure fonctionnelle: un pas décisif vers la maîtrise de la qualité", Bulletin de la Société Suisse de Chronométrie, Switzerland, Avril 2007

16. F. Bourgeois, Y. L. de Meneses, S. Koelemeijer Chollet, P-A. Adragna, M. Pillet, J. Jacot, "Tolerancing strategy for microsystem assembly", submitted to Precision Engineering the 5th of July 2006

17. J. Gfeller, S. Koelemeijer Chollet, F. Bourgeois, J. Jacot, "Functional tolerancing and testing increase microassembly yield", IEEE International Symposium on Assembly and Manufacturing (ISAM07), Michigan, July 2007

18. F. Bourgeois, Y. L. de Meneses, S. Koelemeijer Chollet, J. Jacot, "Defining assembly specifications from product functional requirements using inertial tolerancing in precision assembly", IEEE International Symposium on Assembly and Task Planning (ISATP05), Montreal, July 2005

IMPACT FORCES REDUCTION FOR HIGH-SPEED MICRO-ASSEMBLY

Ronald Plak, Roger Görtzen, Erik Puik
TNO Science & Industry, Eindhoven, the Netherlands

Abstract During the placement of components in micro-assembly, high impact forces occur. The current approach is to reduce these impact forces by coupling the gripper to the drive unit of the placement device with 5 DOF, wherein the gripper that contacts the component has a relatively low mass. To prevent the gripper from bouncing back at the end of the placement collision a force must be exerted between gripper and drive unit, which can significantly increase the impact forces. A solution has been found to realise an adequate force build-up between gripper and drive unit such that a rebounce of the gripper is prevented without significantly increasing the impact forces. This solution can be implemented relatively easily by placing a spring between gripper and drive unit combined with a force limiter.

Keywords assembly, collision, force reduction, gripper, impact, micro system, rebounce

1 Introduction

In micro-assembly, placement devices are used for assembling components or placing components on a substrate, e.g. a printed circuit board. During the placement motion it is generally not known exactly when the final position of the components will be reached due to tolerances of the component heights. Therefore components collide with a certain speed, leading to unwanted impact phenomena such as high contact forces and placement inaccuracies.

For the placement devices it is desired to have a high number of pick & place actions per minute, around 600 placements per minute averaged is state of the art. Second, low impact forces are crucial to avoid damage of the parts. This value is strongly depending on the component; state of the art is in the range of 0,10 to 8,00 [N]. Thirdly, high placement accuracies are needed (3-sigma better than 3microns). However, increasing the placement speed tends to increase the contact forces exerted by the gripper on the component or on the substrate. The current approach is to reduce these impact forces by moveably coupling the gripper to the drive unit of the placement device, wherein the gripper that contacts the component has a relatively low mass [1]. In such a setup the drive unit must exert a force on the gripper to prevent the gripper from bouncing back at the end of the placement collision. A rebounce is undesired as it causes a second impact, which increases the risk of:

a) Damaging the component and/or substrate

b) Increasing pollution of component and its environment

c) Introducing vibrations in the placement device that cause significant inaccuracies in the placement of the component.

The impact forces can be reduced by lowering the impact speed, minimising the gripper mass or optimising the force between gripper and drive unit. In this article a method will be presented to reduce the impact forces by optimising the necessary force exerted on the gripper for preventing rebounce of the gripper.

2 Problem analysis

The primary goal of a Pick and Place robotic system is to efficiently mount components on the target surface, usually a Printed Circuit Board (PCB). This means there are time constraints, as well as accuracy constraints that have to be met. The problem is that these requirements are contradictive, a time optimised system lacks competitive accuracies and vice versa [2].

During the collision, when the gripper holding the component, is hitting the target surface, energy is transferred in elastic and plastic deformation. The elastic energy is not absorbed in the system, but only stored temporarily, like energy stored in a spring. Instead of storing the kinetic energy of component and gripper in the flexible structure, some parts of the system suffer from permanent deformation. These parts, usually the material around the contact points between target structure and components, act as a damped spring system. The energy that is not transformed to elastic deformation will be absorbed by the components. This could lead to damage of the component. The failure mechanism, caused by the absorbed energy is not subject of investigation in this publication but will be addressed in a separate paper.

The elastically stored energy (stored spring energy) will be partially converted back to kinetic energy of the gripper at the end of the collision. The gripper will therefore tend to bounce up again. In a test setup, this could be registered with a high-speed camera (Figure 1). Multiple bounces, as much as ten times would occur before the component showed no movement.

The obvious solution for preventing the gripper and the component from bouncing is to continuously exert a force of a larger magnitude between gripper and target surface. In state of the art equipment, this is done with a pretensioned spring with flattened spring-characteristic. Now we come to the real difficulty; the maximum reaction force of the elastically deformed component will determine the load of this pretensioned spring. This load however will increase the impact force thus leading

Fig. 1. Experimental setup for impact analysis including a z-stage for controlling the placement motion, a high-speed camera for observing the impact phenomena and a scope for measurements. The z-stage is build-up from a linear motor which can reach accelerations of up to 60g. The high-speed camera with additional set of microscope lenses is able of filming with a frame rate of 10.000 fps with a resolution of less then 2 micrometer per pixel.

on its term to a higher reaction force again. The result would initially lead minimal reduction of rebounce.

The collision process is typically non-linear [3-6], so increase in collision force will lead to a higher amount of plastic deformation and elastic amount of deformation remains of the same order. If the extra downwards directed load is strong enough, elastic deformation, that causes the rebounce has become relatively low so the contact force will be maintained during all stages of the collision and there will be no rebouncing. However, because this extra force is significantly higher than the necessarily placement force, there will be stronger impact, more plastic deformation and higher risk of damaging the components and/or the substrate.

3 Research and development

The hypothesis was that an adequate force build-up on the gripper for preventing rebounce could be realised by placing a small spring **in the gripper, as near to the component as was mechanically possible**. The spring was combined with a force limiter, to limit placement forces when board height suffers from tolerances (when component touches the board before or after the expected moment) [7-10].

By tuning the spring constant, the force on the gripper can be built-up in a controlled way during the placement collision such that the impact forces are not increased significantly. The force limiter, e.g. a pretensioned spring with low stiffness, prevents the force on the gripper of becoming too big when the drive unit continues moving relative to the gripper during the placement collision.

For optimising the force to be exerted on the gripper to prevent rebounce, the collision of the gripper with component on a substrate has been modeled as a single degree of freedom damped mass-spring system. The equivalent parameters of the model have been determined by conducting a series of experiments (Figure 1) with a gripper, with 1 DOF remaining, mounted to the drive unit without any extra force added between gripper and drive unit. During the experiments the impact speed, rebounce speed and contact duration of the first collision has been measured. The experiments showed that within the measurement range of the collision speed (50-350) mm/s, the contact duration was not significantly influenced by the collision speed. The collision duration was also found out to be almost linearly proportional to the mass of the gripper. Therefore a spring with constant spring stiffness is used to represent the deformations of gripper, component and substrate leading to the following equations of the lumped parameter model [11, 12]:

$$m_{collision} \cdot \ddot{z} + c_{collision} \cdot \dot{z} + k_{collision} \cdot z = F_{drive\_unit} \quad (1)$$

$$m_{collision} = m_{gripper} + m_{component} \quad (2)$$

$$k_{collision} = m_{collision} \cdot \left(\frac{2\pi}{t_{contact}}\right)^2 \quad (3)$$

$$c_{collision} = \frac{2 \cdot \ln\left|\frac{v_{rebounce}}{v_{impact}}\right| \cdot m_{collision}}{t_{contact}} \quad (4)$$

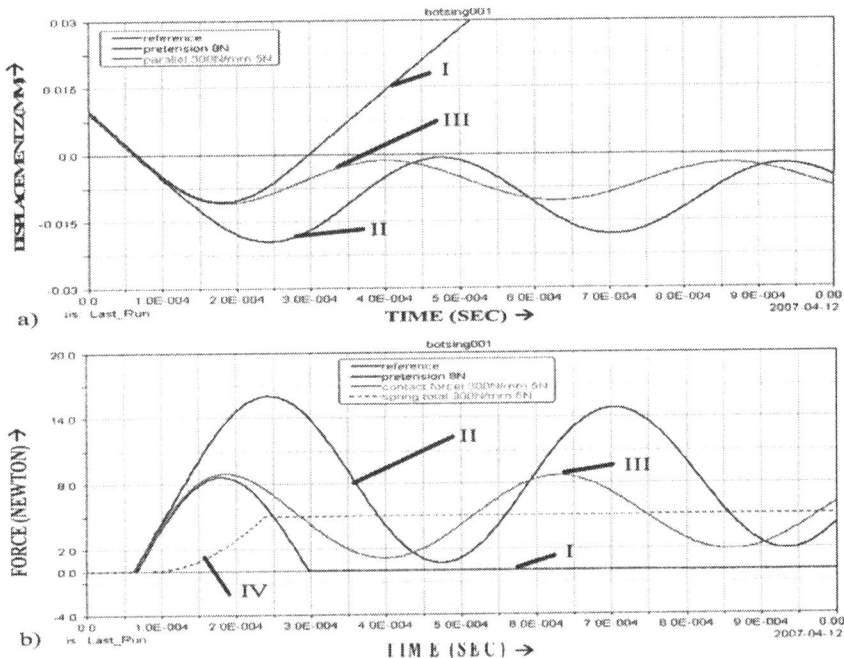

Fig. 2. Simulation result of collision between gripper and base structure, mass gripper 4.4 g, equivalent collision stiffness 821 N/mm; *a)* displacement curve of gripper, zero height is top surface of base structure, negative length represents deformation of material around contact points, *b)* contact force curve, I/ free moving gripper guided by drive unit, II/ gripper guided by drive unit and pretensioned spring between gripper and drive unit with pretensioned force just big enough (8N) to prevent rebounce of gripper, III/ gripper guided by drive unit and spring combined with force limiter between gripper and drive unit, stiffness spring 300 N/mm, pretension force of force limiter 5N, IV/ resultant force between gripper and drive unit for spring combined with force limiter.

in which m is mass, k is spring stiffness, c is damping, z is distance between gripper/component and substrate (negative value represents a deformation of the material around the contact points) and $t_{contact}$ is time between first moment of contact between gripper/component and substrate and the moment at which the contact is broken due to a rebounce of the gripper and component. In the model an extra spring between gripper and drive unit can be added of which the stiffness can be tuned such that the force on the gripper builds up just fast enough to prevent rebounce and subsequently the force build-up can be limited as much as possible by adding a force limiter (see Figure 2a).

The simulations showed that in the new solution (with spring) the impact forces could be reduced to almost half of the impact forces that occur with the general solution of exerting a constant force to prevent rebounce (see Figure 2b).

4 Design

The spring and force limiter must be combined such that at the start of the placement impact the force build-up is controlled by the spring making the force depended on the distance between gripper and drive unit. After the force has reached a predetermined value the force limiter must make the force independent of the distance between gripper and drive unit. The force limiter is needed to prevent the force on the gripper of becoming too high when the drive unit continuous moving towards the substrate while the gripper/component is already in contact with the substrate. The spring and force limiter can either be placed in series or parallel. The parallel configuration is preferred because then less mass is involved in the collision since both spring and force limiter are connected to the drive unit. In this configuration both spring and force limiter exert a force on the gripper but in opposite direction. At the start of the placement collision, both elements exert the same force leading to no resultant force on the gripper. During the collision the spring will be depressed making the spring force smaller then the force of the force limiter, resulting in a higher force on top of the gripper. After the spring is completely depressed until it is zero length, it will loose contact with the gripper making the force on the gripper depend only on the force limiter. An air spring has been chosen as force limiter because it has a relative low stiffness and the force can be controlled easily. For the spring a cupped spring washer has been used to make it possible to change the stiffness easily. For the guidance between gripper and drive unit a membrane has been used giving a low weight symmetrical design. The design of the gripper prototype is shown in Figure 3. In a further design the gripper has been equipped with a vacuum needle and the spring and vacuum supply have been integrated in the holder structure.

5 Results

A prototype of the developed gripper has been build and tested (see Figure 3b). In the prototype the force build-up could be tuned by changing a cupped spring washer and by regulating the pressure on the membrane. When a cupped spring washer with the highest available stiffness is used a situation is created similar to the general solution of using a constant force between gripper and drive unit to prevent rebounce. The test results showed that when the cupped spring washer with the highest available stiffness was replaced by one with the stiffness found with the simulation, the pretension force on the membrane could be reduced with 35%. With a cupped spring washer, with a stiffness that was too low, rebounce of the gripper could not be prevented at all. These test results were in close accordance with the simulation results (see Figure 2) showing that by simply adding a spring with the right stiffness, the impact forces can be reduced with 50% compared to the general solution of having a constant force between gripper and drive unit.

With the further developed gripper also a low mass (less then 1 g) of the parts involved in the collision (e.g. component, gripper needle, guidance, vacuum supply)

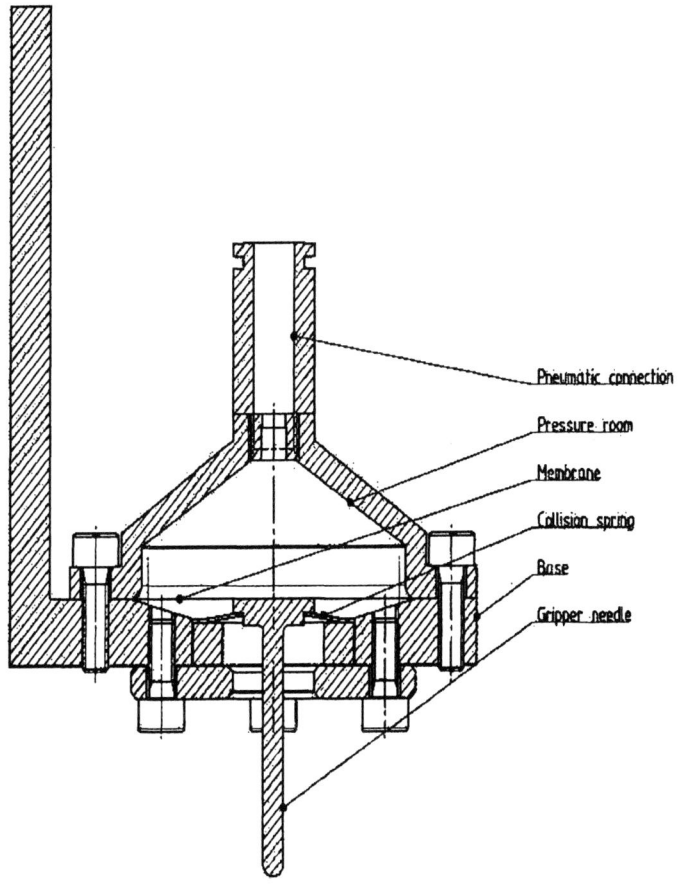

Fig. 3. Prototype of developed gripper for micro assembly: a) drawing gripper prototype, b) realized gripper prototype. Note the "collision spring", which takes care of preventing the rebounce. The mass of the gripper needle can be reduced more; this will lead to further increase of performance.

has been realised. Additionally the placement accuracy has been improved by adding a tilting member in the gripper, which prevents the gripper of becoming overconstrained during the placement collision.

Patents are pending for the solution of "reduced impact for micro-gripper" and for the designed "statically determined gripper".

5 Conclusion

The impact forces during component placement can be reduced significantly by optimizing the force exerted on the gripper for preventing rebounce of the gripper. An adequate force build-up can be realised by placing a spring between gripper and drive unit combined with a force limiter. In this way a real time mechanical solution can be realised in a relative simple way for preventing rebounce of the gripper while keeping the impact forces minimal.

Acknowledgments

This work was supported in part by the MicroNED program, project number 3-B-1: Concept design of miniaturized unit for micro part joining and assembly.
R. Plak was with TNO Science and Industry, department of Micro Device Technology, Eindhoven, The Netherlands.
R. Görtzen is with TNO Science and Industry, department of High-end equipment, Eindhoven, The Netherlands (phone: +31-40-265-0928; fax: +31-40-265-0305; e-mail: Roger.Gortzen@tno.nl).
E. Puik is with TNO Science and Industry, department of Electronic System Integration, Eindhoven, The Netherlands. He is professor, chair of Micro-Systems-Technology at university of applied sciences Utrecht, The Netherlands (e-mail: Erik.Puik@tno.nl).

References

[1] E. Bos, Euspen (2007).
[2] S. K. Nah, Z. W. Zhong, Sensors and Actuators A: Physical 133 (2007) 218-224.
[3] J. H. Oh, D. G. Lee, H. S. Kim, Composite Structures 47 (1999) 497-506.
[4] I. Mizuuchi, M. Inaba, H. Inoue, Robotics and Autonomous Systems 28 (1999) 99-113.
[5] H. Choi, M. Koc, International Journal of Machine Tools and Manufacture 46 (2006) 1350-1361.
[6] T. Klisch, Mechanism and Machine Theory 34 (1999) 665-675.
[7] S. S. Rao, P. K. Bhatti, Reliability Engineering & System Safety 72 (2001) 47-58.
[8] H. Bremer, F. Pfeiffer, Control Engineering Practice 3 (1995) 1331-1338.
[9] J. Yanof, C. Bauer, B. Wood, International Congress Series 1268 (2004) 521-526.
[10] J. De Schutter, D. Torfs, H. Bruyninckx, Robotics and Autonomous Systems 19 (1996) 205-214.
[11] Goldsmith, Impact, 2001.
[12] Johnson, Contact Mechanics, 1987.

PART IV

Development of Micro-Assembly Production Systems

Chapter 9

Design of Modular Reconfigurable Micro-Assembly Systems

STRATEGIES AND DEVICES FOR A MODULAR DESKTOP FACTORY

Arne Burisch, Annika Raatz, Jürgen Hesselbach

Technical University Braunschweig
Institute of Machine Tools and Production Technology (IWF)
Langer Kamp 19b, 38100 Braunschweig, Germany

Abstract The first part of the paper describes different concepts for micro production. The potential of miniaturised size adapted robots is pointed out based on their flexibility. Flexibility thereby is discussed in a detailed way that leads to two different ways of robot integration. The second part of the paper deals with the desktop factory concept. Beginning with a short state of the art for desktop factories, the preconditions for future concepts are discussed. Finally, with regard to a module for a desktop factory, the miniaturised precision robot Parvus, a micro gripper, a gripper changer and a robot control are presented.

Keywords desktop factory, flexibility, miniaturised robot, micro gripper, gripper changer, compact robot control

1 Introduction

Today, a trend of miniaturisation with regard to product development in several industrial sectors can be observed. Based on the *Nexus III* market study [1], it can be assumed that the market of millimeter-sized MST-products will grow by 16% per year. However, the gap between the dimensions and costs of the products and the production systems used is increasing. Assembly lines and clean rooms for millimeter-sized products often measure some tens of meters and are mostly too expensive for small- and medium-sized businesses. Therefore, many micro-products are assembled by hand, which causes high assembly costs that amount to 20% to 80% of the total production costs [2].

2 Potentials for miniaturised robots in micro production systems

In recent years, the reduction of size and costs of micro production systems has been widely discussed in various papers. Most of these concepts relate to one of the two general groups explained in the following.

The first group consists of piezo driven, small walking micro robots and handling machines. These autonomous robots are suitable for positioning small objects such as the *MINIMAN* of *Fatikow* [3], a handling device for samples in a scanning electron microscope. On the one hand, these micro robots are very promising for new trends such as nano assembly. On the other hand, by using autonomous robots difficulties occur with coordination and interaction of these robots, movement on rough surfaces and energy supply.

The second group describes cost-efficient, size adapted handling devices, which fill the gap between piezo driven, small walking micro robots and conventional robots. A possible solution for this strategy is to find out the highest degree of miniaturisation of conventional robot technology, using innovative, miniaturised machine parts. With these size adapted handling devices, in the range of several centimeters to a few decimeters, easily scalable and highly flexible production technology can be achieved.

In the following, the potentials of flexibility of size adapted production machines will be discussed in a general view. First of all, flexibility can be subdivided into *flexibility of function*, *flexibility of production volume*, *flexibility of placement* and *flexibility of property*.

- Flexibility of function

    Compared to conventional industrial robots, miniaturised robots can nowadays be equipped with a similar range of functionalities (e.g. degrees of freedom, sensors and tools). The combination of such miniaturised machines with freely programmable control systems, miniaturised drive systems and micro grippers can lead to the development of a micro production system with a high range of functionalities and motion-sequences.

- Flexibility of production volume

    Less required space for miniaturised production machines makes it possible to replace a conventional robot by several miniaturized robots with increased productivity with regard to space. These concepts, known as the desktop factory concepts, increase the availability and the production volume. Hence, the manufacturer is more flexible in the number and variety of produced pieces.

- Flexibility of placement

    Miniaturised production systems with a high density of functionality are flexible regarding their location. They can be placed in conventional clean rooms as well as in local clean room cells. Furthermore, companies profit from a high flexibility of placement with regard to the development of processes as well as the expansion and relocation of the company.

- Flexibility of property

    Most conventional production systems are not sufficient for future applications. For most companies, a complete substitution of these production systems is impossible, as the newly acquired expensive systems have to be paid off. Thanks to miniaturised production machines, already existing larger production systems can be equipped with extended functionalities. Due to the smaller amount of space, miniaturised robots can be integrated into larger machines. The symbiosis of conventional technology and micro technology leads to improved flexible properties.

Based on the aforementioned assumptions, several of the current problems in micro assembly or the semiconductor industry can be solved by integrating a miniaturised robot pursuing two different strategies (Fig. 1):

a) As a component for miniaturised production systems such as visionary desktop factories, e.g. in micro production or micro assembly industry.

b) As a miniaturised production machine integrated into a conventional bigger machine, e.g. in testing machines for conductor boards.

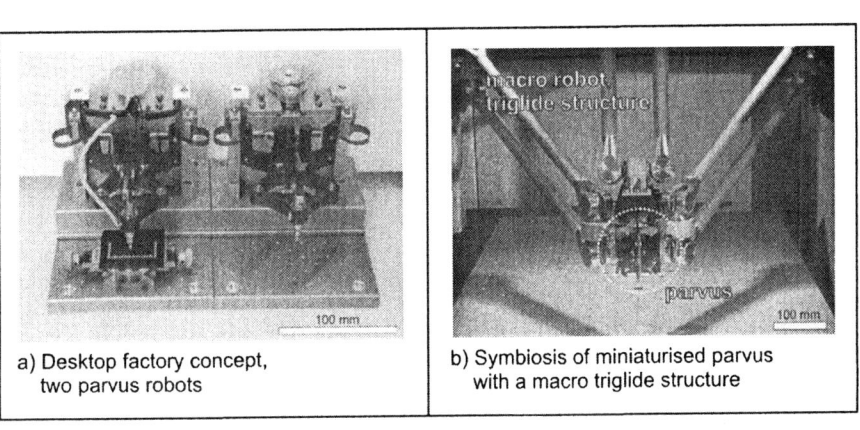

a) Desktop factory concept, two parvus robots

b) Symbiosis of miniaturised parvus with a macro triglide structure

Fig. 1. Two strategies of system integration for the miniaturized robot *Parvus*

In the following, the paper will only deal with the aforementioned strategy (a) - the integration of miniaturised handling devices into a visionary desktop factory.

3 The modular Desktop Factory

As for visionary desktop factories, it is necessary to develop highly modular systems by playing on the potentials of size adapted flexible handling systems, as presented in the previous chapter. Therefore, miniaturised components and handling devices with high *flexibility of placement* have to be developed. In addition, Breguet et al. [4] proclaimed in their paper *Toward the Personal Factory* that personal desktop factories should not be all-purpose but highly modular systems. In conventional automation technology, examples of modular production cells such as the system of *IMSTec* [5] can be found. However, there are only a few examples in research and industry of concepts for modular desktop factories. The concept of assembly modules mounted around a fixed platform is followed up in research projects by *Gaugel et al.* [6] and *Rochdi et al.* [7] and by the industrial manufacturer *MiLaSys* [8]. Many other concepts follow up the idea of a fixed production cell equipped with a main handling device and several subsystems, such as e.g. *Uusitalo et al.* [9] in research and *Klocke Nanotechnik* [10] as a manufacturer. In particular, these concepts and the aforementioned system of *IMSTec* point out that the size of the whole production system is limited by the size of the conventional precision robot used. These miniaturised production systems obviously require high-miniaturised conventional precision robots and a miniaturised environment such as the concepts for modular *Microboxes* with the miniaturised precision robot *Pocket-Delta* of *EPFL* and *HTI-Biel*, Switzerland [11], [12].

Motivated by the aforementioned necessity of miniaturised components the *Institute of Machine Tools and Production Technology (IWF)* of the *Technical University Braunschweig* worked in cooperation with *Micromotion GmbH* to develop a completely miniaturised precision robot with micro motors, micro encoders and high precise *Micro Harmonic Drive* gears. Furthermore, in cooperation with other institutes the environment for the robot will also become miniaturised.

3.1 The functional model Parvus

The challenge of the functional model *Parvus* was to develop a miniaturised precision industrial robot with the full functional range of larger models.

The robot consists of a typical parallel structure, driven by *Micro Harmonic Drive* [13] gears combined with *Maxon* electrical motors. This plane parallel structure offers two translational DOF in the x-y-plane. The z-axis is integrated as a serial axis in the base frame of the robot. The easy handling of the whole plane parallel structure driven in z-direction is possible because of its minimised drive components and light aluminum alloy structure. The rotational hand axis Ψ was designed as a hollow rotational axis as the Tool Center Point (TCP) of the parallel structure. This allows media such as a vacuum to be passed along the hand axis. This axis, with a diameter of 2.5 mm, can be equipped with several vacuum grippers. Further details are given by the technical specifications in Table 1.

The development of the *Parvus*, its fundamentals, the miniaturised drive systems and the robot design approaches were already described in several previous papers [14], [15].

Criterion	Value	Unit
Workspace (xy, absolute)	4658	mm²
Workspace (max. cubical)	60x45x20	mm3
Footprint	100x53	mm²
Robot cell	130x170	mm2
Resolution max. (xy-plane)	< 1	µm
Repeatability (best, worst)	5.9, 14.1	µm
Linear speed	> 100	mm/s
Rotational speed (Ψ axis)	187* / 60**	rpm
Angular resolution (Ψ axis)	0.022* / 0.007**	°
Payload	50	g
*ratio 160:1 / **ratio 500:1		

Table 1. Technical specifications of the first prototype

3.1.1 Pick and Place applications with the *Parvus*

For demonstration applications, the robot is equipped with a standard vacuum gripper (Fig. 2) to pick and place glass balls with a diameter of about 1.5 mm.

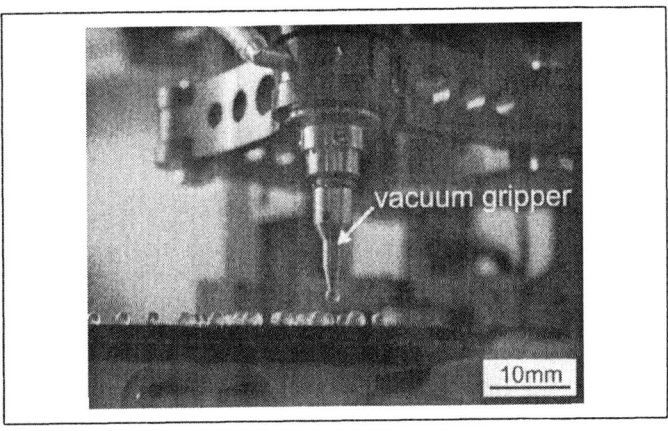

Fig. 2. Pick-and-place application of glass balls (Ø 1.5mm) with *Parvus*

For gripping much smaller objects, a recently developed pneumatic micro gripper driven by a single channel for vacuum and pressure was integrated into the hand axis of the *Parvus*. The *Institute for Microtechnology (IMT)* of the *Technical University Braunschweig* developed this two-jaw micro gripper based on a micro pneumatic actuator manufactured in micro technology [16]. In cooperation between the *IMT* and the *IWF* the micro gripper was adapted for the *Parvus*. Experiments showed that it is possible to grip small ruby balls with 200 μm in diameter (Fig. 3).

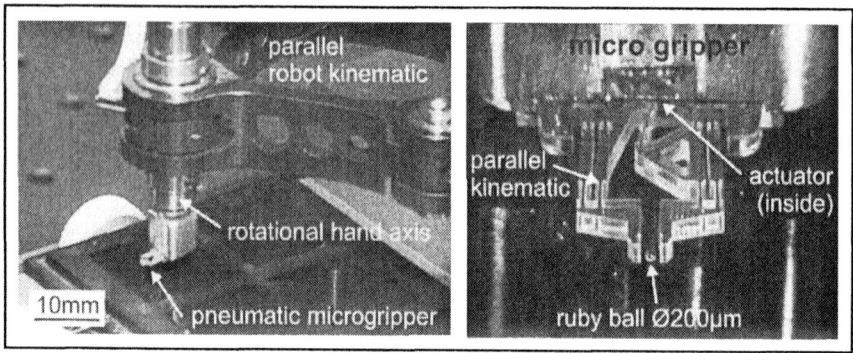

Fig. 3. Parvus equipped with a pneumatic micro gripper

To enable different flexible pick-and-place applications the *Parvus* will be equipped with a miniaturised tool changing station. The basic design of the tool changer consists of a rotating wheel with six tool seats. The different grippers can be disconnected and connected with the *Parvus* by a pneumatic working principle. Figure 4 shows the conceptual design of the gripper changing station, which is actually under construction.

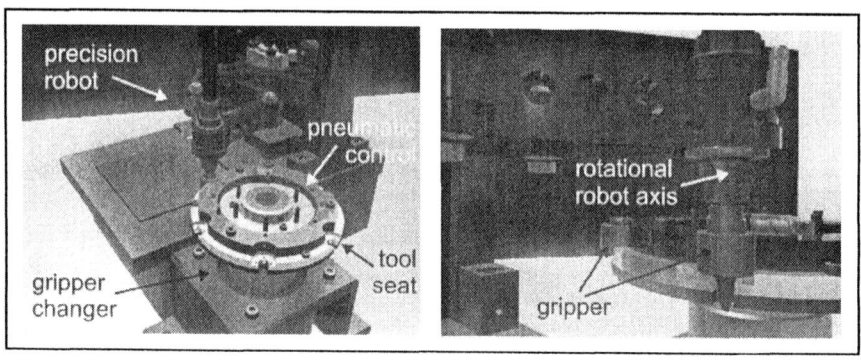

Fig. 4. Concept for pneumatic gripper changing station for the Parvus

3.1.2 Compact industrial robot control

The first robot control for the *Parvus* was developed on a rapid prototyping system *dSPACE* in *Matlab Simulink*. To prepare a robot for the desktop factory concept, a compact industrial robot control is necessary. In cooperation with the company *B&R* [17] the *Parvus* recently was equipped with a PC based I/O System using a *Celeron 650* processor. This system enables different task classes to run simultaneously. The transformations of the robot drives q and the world coordinates p can be calculated in real time. The central unit can be easily upgraded and operate with several I/O modules and controllers connected with the robot drives and sensors or for communication with a superior field bus. By a VNC-Viewer, a graphical CNC Interface running on any PC connected via TCP/IP to the control system, the robot can be operated.

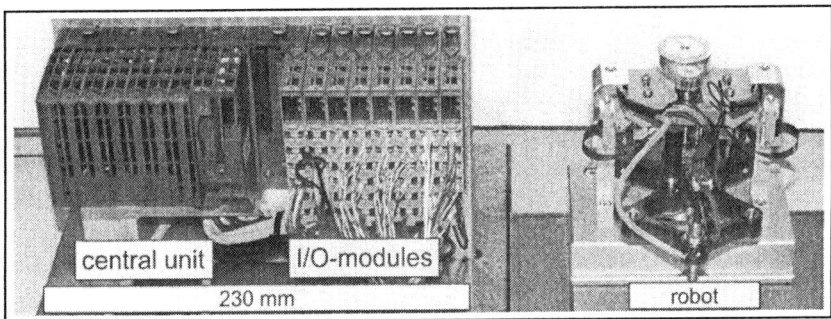

Fig. 5. Compact robot control of the Parvus

4 Conclusion

The paper points out that flexible desktop factories necessitate highly modular systems which have to be entirely miniaturised. Therefore, the development of such desktop factories needs a lot of technical expertise in several disciplines. The *IWF* researches for miniaturized handling devices in cooperation with institutes of the *Collaborative Research Center 516* and industrial partners. A result of such interdisciplinary research is the miniaturized precision robot *Parvus* equipped with micro gears, a micro gripper and gripper changer as a possible module for a desktop factory.

To achieve a highly modular desktop factory it is necessary to solve further challenges concerning the size and accuracy of many other handling devices. For instance, precise and miniaturised feeding systems or miniaturised magazines have to be developed in the future.

References

[1] Wicht, H.; Bouchaud, J.: NEXUS Market Analysis for MEMS and Microsystems III 2005-2009. In: MST-news, Verlag VDI/VDE Innovation+Technik GmbH, Vol. 5, pp.33-34, 2005

[2] Koelemeijer, S.; Jacot J.: Cost Efficient Assembly of Microsystems. In: MST-news, Verlag VDI/VDE Innovation + Technik GmbH, Vol. 1, pp. 30-32, 1999

[3] Fatikow, S.: Miniman. In: Mikroroboter und Mikromontage, pp.277, ISBN 3-519-06264-x, Teubner Verlag, Stuttgart – Leipzig, 2000

[4] Breguet, J.-M.; Bergander, A.: Toward the Personal Factory?. In: Proc. of SPIE, Microrobotics and Microassembly III, Vol. 4568, pp. 293-303, 2001

[5] IMSTec. http://www.imstec.de, 08.11.2006

[6] Gaugel, T.; Bengel, M.; Malthan, D.: Building a Mini-Assembly System from a Technology Construction Kit. In: Proc. of Intern. Precision Assembly Seminar (IPAS'2003), Bad Hofgastein, Austria, pp. 137-142, 2003

[7] Rochdi, K.; Haddab, Y.; Dembélé, S.; Chaillet, N.: A Microassembly Workcell. Proc. of Intern. Precision Assembly Seminar (IPAS' 2003), Bad Hofgastein, Austria, 2003

[8] MiLaSys technologies GmbH. http://www.milasys.de, 08.11.2006

[9] Uusitalo, J. J.; Viinikainen, H.; Heikkilä, R.: Mini assembly cell for the assembly of mini-sized planetary gearheads. In: Journal of Assembly Automation, Vol. 24/1, pp. 94-101, 2004

[10] Klocke, V.; Gesang, T.: Nanorobotics for micro production technology. In: Proc. SPIE Conference Photonics Fabrication Europe, Brugge, Belgium, Vol. 4943, pp. 132-141, Oktober 2002

[11] Codourey, A.; Perroud, S.; Mussard; Y.: Miniature Reconfigurable assembly Line for small Products. In: Proc. of the Third Intern. Precision Assembly Seminar (IPAS'2006), Bad Hofgastein, Austria, 19-21 February 2006, S. Ratchev (Edt.), ISBN 0-387-31276-5, pp. 193-200, 2006

[12] Verettas, I.; Codourey, A., Clavel, R.: "Pocket Factory": Concept of Miniaturized Modular Cleanrooms. In: 1st Topical Meeting of Desktop MEMS and Nanofactories (TMMF2005), Tsukuba, JP, October 17-19, 2005

[13] Slatter, R.; Degen, R; Burisch, A.: Micro-Mechatronic Actuators for Desktop Factory Applications. In: Proc. of Intern. Symposium on Robotics ISR/Robotik 2006 München, pp. 137, VDI Wissensforum IWB GmbH, Düsseldorf, ISBN 3-18-091956-6, 2006

[14] Burisch, A.; Wrege, J.; Soetebier, S.; Raatz, A.; Hesselbach, J.; Slatter, R.: "Parvus" A Micro-Parallel-SCARA Robot for Desktop Assembly Lines. In: Proc. of the Third Intern. Precision Assembly Seminar (IPAS'2006), Bad Hofgastein, Austria, 19-21 February 2006, S. Ratchev (Edt.), ISBN 0-387-31276-5, pp. 65-74, 2006

[15] Burisch, A.; Wrege, J.; Raatz, A.; Hesselbach, J.; Degen, R.: PARVUS – miniaturised robot for improved flexibility in micro production. In: Journal of Assembly Automation, Emerald, Vol. 27/1, pp. 65-73, ISSN: 0144-5154, 2007

[16] Bütefisch, S.; Brand, U.; Leester-Schädel, M.; Hoxhold, B.; Büttgenbach, S.: Characterisation of Pneumatic and SMA Micro-Actuators with Short Response Times and Large Exerted Forces and Deflections. In: Proc. of Actuator 2006, Bremen, Germany, pp.483-486, 2006

[17] B&R Industrie-Elektronik GmbH. http://www.br-automation.com, 20.06.2007

A DECISION MAKING TOOL FOR RECONFIGURABLE ASSEMBLY LINES - EUPASS PROJECT

F. Wehrli, S. Dufey, S. Koelemeijer Chollet, J. Jacot

Laboratoire de Production Microtechnique, Institut de Production et Robotique,
Ecole Polytechnique Fédérale de Lausanne, SWITZERLAND
Laboratoire de Production Microtechnique
EPFL
Station 17
1015 Lausanne
Switzerland
Tél : ++ 41 21 693 5997
Fax : ++ 41 21 693 3891
frederic.wehrli.epfl.ch

Abstract A decision making tool is proposed to evaluate assembly costs of micro-products (< 1 dm^3) and to compare different assembly strategies for a given product or product family. The tool takes into account equipment costs as well as running costs, for manual and automatic assembly. Its specificity lies in its ability to deal with different production mixes and production volumes over the life cycle of the product family. The underlying concept of the tool is that the cost of a given assembly function (or process) is constant for different technical solutions. This paper describes a cost evaluation for reconfigurable, manual, automatic or semi-automatic assembly lines, and shows its easiness of use through a simple test case.

1 Introduction

Assembly is a crucial issue in production, and a main cost driver for the total production cost (up to 80%) [1, 2]. Assembly is often more expensive than predicted, due to the lack of quality of components, production stops, and a lower than expected productivity. Companies are furthermore faced with the problem of short product life cycles, and uncertainty in the evolution of the demand. Industrials hesitate to invest in expensive assembly equipment that takes too much time to reach full productivity, and is too expensive for a pay-back within the product life cycle. The risks are too high. They would prefer to start with manual assembly for low production volumes, and be able to upgrade their assembly system to a more automated one very quickly when needed [3].

The purpose of the EUPASS project is to respond to this demand by offering a set of (re)configurable generic assembly modules that can very quickly be set up together to form a functional assembly system. The challenge of reconfigurable assembly systems is to offer economic profitable solutions. The aim is to keep manufacturing industries in Europe by providing flexible solutions that can compete with the low wages in some countries such as China [3]. Keeping manufacturing in Europe is not only important from a workforce and employment point of view, but is essential to keep the know-how and the added value to products. It is thus important to also consider logistics, and specific costs related to outsourcing (such as costs due to quality problems and a less trained workforce) to get a complete picture of the productions costs. Cost modelling and evaluation is thus an important task during the development and planning process of a new product and its manufacturing environment.

One task of the EUPASS project is to propose a configuration tool for reconfigurable assembly systems, and the cost evaluation tool is part of this. The specificities of fast evolving demand and the possibility of fast changing physical configuration of the equipment have to be taken into account.

2 State of the art

There are very few ready to use cost models on the market. Most of the time, engineers do cost calculations by their own estimations and by adding different parts of equipment cost. Often, costs are underestimated by a lack on data on significant parameters such as yield, idle times and real productivity values. Existing cost models such as the model for automated assembly lines proposed by Oulevey et al. [4], do not take into account evolving or reconfigurable assembly systems or equipment during the product life cycle. They simply consider a fixed cost calculated on the total units produced by the equipment during the life cycle of the product. Furthermore, the possibility to mix manual and automated assembly is not integrated in the model. There is a need for a simple to use cost evaluation tool, which handles changing demand over the product life cycle, as well as evolving a mix of manual and automated assembly, and evolving equipment.

3 The cost evaluation tool

3.1 Concept for process cost

The concept lying behind this tool is that a set of standard parameters responsible for the assembly cost of a product can be defined for each generic assembly process. Those parameters are typically the manual assembly time, the equipment cost, the assembly yield, the time an operator is needed to watch an automatic process, etc. Those parameters are mean values, and thus not related to a particular technical solution that may be chosen. They are stored in an internal database of the tool. On the shop-floor, a given process will be more or less complex or tricky, depending on the design of the components. This has a direct influence on the set of parameters. To estimate the assembly cost of a product, one has thus to define the required generic processes, and their complexity level. Major factors are the precision required and the geometric dimensions. This information can either be found in the product specifications, or by analyzing the components to assemble. A set of guidelines is provided to the user of the cost tool to help him/her in defining the complexity level related to each process and to each component. The user of the tool should be an expert, which means that he should have strong assembly skills.

Complexity Level	Criteria
Very easy	- Symmetrical components - Very asymmetrical components - Already oriented (trays) - Length and width > 6mm - Thickness > 2mm - Easy to grasp with one hand
Easy	- Length or width < 6mm - Can be grasp with one hand - Flexible components, but with length and width > 6mm and thickness > 2 mm
Medium	- Components few asymmetrical - Length and width < 6mm - Need of grasping aids (tweezers, binocular...) - Difficult to grasp, but with length and width > 6mm and thickness > 2 mm - Thickness < 0.25mm
Difficult	- Combination of medium factors - Delicate, sticking, cutting or slipping components - Intermingled components in bulk
Very Difficult	- Flexible sub-assembly - Combination of many difficult factors - Need of specific tools for manipulation

Table 1. Complexity level evaluation for manual feeding and orientation

An important parameter is the manual assembly times for small and precision products. The method used is adapted from Design for Assembly of Boothroyd [1]. As Boothroyd's method covers the assembly of macro-products, it was important to ensure that both the parameters and the values were valid for the assembly of small parts. Times have been adapted, re-estimated and then tested in a company assembling micro-switches. An extract can be found in Tables 1 and 2. Furthermore, the presentation of the tables has been simplified, so that they are easier to use.

Another point behind the concept is that a few generic assembly processes cover most of the requirements. Other processes can be related to one of those generic processes, and there is thus no need to handle an exhaustive list and database. The most common processes will be proposed as standard EUPASS modules, or exist as internal standards of equipment providers.

Operation	Complexity Level	Time [s]
Feeding (manipulation – orientation)	Very easy	1.5
	Easy	3
	Medium	4.5
	Difficult	6
	Very difficult	8
Insertion	Very easy	2
	Easy	4
	Medium	6
	Difficult	8
	Very difficult	10
Screwing	Easy	5
	Medium	8
	Difficult	10
Measurement	Easy	2
	Medium	5
	Difficult	8

Table 2. Manual assembly times for a set of relevant processes

3.2 Business related parameters

Other significant parameters for the cost are related to the production volume over the time, the number of variants, the batch size, the productivity, and the cost of manual labour. This information is typically related to the market demand and the business strategy, and can be found in the business specifications. One of the objectives of the EUPASS project is to be very reactive, and thus to respond very closely to the production changes. Variations in product mix and production volume are two of the major and most frequent changes. The cost tool handles this by calculating the costs per period, a period being a duration during which there are no major changes in the production mix or volume, or in the physical configuration of the line.

The set of standard values thus also includes parameters such as mean setup time, mean configuration time, mean intervention time, or mean maintenance time.

The tool provides a set of pre-selected outputs that helps the expert user in the choice of a configuration. One of the outputs is an indication of the most cost effective solution between automation and manual assembly for each process and each period. Other outputs are the total cost per period, the unit cost per period, the productivity per period. The user may also be interested in other data, such as the set-up cost, the operator intervention cost, etc, which he may directly consult in the tool. The evolution of the cost over the different periods helps the designer of the assembly line to choose the most interesting solution in a cost point of view. If, for example, the tool indicates that a manual process is more suitable for a pick and place, and an automatic process more profitable for a control on a given period, he will chose a semi-automatic system. If for a later period with a higher production volume, the tool indicates that the automatic solution becomes more interesting also for the pick and place, he can plan the reconfiguration of his assembly system at the right moment.

4 Case study

In order to validate the cost evaluation tool and to show its working, an example is illustrated. The case study handles with the insertion of an inner nut in a cog wheel. As shown below (Figure 1), there are three kinds of material for the cog wheel (PVC, steel, brass), all of them are the same size.

Fig. 1. Case study: three types of cog wheels with one type inner nut

The assembly sequence is the following:

1. Feed cog wheel from bulk
2. Place cog wheel on a support
3. Vision test
4. Transfer
5. Feed nut from bulk
6. Place nut in the centre of the cog wheel
7. Insert the nut into the cog wheel
8. Transfer of the assembly
9. Evacuation of the assembly

4.1 Scenario

The life cycle is presented in Figure 2 and Table 3, which contains the production volume and the batch size for the three variants (PVC, steel and brass) and for each period. The production duration being 2 years and 6 months and the period duration being 3 month, the whole production is divided into 8 periods.

The batch size is more or less proportional to the production volume, except for period 5 and 6. The production volume remains constant, but batch size 1'000 in period 5 and 10'000 in period 6.

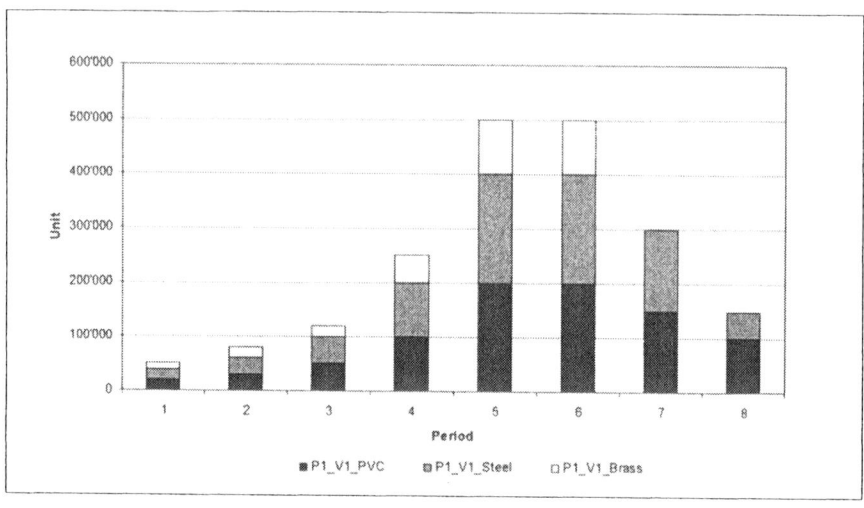

Fig. 2. Production volume for each product and each period

Product to be assembled	Batch size P1	Batch size P2	Batch size P3	Batch size P4	Batch size P5	Batch size P6	Batch size P7	Batch size P8
PVC-variant	200	200	500	1'000	1'000	10'000	500	500
Steel-variant	200	200	500	1'000	1'000	10'000	500	500
Brass-variant	50	50	100	200	1'000	10'000	0	0

Table 3. Batch size for each product and each period

4.2 Production mode comparison

The tool provides the assembly cost for the recommended optimal mode (mix of manual and automatic), as well as for full automatic and for manual production (Figure 3). The optimal solution is not necessary the cheapest way to assemble in one period, but the cheapest global solution, because configuration costs have to be taken into account.

Process list	Period 1	Period 2	Period 3	Period 4	Period 5	Period 6	Period 7	Period 8
Cog wheel feeding	M	M	M	A	A	A	A	M
Positionning cog wheel	M	M	M	M	A	A	M	M
Vision test	M	M	M	A	A	A	A	M
Transfert	M	M	M	M	A	A	A	M
Feed nut	M	M	A	A	A	A	A	A
Positionning nut	M	A	A	A	A	A	A	A
Insert nut	M	A	A	A	A	A	A	A
Transfert	M	A	A	A	A	A	A	A
Evacuation	M	M	M	M	A	A	M	M

Table 4. Optimal production mode recommended by the tool for each process and each period

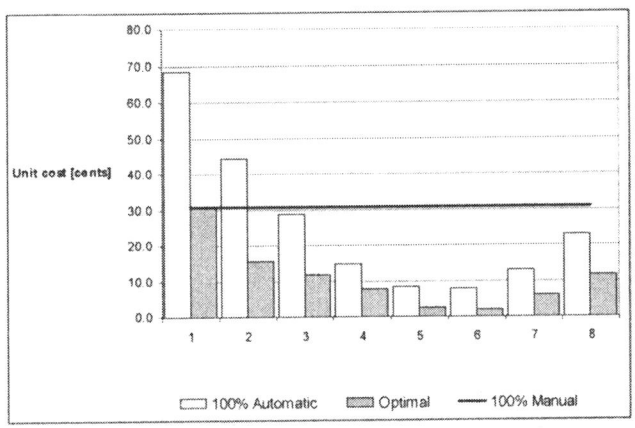

Fig. 3. Costs comparison between a full automated line, a full manual line and the optimal solution proposed by the tool

5 Conclusion and further work

The specificity of this tool is that all costs are calculated for each production period during which the production volume, the production mix and thus the configuration of the line is fixed. This allows taking into account the product life cycle, which is the main added value of this tool. Furthermore, the tool allows comparing the cost of automatic and manual assembly.

Further work to improve the tool will be to complete the internal database, and to check the tool on a real case. The integration of shipping cost, which is another logistics aspects that will help to compare the cost in different countries (mainly European or occidental versus low wage countries).

Acknowledgements

The manual assembly times and parameters have been elaborated and tested with M. Vito Carrea of the company MICROPRECISION ELECTRONICS SA, in Vouvry, Switzerland.

References

1. G. Boothroyd and P. Dewhurst, Product Design for Assembly, Boothroyd Dewhurst, Inc., Wakefield, 1991

2. S. Koelemeijer, F. Bourgeois, C. Wulliens, J. Jacot, "Cost modelling of microassembly", Proceeding of IPAS, First International Precision Assembly Seminar, Bad Hofgasten, Austria, march 2003

3. M. Onori, J. Barata, F. Lastra, M. Tichem, European Precision Assembly Roadmap 2010, Doc ref no: DO2c-2003-09-15-KTH, Assembly Net G1RT-CT-2001-05039

4. M. Oulevey, P. Roduit, S. Koelemeijer Chollet, J. Jacot, O. Ryser, K. Taferner, "A cost model for flexible high speed assembly lines", Proceeding of IPAS, International Precision Assembly Seminar, Bad Hofgasten, Austria, 2004

STANDARDISED INTERFACE AND CONSTRUCTION KIT FOR MICRO-ASSEMBLY

Matthias Haag, Samuel Härer, Andreas Hoch, Dr. Florian Simons

SCHUNK GmbH & Co. KG, Spann- und Greiftechnik,
Bahnhofstr.106-134,
D-74348 Lauffen/Neckar, Germany.

Abstract Automated assembly of microsystems is often characterised by small batch sizes combined with high product diversity. This characteristics demand a large number of endeffectors. In the last years in particular grippers have been developed for automatic assembly of hybrid microsystems. These have been optimised for a specific task and have one thing in common: there is no compatible interface. From the economic point of view the application of a construction kit is a sensible approach enabling both:

- **quick system configuration** as well as initial operation of customised automated solutions and
- **quick system reconfiguration** for adapting the customised handling system to different applications even in the process. In this context standardisation of interfaces is obligatory for universal application and expandability of the system.

However, up to now the field of microsystems assembly is also characterised by strict specialisation of automation solutions to specific tasks and consequently low modularity and rare standardisation of interfaces.

1 Introduction

In general construction kits for robotics have existed for 15 years. They consist of:

- a set of endeffectors,
- an optional tool change system (see Figure 1 left) enabling fast reconfiguration for different applications,
- a positioning system (e.g. swivel units as one representative component, see Figure 1 middle and right).

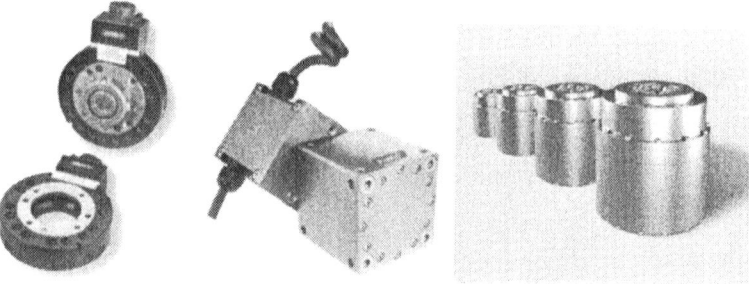

Fig. 1. *Left:* Gripper Change System (diameter >50mm)
Middle: PowerCubes® since 1991
Right: Swivel Units

To obtain an adequate construction kit for the world of fine mechanics or micro technology requires downscaling the existing modular solutions mentioned above. However, the downscaling process is coupled with the request for increasing precision and energy density. Thus modular solutions in robotics have rarely as yet been transferred into the field of precision engineering or microtechnology.

This paper presents a construction kit for micro robotic applications containing the following modules:

- **Endeffectors** (example see Figure 2 left)
  The endeffectors currently consist of different grippers and will be extended by alignment modules, fine positioning units, compliance modules, etc.
- **Micro tool change system** (see Figure 2 right)
  Beyond the mechanical coupling of the grasping module to the handling system, the micro tool change system comprises also electrical and fluidic lead throughs as well as a central aperture.
- **Positioning system** (see Figure 3)
  The positioning system consists of miniature swivel units, rotation units and base profiles variable in length. They are characterised by watertightness, extremely compact build up and integrated electrical and fluidic lead through.
-

Fig. 2. Left: Endeffector: Parallel Micro Gripper (with integrated standardised interface)
Right: Micro Tool Change System

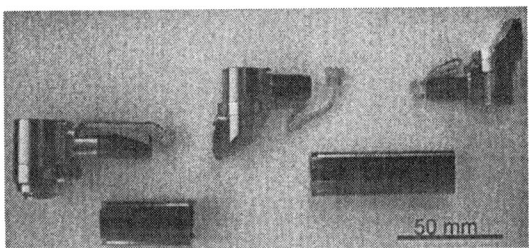

Fig. 3. Positioning System Components

The micro tool change system and the endeffectors conform to the new standard "DIN 32565: Interface between Endeffector and Handling Device". The standardisation of the interface plays a central role in mounting customised handling systems. It assures straightforward exchange of different modules. This is a precondition to perform complex assembly tasks at small batch sizes economically. The standard can be introduced to the international audience of the IPAS seminar as a domain base for micro assembly.

2 Research methodology

Cartesian systems have a good ratio of workspace to overall size and can be realised for high load, high precision and high stiffness but are consequently heavy and cost intensive (like granitic-bed in temperate surrounding). Innovative delta kinematics are absolutely precise and enable short cycle times. However, as they are usually installed overhead, these kinematics require high installation effort and adequate installation space. Additionally both systems require protection devices.

For micro- and nanopositioning in the laboratory environment, e.g. piezo electrical actuator based positioning systems, operating a range of operation of some millimetres, have been increasingly applied in recent years. Currently, systems containing such high precision actors are extending from scientific to industrial use. However, requiring powerful control units, they are cost-intensive and therefore are only used for assorted applications.

An economic use of fixed handling systems requires adequate quantities and capacity utilisation. In the case of moderate lot sizes, automated handling can only be carried out with accordingly lower investment costs. Additionally, in order to maximise capacity utilisation the system should be easily portable and configurable to different kinematic and geometric variations.

Recapitulatory, a system with the following properties will be introduced:

- Reduced costs of investment and installation to enable economic operation even **with low capacity utilisation** on the one hand and
- configuration variability and portability to **maximise capacity utilisation** on the other hand.

An actual industrial initiative is in progress for miniature robotic handling with the goal of optimised required space, portability and high configuration variability. As a consequence new solutions have to be found to receive the required preciseness. A level of precision suitable to the application can be reached economically using an optical feedback in combination with closed loop controlled fine positioning axes at the front-end. A key characteristic here is a central aperture through all modules of relevance like the gripper, its interface and the rotary unit in order to realise direct process monitoring.

The handling system exemplarily introduced in this paper is based on the "Scara" kinematics (see Figure 4) which is typically known as a compact unit for pick and place applications, where the linear z-axis is located at the robots front end. This configuration is motivated by the minimisation of moments at the linear bearing. However, for realising the central aperture at the robots front end here it was reasonable to locate the linear axis at the base of the kinematics (first axis).

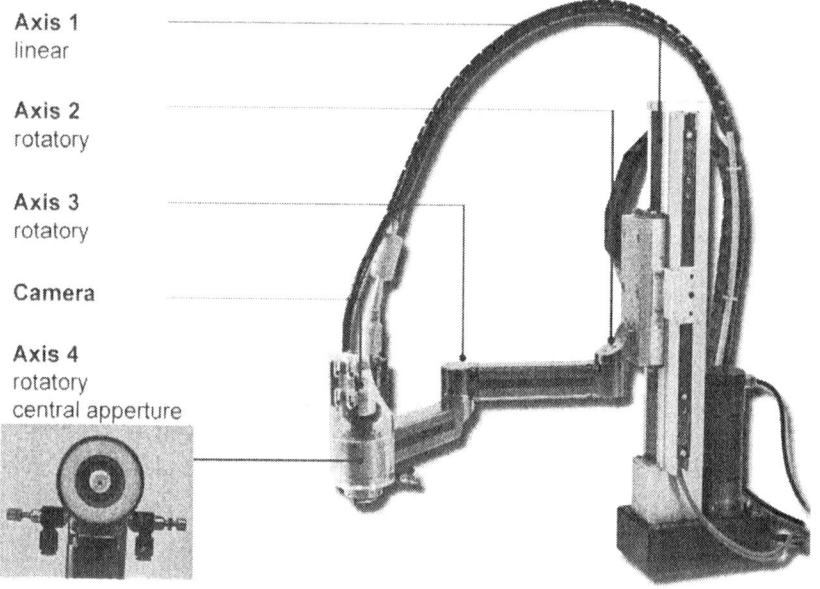

Fig. 4. Miniature Handling System based on Scara kinematics

Making use of the standardised interface DIN 32565, such components guaranty a quick exchange of manufacturer spanning compatibility. Therefore in a first step the micro tool change system mentioned above has been developed. Users and manufacturers of assembly devices as well as research institutes have been cooperating since 2000 for the initiation of the standard and the extension of the construction kit.

Beside the economic requirements, the development of the micro tool change systems includes the following technical requirements:

- Preciseness: In this context the repeat accuracy when changing the endeffectors is of special interest.
- Stability: High load capacity combined with high acceleration requires low deformation at high radial moments.
- Usability: Besides the installation of the interfaces a simple and reliable manual connecting as well as automated connecting and disconnecting has to be ensured.

The main challenges of the development have their seeds in the effects of the downscaling process: while stability can be changing exponentially, the elastic strength is changing linearly by variation of dimension. In consequence new tolerances had to be defined and mechanical principles had to be modified accordingly.

Several research projects like Briolas or Profam (publicly funded by German Federal Ministry of Education and Research), Eupass (publicly funded by European Union) and companies in the sector of system integration like Rohwedder® or MiLaSys technologies® make use of the standardised interface as a design base (see figure 5). So the essential further development is embedded and pushed ahead by the practical demands. Components suitable to the interface, like fine positioning units, optical devices, or whole assembly units have been designed by different applicators.

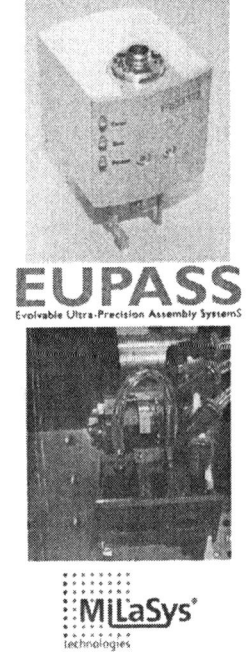

Fig. 5. Examples of first applications of the standardised micro tool change system

3 Major results

One overriding purpose of the paper is to demonstrate the advantages of standardisation in micro technology at large. A basic construction kit for the field of microsystems assembly has been developed with partners from industry and research in the course of several projects. In lieu of solutions specified to the application the construction kit gives rise to economical and modular assembly of microsystems. The standardised interface here is an important step towards cross manufacturer compatibility. The interface between endeffector and machine however is only part of an overall master plan.

Currently the new standard DIN 32565 is applied. It specifies requirements on an interface between endeffector and handling system or between modules. On the front end side of the mentioned micro tool change system several tools can be mounted. An automatic interchange of these tools can be realised provided by a micro change magazine. As a superior system swivel and rotary units have been designed. In a sequence a system will be completed for build up of customised "on table handling" solutions.

The standard has been worked out by the committee "Manufacturing Equipment for Microsystems" in the DIN NAFuO (Optics and Precision Mechanics Standards Committee), the standardisation body for precision engineering and optics. Currently DIN 32565 is transferred to ISO. The committee TC39/WG16 plans to replace the German standard to a world-wide standard in October 2007.

Apart from the mechanical interface the new standard defines also position and specification of lead through for fluidic and electric coupling units. A distinctive feature is the central aperture for optical applications (e.g. laser, camera, etc) or for feeding of parts or media. Only the adapter plate is specified in the standard. This way the realisation of the head plate is up to the equipment manufacturer. The goal of this standard is exchangeability of arbitrary manufacturers. Standardisation is a precondition for a wide dissemination of this interface.

References

[1] Presentation of the efforts of standardisation and a first embodiment of a change system based on this: Micro.tec congress in October 13-15, 2003 Munich (Germany): Proceedings, page: 279-284, ISBN 3-8007-2791-9 VDE Verlag GmbH
[2] DIN Standard: DIN 32565 „Schnittstelle zwischen Endeffektor und Handhabungsgerät" (interface between endeffector and handling device), Deutsches Institut für Normung e. V., Berlin, August 2003
[3] Presentation of first SCHUNK products of a modular system based on mentioned DIN standard: Minat fair in June 12.-14, 2007 Stuttgart (Germany).: Expert forum: Normen und Standards in der Mikromontage - Chance oder Risiko für KMUs? (standards in micro assembly - chance or risk for SME?)

TOWARDS A PUBLISH / SUBSCRIBE CONTROL ARCHITECTURE FOR PRECISION ASSEMBLY WITH THE DATA DISTRIBUTION SERVICE

Marco Ryll and Svetan Ratchev

School of Mechanical, Materials and Manufacturing Engineering,
The University of Nottingham, UK
[epxmr3 | svetan.ratchev] @nottingham.ac.uk.

Abstract This paper presents a comprehensive overview of the Data Distribution Service standard (DDS) and describes its benefits for developing robust precision assembly applications. DDS is a platform-independent standard released by the Object Management Group (OMG) for data-centric publish-subscribe systems. It allows decoupled applications to transfer information, regardless of what architecture, programming language or operating system they use. The standard is particularly designed for real-time systems that need to control timing and memory resources, have low latency and high robustness requirements. As such, DDS has the potential to provide the communication infrastructure for next-generation precision assembly systems where a large number of independently controlled components need to communicate. To illustrate the benefits of DDS for precision assembly an example application is presented.

1 Introduction

Current trends in manufacturing such as modularisation of processes require modern assembly systems which integrate a large number of subsystems such as assembly stations, fixturing devices, Human Machine Control (HMI) etc. This leads to increased information exchange, together with ever more demanding requirements for flexibility and reconfigurability.

Due to these challenges, distributed industrial control systems have attracted a vast amount of research interest. Campelo et al. [1] have proposed a distributed control architecture for manufacturing systems which addresses the problem of fault-tolerance to ensure recovery when components fail. Other examples for fault-tolerant, distributed architectures with real-time characteristics include the DACAPO system [2, 3] and DELTA-4 [4]. Delamer et al. [5-7] have described the CAMX framework (Computer-Aided Manufacturing using XML) which is a message-oriented middleware providing event-distribution in form of standardised

XML-messages with a publish-subscribe approach. Another message-oriented publish-subscribe middleware is the Java Messaging Service (JMS). Although, these systems allow robust communication between many nodes, it appears that message-oriented communication consumes larger amounts of the processing time for the translation of the messages in the nodes. In particular, JMS is limited to the Java programming language and therefore lacks platform-independence. Moreover, JMS does not support dynamic discovery of new components in plug & produce manner, since application discovery is administered and centralised [8].

With the Data Distribution Service (DDS) [9], the Object Management Group (OMG) has recently released a platform-independent standard for a data-centric publish-subscribe middleware that is specifically targeted towards the needs of real-time applications with limited resources. Unlike the previously mentioned approaches, DDS does not exchange data encapsulated within messages. Instead, the data structures to be exchanged are modelled with the formal Interface Definition Language (IDL). On the basis of this data definition all source code to accurately transmit and receive information is generated automatically. This way, information exchange can be achieved significantly faster.

This paper aims at providing a comprehensive overview on the DDS standard and its potential benefits for the precision assembly domain. In the first section, the basic concept of a DDS-based application is described and an architectural overview of the entities that establish data transmission is provided. This is followed by a discussion of three features that can facilitate the development of robust precision assembly platforms. In the last part of the paper, a simple example application is described to illustrate some of the advantages of the DDS middleware.

2 Data Distribution Standard

DDS provides common application-level interfaces which allow processes (so-called participants) to exchange information in form of topics. The latter are data-flows which have an identifier and a data type. A typical architecture of a DDS application is illustrated in Figure 1. Applications that want to write data declare their intent to become "publishers" for a topic [9]. Similarly, applications that want to read data from a topic declare their intent to become "subscribers". Underneath, the DDS middleware is responsible to distribute the information between a variable number of publishers and subscribers. It manages the declarations, automatically establishes connections between publishers and subscribers for a matching topic and dynamically detects new participants in the system. Additionally, DDS allows for precise configuring of the properties of the information exchange with the Quality-of-Service concept (QoS) which is particularly important for real-time systems with limited resources, and for overcoming the "impedance mismatch" between systems with different communication requirements.

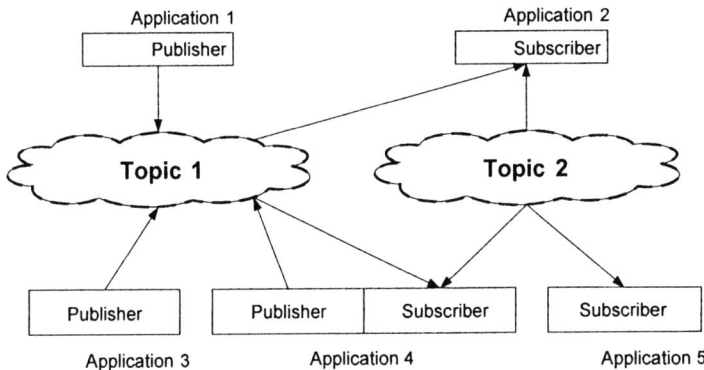

Fig. 1. Decoupling of publishers and subscribers with the DDS middleware

As a result, the utilisation of DDS reduces the complexity of data distribution to simple calls of API-functions which are provided by the standard. This saves development time and is less error-prone than the development of own proprietary data exchange infrastructures. Also, the utilisation of a platform-independent standard allows for better integration with other applications running on different architectures. The middleware hides all details like location of the participants, type of the network or what programming language has been used to implement the participants. Since publishers and subscribers are only connected to topics, communication between them is decoupled. This allows the establishment of dynamic communication channels without reconstructing the source code.

2.1 Architectural Overview

The core of the DDS specification is the Data Centric Publish Subscribe model (DCPS) which organises the data exchange between the communicating participants. Figure 2 presents a simplified view of this model. As can be seen in the diagram all classes that accomplish communication are extended from the central class *Entity*. This class provides the ability to be configured with Quality-of-Service parameters, be notified of events by listener objects, and be attached with conditions that can be waited upon by the application. All child classes of *Entity* have a specialised set of *QoSPolicies* which provides the ability to fine-tune the data exchange.

The publishing side of the communication is represented by the association between a *Publisher* and one or more *DataWriter* objects. The *Publisher* is internally used by the DDS middleware to issue data to be sent. A *DataWriter* acts as a typed accessor to the *Publisher* for the application and is specific for the data type to be sent. The application must use this object if it wants to send data of a given type, which then triggers the *Publisher* to issue the data according to the

Quality-of-Service settings. The subscribing end of the communication is similarly structured. The *Subscriber* is internally responsible for receiving published data and making it available according to the QoS-settings. The application can access the received data by *DataReader*-objects generated for each data type.

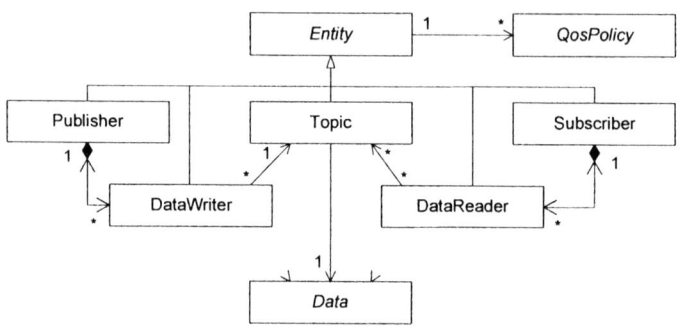

Fig. 2. Overview of the DCPS model (based on [9])

The association between publications and subscriptions is accomplished by means of *Topic* objects. A *Topic* associates a unique name, a data type and Quality-of-Service settings related to the data itself. DDS middleware implementations provide tools that automatically generate the code for these classes on the bases of the data type definitions. The application must use these classes and call the provided methods in order to achieve communication.

2.2 Relevant features for the development of precision assembly platforms

As mentioned before, precision assembly platforms are increasingly implemented as dynamic distributed systems where a number of components need to exchange data in a robust, platform-independent and deterministic way. According to Joshi [10] the key technical challenges for the development of such systems are: (1) impedance mismatch; (2) dynamic real-time adaptation and (3) incremental and independent development. The first challenge refers to the problems with the integration of applications that impose different requirements for the data exchange, such as data volume, data rates or timing constraints. Dynamic real-time adaptation addresses the need to discover topology changes as components are added or removed from the platform. The third challenge refers to the fact that the various subsystems of an assembly platform are often developed by independent parties which upgrade their components incrementally whilst utilising different operating systems and programming languages. The following sections give an overview on aspects of DDS that can reduce the impact of these challenges.

2.2.1 The Quality-of-Service Concept

With the Quality-of-Service concept, DDS provides a sophisticated means to dynamically tailor data distribution according to the application requirements. In particular, QoS provide the ability to control and limit the use of resources like network bandwidth or memory as well as reliability, timeliness and persistence of the data transfer.

The DDS QoS model is a set of classes which are derived from *QosPolicy*. Each of them controls a specific aspect of the behaviour of the service and as shown in Figure 2 they can be attached to all objects that accomplish communication. The middleware automatically checks if the settings for the publishing and subscribing side are compatible. Communication is only established if the offered communication properties of the publisher meet the requested behaviour of the subscriber. The complete specification can be found in [9].

The QoS concept solves a number of typical problems in the development of distributed applications. For example, with the TIME_BASED_FILTER QoS a subscriber can precisely define how often it needs to be updated with new data. This prevents it from flooding when it is integrated with systems that produce data at much higher rates (impedance mismatch). The combination of OWNERSHIP and OWNERSHIP_STRENGTH provides an easy way of integrating redundancy and a seamless failover to backup systems. The HISTORY QoS alleviates the challenge of providing late-joining processes with already sent (so-called historical) data. These features are demonstrated in section 3.2.

2.2.2 Automatic Discovery and dynamic real-time adaptation

In order to react to changing product or process requirements it must be possible to dynamically add or remove components without the need to rebuild the control software. For example, an assembly line might be extended by modules with additional functionalities or monitoring devices are plugged and unplugged during the production process without affecting the overall process.

DDS addresses this aspect with its decoupled publish-subscribe architecture approach and the automatic discovery of participants. Since publishers and subscribers are only connected to topics, communication between them is decoupled. This allows the establishment of dynamic communication channels without reconstructing the source code. New publishers and subscribers for a topic can appear at any time and the DDS middleware interconnects them automatically. Similarly, the middleware provides mechanisms to inform the application when participants have been removed.

2.2.3 Platform-independence and independent development

The components of modern manufacturing systems are often developed by different vendors. As a consequence, these subsystems might be implemented on the basis of different hardware architectures, operating systems and programming languages. This increases the challenge of integrating them into a working system. Also, each component might be subject to incremental changes or upgrades.

DDS is defined by a Platform Independent Model (PIM) and thus can be implemented for any combination of processor architecture, programming language and operating system. Commercially available DDS middleware implementations offer solutions for the programming languages C, C++ and Java and a wide variety of operating systems such as VxWorks, Windows, Lynx and Unix derivates. Since DDS hides the communication aspects from the application code it allows the seamless integration of an application written in C running on VxWorks with an application developed with Java running on a Windows PC. Hence, the usage of DDS as a backbone for the communication within the assembly system can significantly reduce the difficulties of the system integration task. Additionally, the data-centric publish-subscribe paradigm introduces less dependencies between the applications than conventional object-oriented or client-server approaches [10]. This is because in a data-centric architecture, applications are only connected by the data model and do not expose behaviour. Since the data model is usually the most constant aspect of an application, DDS supports the incremental and independent development of subsystems.

3 Example Application

This section aims to illustrate some of the previously mentioned features of DDS with a simplified example application. It was developed using the commercially available DDS implementation from Real-Time Innovations, Inc., called RTI DDS 4.1e. The example uses the DDS standard for the implementation of a sensor-based active fixturing system. A simplified overview of the proposed system is shown in figure 3.

The system consists of a variable number of physical fixture modules, a fixture control software, a variable number of Human Machine Interfaces (HMI). For the sake of simplicity, each module consists of one linear actuator and three sensors. The former acts as the locating and clamping pin against the workpiece, while the sensors feedback reaction force, position and temperature of the contact point. The fixture modules are implemented as smart devices with local control routines for their embedded sensor/actuator devices. It is further assumed that each fixture module is configured with a unique numerical identifier and with meta information about its sensors and actuators. In this way, the module is able to convert the signals coming from the sensors (e.g. a voltage) into meaningful information (e.g. reaction force in Newton) which is then published via DDS. The fixture control implements the global control routines of the fixture. It processes the data coming from the various modules and controls the movement of the actuators by publishing their desired status.

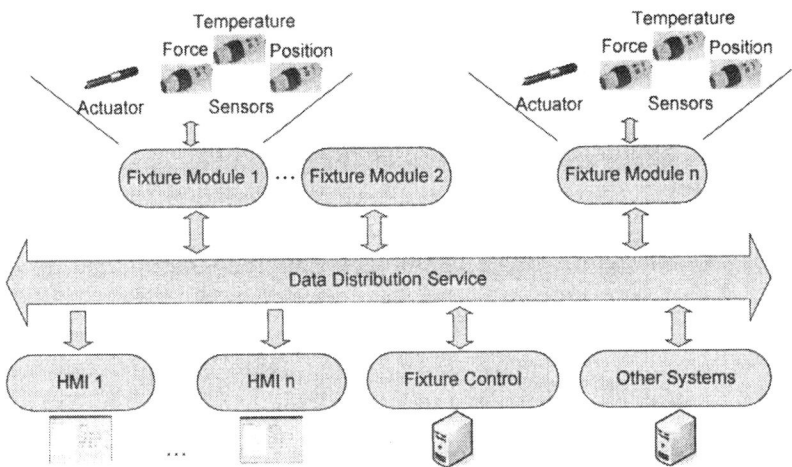

Fig. 3. Overview on the example application

In order to demonstrate some features of DDS we define the following requirements for the overall system:
1. At the initial start-up each fixture module publishes meta information about its sensors and actuators in order to allow subscribing applications to interpret the sensor data. Late-joining applications shall receive this information automatically.
2. HMI applications shall receive force sensor readings every 500ms and temperature readings every 1000ms, regardless how fast the modules publish this information.
3. The fixture control and connected HMI applications only consider the temperature readings of 1 fixture module at a time. If no data from this most-trusted sensor source is received within 4000ms, temperature data shall automatically be received from another module, allowing a seamless failover.

Details about the control logic of this application are not described within this paper. Instead the description concentrates on the data modelling and the definition of the QoS-settings to meet the requirements.

3.1 Design of Data Structures

The development of a data-centric application starts with the definition of the data structures that shall be exchanged between the applications. In our simple example, we create two data structures for each sensor type. The first data structure contains the actual sensor data to be transferred during the assembly process. It contains fields to uniquely identify the sensor and a field for the current sensor reading. In order to interpret the sensor values correctly, subscribers need additional meta-information. These details are modelled in an additional structure and only need to be published when the application starts or if sensors are exchanged. This way, sensor data and its interpretation are separated and network load can be reduced significantly. Subscribing applications can be dynamically reconfigured to accurately interpret incoming data from different sensors. As an example, the data structures for force sensors are shown below according to the Interface Definition Language (IDL). The structure *Force* is used to transmit the sensor readings, whereas *ForceSensorInfo* contains meta-information to interpret the sensor data correctly. For the other sensor types the data structures are defined accordingly.

```
struct Force{                    struct ForceSensorInfo{
    long module_id;                  long module_id;
    lond device_id;                  long device_id;
    double value;                    MeasuringUnit unit;
};                                   double maxForce;
                                     double minForce;
                                     double resolution;
                                     double sampleRate;
                                 };
```

Based on these data type definitions, the source code for all the subscribers, publishers, data readers and data writers are automatically generated in the specified programming language. These classes have to be used by the application programs that implement the fixture modules, the fixture control and any other participating system like monitoring applications.

3.2 Selection of Quality-of-Service Parameters

To tailor the data transfer according to the requirements, the QoS parameters for each data reader and data writer need to be configured in the application source code. Table 1 shows for each requirement the QoS settings for the data readers on the subscribing side and data writers for publishing applications.

When a fixture module executes its initialising routine it retrieves the meta-data for each sensor from a configuration file and creates the corresponding data structures. The data writers for these data types are configured according to table 1 and each fixture module publishes this information only once. After this publication it can initialise the publishers for the actual sensor data transmission and start publishing the sensor reading. Subscribing applications like HMIs or the fixture control first initialize the subscribers for the meta information and hence retrieve these data before they start subscribing to the actual sensor data. Since the QoS parameters are set to HISTORY.depth = 1 and DURABILITY.kind = TRANSIENT_LOCAL it is assured by the middleware that even late-joining applications will get the necessary information to interpret all sensor readings. This approach significantly saves network resources, since meta information is only transferred when it is necessary. DDS allows the implementation of this feature with only minor programming effort, whereas in traditional distributed systems providing late-joining applications with historical data is an error prone and complex task. The same argument applies for the second requirement. Monitoring applications can limit the number of sample readings simply by setting the time-based filter QoS on their data readers. This way they are protected from being flooded with too much data and can spent more resources for the graphical user interface. In traditional client-server based applications this "impedance mismatch" is a major problem. DDS overcomes this problem by the simple setting a QoS parameters.

Requirement	Data type	QoS DataReader	QoS DataWriter
1	ForceSensorInfo, TemperatureSensor Info, DisplacementSensorInfo	HISTORY.depth = 1 RELIABILITY.kind = RELIABLE DURABILITY.kind = TRANSIENT_LOCAL	HISTORY.depth = 1 DURABILITY.kind = TRANSIENT_LOCAL
2	Force	TIME_BASED_FILTER = 500ms	-
	Temperature	TIME_BASED_FILTER = 1000ms	-
3	Temperature	OWNERSHIP.kind = EXCLUSIVE OWNERSHIP_STRENGTH. value = module_id of the fixture module DEADLINE.period.sec = 4	OWNERSHIP.kind = EXCLUSIVE DEADLINE.period.sec = 4

Table 1. Qos Settings for the Applications

The third requirement allows subscribing applications to get temperature readings only from one most-trusted sensor. If this sensor stops working because of damage or other reasons, the applications shall automatically use the readings from a temperature sensor of another fixture module. This automatic and dynamic failover to a backup sensor would require enormous programming effort if implemented manually. With DDS we have to set the OWNERSHIP.kind parameter to "exclusive" to ensure that readers will only receive data from a single sensor. Additionally, each temperature data writer is configured with the identifier of its corresponding fixture module, resulting in a hierarchic order. The data readers in each subscribing application will only receive temperature readings from the fixture module with the highest identifier. In this context, the DEADLINE QoS specifies that the subscribers will automatically failover to the sensor of the fixture module with second-highest ID if it does not receive data within the specified time period. This way, fault-tolerant distributed applications can easily be developed with the ability to dynamically react to failures in the system.

4 Conclusions

In this paper, a new standard for data-centric publish-subscribe communication has been presented and put in context to the development of next-generation precision assembly platforms. The standard is called Data Distribution Service and is particularly targeting real-time applications which need to manage resource consumption and timeliness of the data transfer. DDS allows platform-independent many-to-many communication and alleviates a number of common problems which are of particular interest for the development of distributed assembly systems. For example, with its sophisticated Quality-of-Service support communication can be tailored according to the system requirements and typical challenges such as the delivery of historical data to late-joining applications is achieved automatically and in an efficient manner. Additionally, DDS automatically discovers when components are plugged in or removed. The example presented in this paper shows how these features can potentially be integrated to develop a plug & produce capable active fixturing system for precision assembly.

Acknowledgments

The reported research is conducted as part of the ongoing European Commission FP6 funded integrated project (FP6-2004-NMP-NI-4) - AFFIX "Aligning, Holding and Fixing Flexible and Difficult to Handle Components". The financial support of the EU and the contributions by the project consortium members are gratefully acknowledged. The authors are also grateful to Real-Time Innovations, Inc. for their support of the research with a software grant for the RTI Data Distribution Service middleware.

References

1. J.C. Campelo, et al., Distributed industrial control systems: a fault-tolerant architecture, Microprocessors and microsystems, Vol. 23 (1999), 103-112.
2. B. Rostamzadeh, et al., *DACAPO: a distributed computer architecture for safety-critical control applications*,IEEE International Symposium on Intelligent Vehicles, Detroit, USA, 1995
3. B. Rostamzadeh and J. Torin, *Design principles of fail-operation/fail-silent modular node in DACAPO*,Proceedings of the ICEE, Tehran, Iran, 1995
4. J. Arlat, et al., *Experimental evaluation of the fault tolerance of an atomic multicast system*, IEEE Transactions on reliability, Vol. 39 (1990).
5. I.M. Delamer and J.L. Martinez Lastra, *Evolutionary multi-objective optimization of QoS-Aware Publish/Subscribe Middleware in electronics production*, Engineering Applications of Artificial Intelligence, Vol. 19 (2006), 593-697.
6. I.M. Delamer and J.L. Martinez Lastra, *Quality of service for CAMX middleware*, International Journal of Computer Integrated Manufacturing, Vol. 19 (2006), pp. 784-804.
7. I.M. Delamer, J.L. Martinez Lastra, R. Tuokko, *Design of QoS-aware framework for industrial CAMX systems*,Proceedings of the Second IEEE International Conference on Industrial Informatics INDIN 2004, Berlin, Germany, 2004
8. J. Joshi, *A comparison and mapping of Data Distribution Service (DDS) and Java Messaging Service (JMS)*, Real-Time Innovations, Inc., Whitepaper, 2006, Available from: www.rti.com.
9. Object Management Group, *Data Distribution Service for Real-Time Systems, Version 1.2*, 2007, Available from: www.omg.org, June 2007
10. J. Joshi, *Data-Oriented Architecture*, Real-Time Innovations, Inc., Whitepaper, 2007, Available from: www.rti.com.

SMART ASSEMBLY - DATA AND MODEL DRIVEN

Juhani Heilala*, Heli Helaakoski[§], Irina Peltomaa[§]

*VTT Technical Research Centre of Finland, P.O.Box 1000, 02044 VTT, Finland
juhani.heilala@vtt.fi
[§]VTT Technical Research Centre of Finland, P.O.Box 3, 92101 Raahe, Finland
heli.helaakoski@vtt.fi, irina.peltomaa@vtt.fi

Abstract The world is changing distinctly and manufacturing is facing significant challenges. Current manufacturing paradigms need to develop towards better agility to meet current market demand. The relevant research has been launched and the resulting new approaches, like smart manufacturing, digital manufacturing and the cognitive factory, are introduced. However, today's real challenge to manufacturing enterprises is building an infrastructure that will enable the sharing of knowledge and information in a manufacturing environment, regardless of time and place. This article reviews some of the latest approaches and research initiatives and suggests the use of semantic technologies in information integration. Use of semantics enables information and knowledge to be represented in a form understandable to both humans and machines, and offers the ability to adapt more flexibly to change.

Keywords smart assembly, semantic interoperability

1 Introduction

Today, enterprises are living in a changing business environment characterised by global competition, rapid development of technology, short lead-time, increased cost pressure and more aggressive demand from customers. In Western countries, manufacturing enterprises are not able to compete with labour costs and they need to move on from resource-based manufacturing to knowledge and skill-driven manufacturing. They need to focus on customised high quality, high value added products that require a growing amount of information and knowledge. Today, there is a gap between current market demand and current manufacturing paradigms. There is a need to develop the information integration of the manufacturing process and manufacturing resources towards a smart agile manufacturing process.

The intelligent use of modern ICT as information-producing and communication-enabling technologies has become crucially important. Characteristic of modern ICT is that they significantly increase the capacity of enterprises to efficiently generate and process knowledge-based information [1]. In this paper, "smart manufacturing" is defined as the efficient use of modern ICT, including semantic technologies, ubiquitous computing, novel interfaces for human-technology interaction, human-centred automation, and affordable and flexible robot automation technology in a traditional manufacturing environment. According to Caie [2], the key elements of smart assembly are:

- Empowered, Knowledgeable People: Multi-disciplined, highly skilled workforce empowered to make the best overall decisions.
- Collaboration: People and automation collaboratively working in a safe, shared environment for all tasks.
- Reconfigurable: Modular, plug and play system components easily reconfigured and reprogrammed to accommodate new product, equipment, and software variations and to implement corrections.
- Model and Data Driven: Modelling and simulation tools enabling all designs and design changes to be virtually evaluated, optimized, and validated before being propagated to the physical plant.
- Capable of Learning: Self-integrating and adaptive assembly systems that prevent repeated mistakes and avoid new ones.

The above definition was created in the USA in 2006. The European Technology Platform Manufuture Strategic Research Agenda [3] has similar definitions, with a vision of the manufacturing of the future. According to Westkämper [4], intelligent manufacturing visions are holistic systems, operating in primer fields of high performance and managed by highly skilled workers. They can be adapted by plug-and-produce mechanisms and are linked in a digital and virtual engineering and management IT. Adaptive manufacturing recombines new and innovative processes, uses intelligent combinations and flexible configuration of products and manufacturing systems to overcome existing process limitations, and transfers manufacturing know-how using completely new themes or manufacturing-related themes. Adaptive manufacturing takes into account intelligent manufacturing technologies and includes the field of automation and robotics. Adaptive manufacturing also includes new solutions of automation by integration of new methods of cognitive information processing, signal processing and production control by high speed information and communication systems. Intelligent manufacturing can be linked to communications technology networks to assure real-time adaptation. In the future, new technologies like RFID, MES, Wireless, Grid Computing and others will make the vision of a real time factory or smart factory a reality.

A similar approach is the cognitive factory, defined by Zäh et al. [5] as a factory environment with its machines, robots, storages, planning processes and human workers, that is equipped with a sensor network and an IT structure to allow the resources and processes to perceive what they are doing, enable them to control themselves and plan further actions in cooperation with other machines and human workers. The overall goals of the cognitive factory can be summarised as:

- Making the knowledge available in factory systems more transparent
- Enforcing the interconnection between men and machines
- Creating manufacturing systems that are more adaptive than existing ones
- Enabling autonomous planning processes.

In all the elements shown here, Caie [2], Westkämper [4], Zäh [5], and also Dencker [6] highlight the role of information and knowledge, how information can be accessed and aggregated by both humans and machines. The real challenge for manufacturing enterprises today is to build an infrastructure that enables knowledge and information sharing in a manufacturing environment regardless of time and place. Micro-products are especially challenging given that they might have been designed and developed elsewhere, and that production personnel are facing the challenges. Agility and customer-driven manufacturing increases the complexity of manufacturing operations; the equipment can be complex as well as the products to be assembled.

2 The role of modern ICT in manufacturing

One consequence of the deep penetration of ICT into daily life is migration from the conventional factory floor to intelligent manufacturing environments built around the AmI (Ambient Intelligence) paradigm. That is, workplaces with emphasis on greater user-friendliness, more efficient service support, user-empowerment, and support for human interaction. Large scale industrial AmI solutions are still rare, even if research efforts are being carried out.

In a manufacturing environment where workforces are surrounded by a collection of reconfigurable production components (physical agents) that include mechatronics, control and intelligence, the challenge is to develop production automation and control systems with autonomy and intelligence capabilities for collaboration, agile and fast adaptation to environmental changes, robustness against the occurrence of disturbances, and easier integration of manufacturing resources and legacy systems.

Today's requirements for high production-process performance, combined with their increasing complexity, represent a real challenge to staff members at all levels (from worker to plant manager) in controlling the production process such that customer orders will be fulfilled to perfection. The concept of digital manufacturing emphasises that the success of manufacturing industries is mainly related to a great diversity and skills of personnel at all levels [4].

The adaptation to changing situations in factory floor operations, in flexible, lean, agile and re-configurable manufacturing, is a challenge. Each product produced can be unique; production personnel must synchronize the material and resources efficiently. The management of information flow is a key issue as shown in Figure 1:

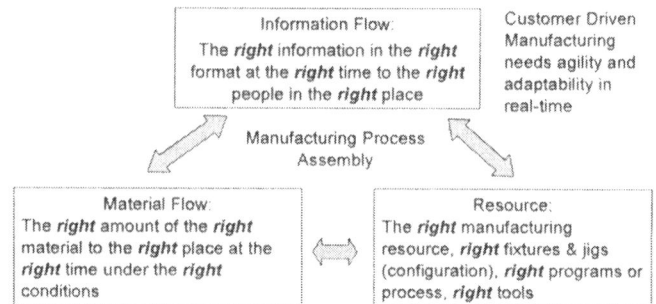

Fig. 1. Information controls material flow and resources

3 Smart manufacturing

In the smart manufacturing approach, the human operator interacts and collaborates with a smart manufacturing environment by exploiting modern ICT. Workers have access to aggregated information and knowledge that is supported by an intelligent assistant delivering the relevant piece of information, whenever and wherever it is needed. This will allow workers at different levels, when relevant, to:
- Improve availability of information, communication and knowledge sharing
- Speed up learning of human operators
- Enhance the quality of products, zero defects
- Improve teaching of automated device
- Shorten production standstills, speed of delivery.

Future manufacturing enterprises will be able to capture individual expertise and experience for efficient reuse and draw on a rich, openly accessible shared base of scientific, business, and process knowledge to make informed, accurate decisions and to ensure the right people get the right information at the right time to do their jobs (knowledge-based enterprises). Improved understanding and shared knowledge of the scientific foundations for material and process properties and interactions will support optimized process design and total understanding of complex transformations and interactions at the micro and macro levels.

Collaborative human-centred automation can be seen as automation of a manual task, mechanisation of physical activities and also as cognitive task, computerisation, information and control activities (Fig. 2). This paper focuses on information and control activities and on how to make information available in ICT systems more transparent.

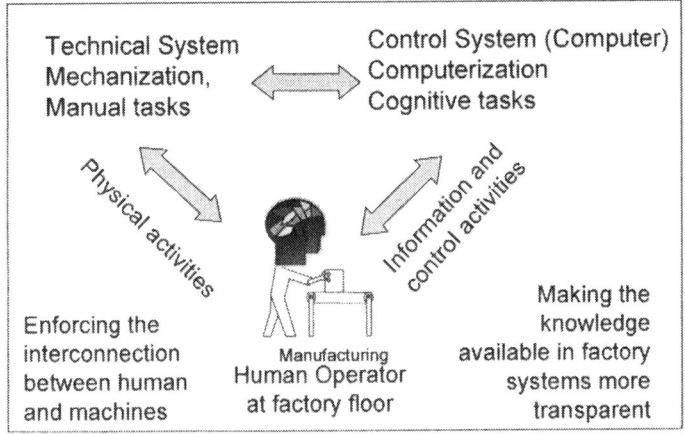

Fig. 2. Collaborative human-centred automation

4 Digital manufacturing in smart assembly

Digital manufacturing uses a wide range of engineering and planning tools, software, and ICT to integrate new technologies into manufacturing processes as quickly and as efficiently as possible. The main area of research is the development of integrated tools for industrial engineering and the adaptation of manufacturing, taking into account the configurability of systems. Digital manufacturing is the most important technology of the future. Digital manufacturing needs [4]:
- Distributed data management
- Tools for process engineering
- Tools for presentation and graphic interfaces
- Participative, collaborative and networked engineering
- Interfaces to reality.

Manufacturing systems, processes and data are growing and becoming more complex. Manufacturing engineering and production management decisions involve the consideration of many interdependent factors and variables. These often complex, interdependent factors and variables are probably too many for the human mind to cope with at one time (Fig. 3). For a human decision-maker it is difficult to locate the relevant pieces of information. That is why users on the factory floor, human operators and production managers all need tools for data mining. The challenge of the man-system interface is to bring context-aware data, as a personal advisor, to the operator.

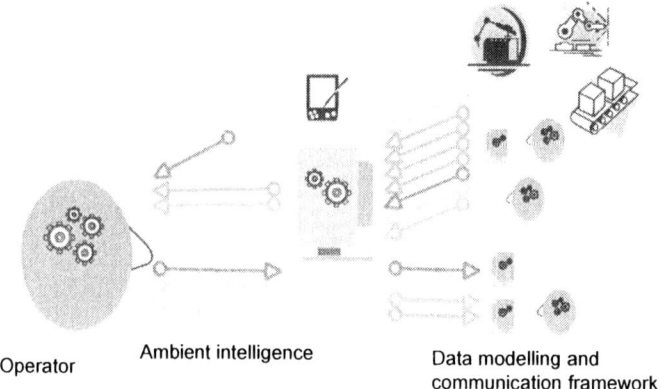

Operator   Ambient intelligence   Data modelling and communication framework

What to show and how to show it: filtering, selecting only relevant information, knowledge sharing and context awareness.

Fig. 3. The amount of information at assembly shops is increasing

Different users require, or are allowed access to, different types of information or the same information presented differently. For instance, a worker, line supervisor and plant manager relies on different levels of data upon which to base decisions — too little or too much data and its utility is diminished if not lost. Time is an important factor in defining how much and what kind of data should be aggregated for the various levels of the organization; a manager or worker cannot afford to be "swimming in data" when making a quick decision. Ultimately, users should be given enough information to enable them to make the decisions necessary to optimize the performance of their job function. One of the challenges is to provide the appropriate granularity of information needed by each class of user.

Vision: *Integrate human and technical resources to enhance workforce performance, safety and satisfaction, help the human to access "instantaneously" information from a vast array of diverse sources into useful knowledge and effective decisions. Bring into the human-machine system all the skills, rules and knowledge necessary to interpret, diagnose and act in a concerted way between machine knowledge and human intelligence.*

Digital manufacturing engineering could be part of the solution. Results presented in a visually attractive way speed up and improve the way they are understood. Virtual reality (VR) and augmented reality (AR) technologies can provide easy to understand visualization. Combined with other production-related tasks AR/VR is an efficient way to create an electronic performance support system (EPSS) for human workers. Examples of how to use AR in assembly work have been given by Sääski et al. [7].

We are able to enable science-based management by workers, if we can provide tools for decision support and information, and the data and models needed in the decision making. Present factory information systems (FIS), like the manufacturing execution system (MES) or automation system capability are geared to support operators and engineers, not planners, financial analysts or upper management.

Enterprise resource planning (ERP) systems have other limitations; they are usually based on static resource models, with unlimited capacity. Currently, manufacturing scenarios cannot be studied efficiently with the ERP system, at least not by factory floor personnel.

Manufacturers have begun the transition from passive data monitoring to conversion of data to information and must attempt proactive, operational and strategic decision making in the future. Simulation and modelling is one of the tools available for proactive planning. Normally production simulation is used for system design analysis, not yet on the factory floor on-line near real-time. Simulation analysis shows future events with given input parameters. Near real-time simulation is one way to get information for decision making and to create a model-driven smart assembly system.

There has been progress in simulation software, and most of the current software is object-oriented with a graphical windows style user interface. At the same time, however, the complexity in manufacturing systems and products has increased. There are development needs in manufacturing system modelling and simulation, such as faster analysis cycle and automated model building. A simulation model and a manufacturing system life cycle should be combined, and the model should be a virtual, concurrently evolving digital image of the real system. Thus a need exists for real-time data coupling from the factory floor and other manufacturing information systems. A future aim could also be hybrid methods —simulation, optimisation and other manufacturing information systems in an embedded application. These could be used by non-simulation experts, even on-line and real-time on the factory floor (plug and simulate), enabling science-based system management and productivity enhancement tools.

4.1 Potential in standardised neutral interfaces

Many standardisation efforts are being made to improve manufacturing system interoperability, but in many cases system integration is still based on custom-built proprietary interfaces. The use of standardised structured manufacturing data in a neutral format (like XML) could clearly increase interoperability between manufacturing information systems. In the digital manufacturing paradigm and use of assembly job shop simulation to improve operations, one of the biggest challenges has been integrating the information system. Work is being done by SISO (Simulation Interoperability Standards Organization (http://www.sisostds.org/)) and CMSD (Product Development Group to create a Core Manufacturing Simulation Data Information Model (CMSDIM)) to standardise a job shop simulation data model. The data elements of the current data model are: Organisations, Calendars, Resources, Skill-definitions, Setup-definitions, Operation-definitions, Maintenance-definitions, Layout, Parts, Bills-of-materials, Inventory, Procurements, Process-plans, Work, Schedules, Revisions, Time-sheets, Probability-distributions, References, and Units-of-measurement [8]. In the future, this development could help lower integration costs, if the SISO CMSD standard is accepted by the industry and by providers of simulation software and other manufacturing information systems. The authors believe that suitable standardisation with semantics modelling, and the use of ontologies, are a future solution. For information sharing and integration, semantics technology is clearly one solution, as shown in the next chapter.

5 Semantics in smart manufacturing

Manufacturing competitiveness is highly dependant on companies' ability to rapidly reconfigure their manufacturing and assembly systems while the variety of products and volumes is changing rapidly [6]. Over the last decade, several manufacturing concepts and high-level strategies aiming mainly at automation or support of human tasks have been developed [5]. However these concepts alone have not been able to meet the demands of rapid reconfiguration enhanced with increased cost pressure, more aggressive demand from customers and precise product delivery.

To meet the challenges of agile business requirements, future smart manufacturing systems must be able to integrate information and knowledge from disparate sources — humans, machines and task-specific applications — and maintain and share information more effectively than before. The interoperability of information can be achieved through the use of semantics, representing the meaning of data in a way that is understandable to both humans and machines.

Semantic interoperability is defined as a dynamic enterprise capability derived from the application of special software technologies that infer, relate, interpret and classify the implicit meanings of digital content without human involvement, which in turn drive adaptive business processes, enterprise knowledge, business rules and software application interoperability [9]. Semantic interoperability enables the integration of data sources developed using different vocabularies and different perspectives on data. To achieve semantic interoperability, systems must be able to exchange data in such a way that the precise meaning of the data is readily accessible and the data can be translated by any system into a form that it understands [10]. Using a model of a given business domain (ontology) as semantic data, the model can rationalise disparate data sources into one body of information. By creating models for data and content sources and adding generic domain information, integration of disparate sources in the enterprise can be performed without disturbing existing applications. The model is mapped to the data sources (fields, records, files, documents), giving applications direct access to the data through the model of a certain domain.

The concept ontology is defined to be an explicit specification of a conceptualisation [11]. Ontology defines a common vocabulary for information sharing in a domain [12, 13] and it includes machine-interpretable definitions of basic concepts in the domain and relations among them [12]. The information infrastructure is created using different ontology architectures and languages. The most efficient languages (RDF [14], OWL [15]) provide semantic and inference support, whereas the lowest level ontology languages like XML [16] only provide information modelling. XML doesn't capture the contextual meaning (or semantics)

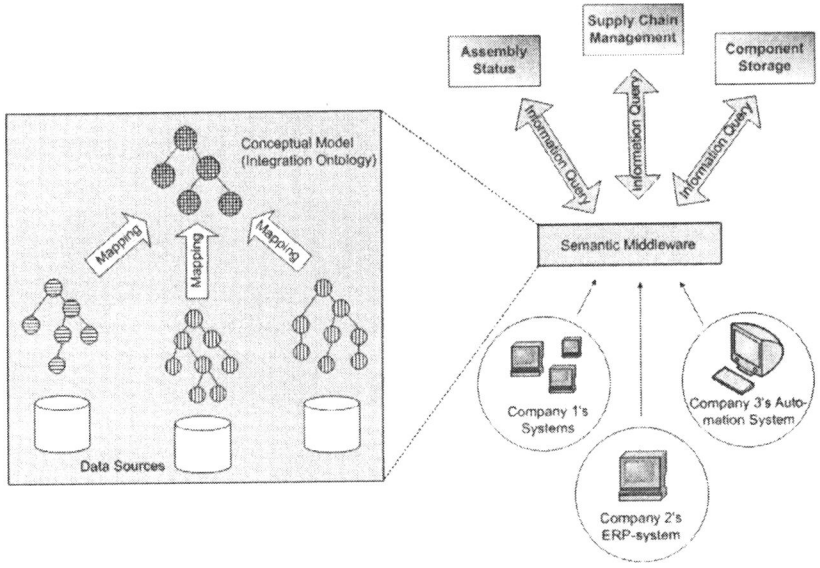

Fig. 4. Through semantic middleware the sharing of mutual information is enabled between a network's companies

of the data since it is a standard for the representation of structured information. The semantics of the XML document is not accessible to machines, only to humans.

The importance of semantics is increasing while companies move away from the traditional view of enterprises as separate, independent entities towards a more collaborative model where the emphasis is on interconnectivity [17]. This development will take over in the near future also in precision assembly. The best knowledge of different expertise fields can be combined together to get competitive advantages in global markets. At this stage the ability to share mutual information and proficiency in interoperability are key assets. All the activities in a company or collaborative networks must be seen as one entity where the same information is advantaged through different applications and through different systems (Fig. 4).

6 Summary

In the area of manufacturing there are big challenges to be met. The rapidly changing situation on the factory floor needs to be addressed, the expertise of individual workers needs to be captured, and the vast amount of data needs to be organized. The importance of information and the requirement to share information is recognized in many of the recent approaches presented in this paper: smart assembly, digital manufacturing, the cognitive factory and proactive assembly.

This paper proposes the use of semantic interoperability as an infrastructure that enables information sharing in the manufacturing environment, including assembly. Semantic technologies are used in business process integration but not yet as commonly in assembly. Semantic technology is quite a new technology and it is seen as an important technology for supporting interoperability between systems and networks. By bringing together the achievements reached in modern manufacturing approaches with the possibilities of semantic technologies, the challenges of assembly could well be met in the future.

Acknowledgements

The authors wish to acknowledge the financial support received from VTT. The development is part of the VTT Theme Complex System Design project KNOWMAN - Knowledge and Skill-Driven Manufacturing of High Added-Value Products.

References

1. M. Castells, *Volume 1: The Rise of the Network Society* (Blackwell Publishers, Oxford & Malden, MA, 1996).
2. J. Caie, NIST Workshop Defines Preliminary Roadmaps for Smart Assembly, ARC Advisory Group ARC Strategies, February 2007, p.16; smartassembly.wikispaces.com/space/showimage/NIST+SA+Workshop+SR.doc
3. The Technology Platform on Future Manufacturing Technologies: MANU*FUTURE*, Strategic Research Agenda, September 2006; http://www.manufuture.org/strategic.html.
4. E. Westkämper, Digital Manufacturing in the global Era, Keynote paper, in: Digital Enterprise Technology, CD-ROM: Proceedings. 3rd CIRP Sponsored Conference on Digital Enterprise Technology, edited by P.F. Cunha (September 18-22, Setubal, Portugal, 2006), p. 11.
5. M.F. Zäh, C. Lau, M. Wiesbeck, M. Ostgathe, and W. Vogl, Towards the Cognitive Factory, in: Proceedings of the 2nd International Conference on Changeable, Agile, Reconfigurable and Virtual Production, CARV 2007 (July 22-24, Toronto, Canada, 2007), pp. 2-16.
6. K. Dencker, J. Stahre, P. Gröndahl, L. Mårtensson, T. Lundholm, and C. Johansson, An Approach to Proactive Assembly Systems - Towards Competitive Assembly Systems, in: Proceedings of the 2007 IEEE International Symposium on Assembly and Manufacturing, ISAM '07 (July 22-25, Ann Arbor, Michigan, USA, 2007), pp. 294-299.
7. J. Sääski, T. Salonen, M. Hakkarainen, S. Siltanen, C. Woodward, J. Lempiäinen. Integration Of Design And Assembly Using Augmented Reality. Fourth International Precision Assembly Seminar, IPAS'2008. Chamonix, France, 10-13 February 2008. Submitted for publication.
8. SISO 2006. Core Manufacturing Simulation Data Information Model, (CMSDIM), PART 1: UML Model. CMSD Product Development Group. Simulation Interoperability Standards Organization. Revision Date: September 13, 2006. www.sisostds.org.
9. J.T. Pollock, and R. Hodgson, *Adaptive Information - Improving Business Through Semantic Interoperability, Grid Computing, and Enterprise Integration* (John Wiley & Sons, New Jersey, 2004).
10. S. Ram, and J. Park, Semantic Conflict Resolution Ontology (SCROL): An Ontology for Detecting and Resolving Data and Schema-Level Semantic Conflicts, *IEEE Transactions on Knowledge and Data Engineering* **16**(2), 189-202 (2004).

11. T. R. Gruber, Towards Principles for the Design of Ontology Used for Knowledge Sharing, *Int. Journal of Human-Computer Studies* **43**(5/6), 907-928 (1995).
12. N.F. Noy, and D.L., McGuinness 2001. Ontology Development 101: A Guide to Creating Your First Ontology, Stanford Knowledge Systems Laboratory Technical Report KSL-01-05, March 2001, p. 25; http://protege.stanford.edu/publications/ontology_development/ontology101.pdf
13. M. Uschold, and M. Gruninger, Ontologies: Principles, Methods, and Applications, *Knowledge Engineering Review* **11**(2), 93-155 (1996).
14. F. Manola, and E. Miller, RDF Primer, W3C Recommendation, 10 February 2004; http://www.w3.org/TR/rdf-primer/.
15. D.L. McGuinness, and F. van Harmelen, OWL Web Ontology Language - Overview, W3C Recommendation, 10 February 2004; http://www.w3.org/TR/owl-features/.
16. T. Bray, J. Paoli, C.M. Sperberg-McQueen, E. Maler, and F. Yergeau, Extensible Markup Language (XML) 1.0 (Fourth Edition), W3C Recommendation, 16 August 2006; http://www.w3.org/TR/REC-xml/.
17. M. Singh, A Review of the Leading Opinions on the Future of Supply Chains, MIT Center for Transportation & Logistics Supply Chain 2020 Working Paper, edited by K. Cottrill, December 2004. p.27; http://ctl.mit.edu/public/opinions_future_supply_chains.pdf.

Chapter 10

Assembly System Integration

MAN-ROBOT COOPERATION — NEW TECHNOLOGIES AND NEW SOLUTIONS

Timo Salmi*, Ilari Marstio*, Timo Malm°, Esa Laine°

*VTT, P.O. Box 1000	°VTT, P.O. Box 1300	˜TUT, P.O. Box 589
FI-02044 VTT	FI-33101 Tampere	FI-33101 Tampere
Timo.Salmi@vtt.fi	Timo.Malm@vtt.fi	Esa.P.Laine@tut.fi
Ilari.Marstio@vtt.fi		

Abstract In this paper we present technologies that enable man-robot cooperation. Several robots are presented that are capable of working safely around humans. Software and sensor-based safety systems are also discussed. We review the regulations and robot standards which determine the construction of today's robot solutions. Finally, we present a few concept man-robot solutions in assembly cells from a safety perspective.

1 Introduction

The first industrial robots in the 1960s and 1970s were developed for tasks that were dangerous or very monotonous to man. Since then, robots have improved and brought flexibility to industrial automation. Compared to traditional rigid automation robots are reprogrammable and adaptable, and compared to man they are more precise and faster.

Nowadays, man and robot are completely isolated from each other for safety reasons. It must be ensured that the robot cannot hit anybody, and that nobody can access the robot without stopping it. Usually this is done by fencing in the robot, and often the enclosure is larger than the robot's workspace to prevent crushing accidents. This requires a lot of space.

Small patch sizes, assembly-to-order ideologies, and short life cycles of products are making large investments in rigid automation unprofitable. Often, manual assembly seems to be the solution to achieving flexibility. At the same time the component sizes and tolerances are getting smaller, causing many assembly tasks to be very challenging or even impossible to man.

In this present situation, combining the robot's accuracy and man's flexibility may be the solution to efficient production. When a worker needs to work very close to a robot it is not necessarily flexible, because the robot has to be stopped before anyone can enter its work area. In this paper we have surveyed different technologies and concept solutions to make man-robot cooperation more flexible,

---

Please use the following format when citing this chapter:

Salmi, T., Marstio, I., Malm, T., Laine, E., 2008, in IFIP International Federation for Information Processing, Volume 260, *Micro-Assembly Technologies and Applications*, eds. Ratchev, S., Koelemeijer, S., (Boston: Springer), pp. 385-394.

mostly from a safety point of view. We do not address what the tasks are that humans and robots do, and the psychological aspect is omitted. [1]

2 Technologies

Safety is a core factor when bringing humans close to a robot. The working environment needs to be safe, yet the work should be flexible and effective. Many new technologies enable this. There are several ways of approaching the problem. Either the robot has to be made such that it cannot carry destructive force, or it has to be controlled by an intelligent controller and/or with safety sensors.

2.1 Robots

In recent years robot technology has undergone immense changes. Control and the user interface have become increasingly PC-based, and calculating power has multiplied. Robot structures have become more monolithic due to higher payload and accuracy demands. At the same time, ways of using robots have changed and therefore safety issues have become more important. Human-robot cooperation and research into it have promoted greater interest in safer robot structures and controlling technologies.

One approach to a safer small robot is the Neuronics Katana robot. The robot itself weighs only 4.3 kilograms and has a payload of 500 grams. It is shaped in such a way that sharp edges and the danger of compression are eliminated. The remaining protruding shapes are protected by bumpers. Safe structures, together with its low weight and slow speeds, are enough to make this robot safe. It can be mounted directly on a desktop without discrete safety devices or safety measures. This robot is a fairly good example of safety achieved simply through the design of the robot arm. The robot is used in real-life precision assembly solutions, where a human is doing the most demanding tasks while the robot performs the monotonous and simple ones. Combining the good characteristics of human and robot, higher production is achieved. [2]

Several approaches give the same result: a light robot that cannot seriously injure a human. Several years ago, Kuka Robotics presented a small robot, the KR 3 SI, which is coated with plastic foam to absorb the kinetic energy in a collision. Inside the foam there is also a proximity sensor, which detects the vicinity of hands or other body parts. When the operator's hand comes close to the robot, the machine stops immediately and continues moving as soon as the hand is moved away. The design of the tool flange also takes safety into account; the flange includes a magnetic switch that flips down if too much force is applied to the tool. [3]

A slightly different approach to safer robotics is the German Aerospace Centre DLRs' Light-Weight Robot (LWR). Compared to the simply structured Katana, the LWR is packed with technology. Their similarity is their light weight; the LWR weighs 13.5 kilograms but can handle over 15 kilograms. The DLR robot's every axis is equipped with force-torque sensors, which can detect collisions along the whole arm of the robot. Alongside the axle are the controlling and other electronics related to axle movement. Upper level control and the brain are in the external computer, which makes decisions based on sensory data. This robot brings to human-robot cooperation solutions extensive collision detection and, more important,

Fig. 1. Neuronics' small Katana robot [2]   *Fig. 2.* DLR's LWR III [4]

options after a collision has occurred. The robot can perform a traditional emergency stop, which is not suitable in a squeeze situation. In such a case, the robot can switch to the gravity compensation mode, in which it compensates for its own weight and can be moved by simply pushing it at any point of the arm. After a collision, the robot can also change its moving direction and withdraw. A certain limit for a force that the robot directs at a collision can be specified. Due to fast servos and sensors, the collision force stays at the set value. [4]

2.2 Software-based safety systems

Reliable control of a robot is essential for safer robotics, especially in human-robot cooperation. To make that possible and feasible, several robot manufacturers have released robot controllers, which provide an auxiliary unit beside the normal robot controller components. The safety controller watches ceaselessly the orientation and speed of the robot and compares it to the programmed and calculated values. If there are any differences between the real-life and predicted movements of the robot, it is stopped immediately. The safety controller makes possible, for example, software-based limit switches, restricting movements in 3D space, safe reduced speeds and safe standstill in any orientation. All these new functions can be used for safety-related purposes. Even though current and even older robot controllers are able to perform some of these actions, they are not safe enough for occupational safety purposes.

2.3 Vision and range sensors

The usual way to ensure the safety of a robot cell is separation. The robot is fenced in, far away from humans. The only way to get near the robot is through a door, which is strictly monitored by safety devices. This approach, however, limits the use of the robots' full potential. Human-robot cooperation is virtually impossible to achieve when distances are small and only fences are used for safeguarding. Standardisation also set strict guidelines for human and robot separation prior to EN ISO 10218-1.

The laser scanner is a commonly known safety device. It is capable of making observations in 2D fields and monitors quite a large area near the dangerous device. Two types of detection zones are warning and danger fields. The warning field warns of an approaching human, and the device is immediately stopped if movement is detected in the stopping field. A few laser scanners can in theory replace the fences and all other presence-sensing devices such as safety light curtains. The downside of the laser scanner is its limited sensing area and only 2D detection field.

A 3D detection field in safety devices has become possible since Pilz GmbH introduced SafetyEYE™. This sensor and detection is based entirely on machine vision. It is meant for human body detection in a freely shapeable 3D field around the supervised machine. Figure 1 clarifies how this device is used. 3D detection is done by three cameras, which observe the vicinity of the robot from above, i.e. the ceiling. The operator can freely modify the size, shape and position of detecting fields within the area of surveillance. Decisions are made by two PCs, which double-check the sensor data with both hard- and software. Actual decisions and signals are made by a programmable logical unit, which can send safety and normal signals depending on the detections made. This sensor is accepted in several safety classes and will be widely available soon. [5]

Machine vision will very likely be the next widely used new technology in safety devices. Research is ongoing into new technologies that could one day be used in safety devices. The Photonic Mixer Device (PMD) is a new type of sensor able to measure distances natively. The device sends a modulated light pulse, the reflections of which are measured by CMOS technology. The sensor gives distance values directly. This technology still has many limitations and problems, such as difficulty detecting a missing signal. Thus it cannot be expected to be viable as a safety device in the near future. [6]

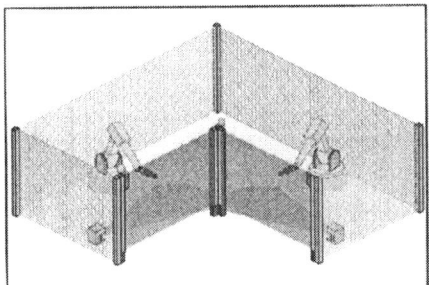 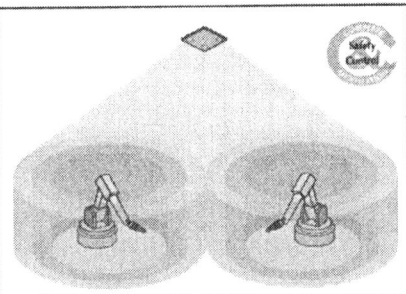

Fig. 3. Example of SafetyEYEs™ usage and advantages. [5]

3 Development of regulations

In Europe the Directives, and especially the Machine Directive, give the basic requirements for new machines. There are about 700 standards supporting the Directive. So-called harmonized standards interpret the requirements of the directives and set the basic level of safety. Harmonised standards do not need to be obeyed to the letter, but the level of safety should not be compromised. This may also need to be proved. Currently the most important safety standard for robots is "ISO 10218-1:2006. Robots for industrial environments -- Safety requirements -- Part 1: Robot". The standard was published 2006 and it gives some new aspects related to man-machine cooperation.

The safety standards aim to support development by determining physical targets for manufactures and integrators. The technical means of how to reach a physical target are open and therefore new technical solutions can appear. The requirements of new standards are quite strict, and in the near future standards will become more detailed and there will be more standards. Therefore the level of requirements will become harder since many details are already determined. This may inhibit technology development in the future, but it guarantees good and safe solutions for the industry. Detailed regulations often ease the work of the integrator, since then the safeguarding principles are well determined in standards.

Robot Standard EN ISO 10218-1 covers several features that are needed in man-machine cooperation.

Some features that the standard covers are as follows:
- Safety related parts of the control system are determined according to ISO EN 13849-1. The requirement specifies in relatively detailed level the safety features of the control system. The most common safety functions are mentioned in EN ISO 10218-1.
- Safety (reduced) speed. Basically, the speed 250 mm/s is considered to be safe enough against impact hazard, but not against crushing, shearing or puncture hazard. Anyway, the idea of the reduced speed is that a person has enough time to avoid an impact.

- High speed for special occasions. Special requirements for e.g. testing the robot at full speed.
- Safety distances according to EN 999 (ISO 13855). This is related to positioning the safety devices and it is possible to apply also for dynamic cases, in which the safety zone changes during operation. If the robot has a slow stopping performance the needed safety distance is also long.
- Hand guiding and related safety requirements are determined.
- The safety can be based on small forces against persons. The planar part of the robot may not initiate greater force than 150 N (static) or 80 W (impact) power against a person. The force is not much and the values can be reached only with relatively light robots and low speed. The stated values give guidelines that robot manufacturers can aim for.
- Soft limits are described. The robot can be safely under control with a programmable system (according to ISO 13849-1). This is a very important factor for safety controllers. Safety controllers make it more convenient, or in some cases make it possible, to change (safely) safety features during the operation. This can greatly help the close cooperation between a human and a robot. Such dynamic safety features must be designed into the system so that the operator cannot change it.

The standard does not cover e.g. equipment besides the robot arm or robots for other than typical industrial use. Considering man-robot cooperation, some requirements are quite hard for current robots (e.g. maximum impact energy against a person and realisation of soft limits), but the requirements set targets for robot manufacturers, and now two important safety targets are light-weight robots with quick performance in case of emergency and sophisticated safety controllers. [7]

4 Concept solutions

In this chapter we present five different ways to orchestrate man-robot cooperation. These are only concepts that give a general idea of the construction and in real cases many more aspects should be considered.

4.1 Robot out of workers' reach

One of the basic solutions to realize man-robot cooperation is to keep the robot out of workers' reach. This is the traditional way and basically it is the same idea as in the automation line, where there are manual workstations in between. The worker does his or her task and passes the product on to an automated machine or robot.

There has to be a courier between the robot and the worker. This can be a circular table as in Figure 4, or a conveyor with or without a palette. If a conveyor without a palette is used, some kind of mechanical positioning system or machine vision system must be used. In the simplest form the courier system can be a gravity-based mechanical slide. If the product needs to return from the robot, a returning system is needed which can also be based on any of the above solutions.

This way both the robot and the worker can carry out their tasks without interference. If there is a divergence in cycle times between the robot and the worker, the courier system can act as a buffer in unison. The downside of this solution is still the distance to the robot. Real cooperation is impossible and the structures may be very expensive.

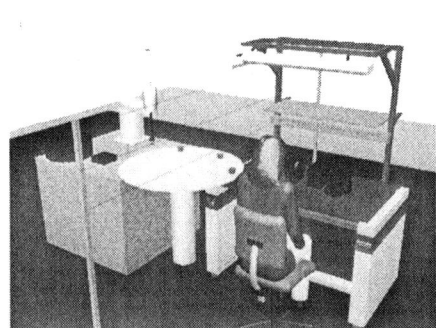 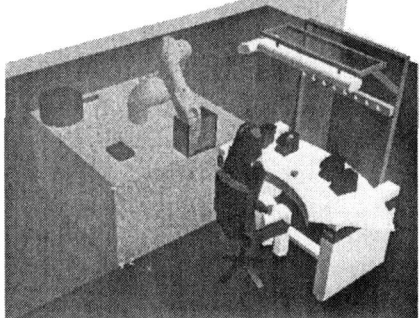

Fig. 4. Robot out of worker's reach. *Fig. 5.* Robot isolated from the worker.

4.2 Access to robot area via a safety-controlled hatch

If there is a plexiglass or other isolation that the robot cannot penetrate between itself and the worker, the robot can be brought closer to the worker. There should be a hatch in the partition, which both the robot and the worker can access so that exchange of material can take place. (Figure 5)

The hatch has to be such that only either the robot or the worker can access the handled material at any one time. This can be arranged mechanically so that only one side of hatchway opens at a time. The other way to do this without complex structures is to use a safety controller and a small simple door. The door is fitted with a safety sensor that sends a signal to the robot if the door is open. When the door is open, a pre-programmed area at least 500 mm from the worker's reach is blocked from the robot by the safety controller. If the robot happens to be in the restricted area when the door is opened, the robot is stopped or switched to safety speed by the safety controller. If the door is large enough for a person to enter, there should be a safety sensor, e.g. optical sensors active when the door is open, and when the sensors detect something the robot should be stopped.

This solution allows a person to work within very close range of the robot, and the latter solution allows humans do this without any massive safety structures.

4.3 Small and slow robot

There are very light and small robots that can cooperate with people without any external safety devices. If no great force or speed is needed in a production task and a robot is justified for the task, one of these light robots can be considered. In assembly tasks this robot can act as a third hand, material handler or precision task executor. (Figure 6)

Even if there is no need for man-robot cooperation, a light robot can save costs because there is no need to isolate it from people.

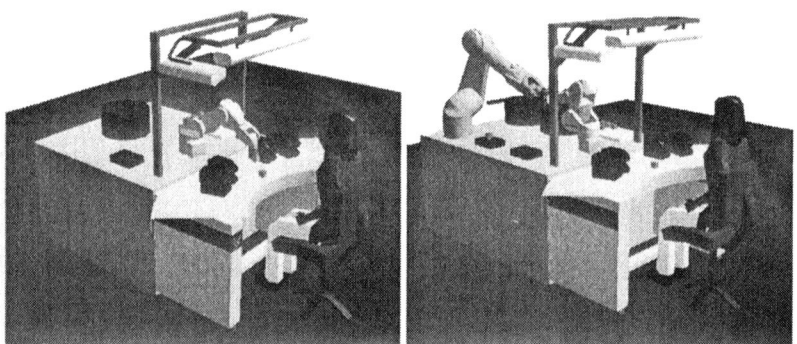

Fig. 6. Safe robot in cooperation with the worker   *Fig. 7.* Safe robot as a courier to a faster robot

4.4 Small and slow robot as a courier

If there is a need for man-robot cooperation and the speed of the safe and light robot is not enough, the light robot can act as a courier to a faster and stronger robot as shown in Figure 7. Even if the courier robot's only work is to act as a material transporter, often it is justified because the costs of a conveyor and its structures can climb pretty high.

In cases where one high-speed robot cannot keep up with the worker's cycle time, adding another (courier) robot that can also do production tasks may be one effective solution. Also when there is a dangerous machine these safe robots act as couriers between the worker and the machine.

4.5 Switching safety areas

If a robot's exact position and the surrounding workers' positions can be detected, it is possible to make a fenceless robot station where workers can work within the robot's reach. The robot's position is easy to obtain from the robot controller, and with a safety controller the position data is safety-classified.

A human's position's acquisition is more complicated. A safety system based on ceiling-mounted vision is in many ways good for monitoring a human's position, but there are still a few drawbacks. The response time in today's vision-based safety systems is around 300 ms, which gives quite long safety distances. But if a safety laser scanner is used which has a response time around 60 ms, safety distances are reasonable to about 500 mm depending on the robot's speed and stopping distance. Also safety pressure mats are very effective in defining a worker's position.

In a safety laser scanner there are warning and stopping areas that can be predefined and changed on-the-run. Safety laser scanners do not give the position of a detected object; consequently it may be necessary to have as many scanners as working areas. When a worker approaches a workstation, entering the warning zone of a laser scanner, the safety scanner sends a signal to the robot's safety controller. If the robot is not working at that workstation, the worker is signalled that he or she is allowed to enter the station, that the station's safety scanner's stopping area is diminished, and that area is removed from the robot's work area by the safety controller. If the robot is working at the station that the worker is entering, he or she is signalled to stop, but if he or she ignores the warnings and enters the safety scanner's stopping area, the robot is immediately stopped. A laser scanner-based safety area switching system is illustrated in Figure 8. In the presented solution, it should be taken into account that the worker does not climb on to the table.

Fig. 8. A laser scanner detects the worker and switches the robot's prohibited areas accordingly.

5 Discussion and conclusions

Ideally, from a safety point of view, robots would be aware of surrounding people's positions and would control their own movements and speed accordingly. In that case no extra fences would be needed and robots could work in a very limited space. However, since getting a worker's exact position in real time in safety-classified ways is not yet a reality, some compromises must be made.

By allowing software-based safety systems into the robot standard, designing these cooperative systems has become possible. Before safety controllers, the whole area around the robot was restricted from humans if the robot was not moving at safety speed. Now, safety controllers solve one problem, which is the robot's position and movements. If safety regulations keep on evolving, safety controllers will be able to create dynamic safety areas that surround the moving parts of the robot.

In addition to knowing the robot's position, the position of the surrounding people has to be known. The most promising solutions are vision-based safety systems. Different systems are being developed but the Piltz system is the only one on the market, and it still lacks several important features needed in man-robot cooperation. For example, the robot's working area has to be muted out from SafetyEYE™ because it cannot yet distinguish a robot from a human.

Even with today's technology it is still a challenge to build a seamless flexible man-robot cooperative unit. The costs of safety system constructions may become very high and other solutions might be more cost-effective. Nevertheless, cooperative tasks like teaching or controlling the robot by moving it by hand and working within the robot's reach are becoming more common. Also the worker's perspective has to be taken into account. Even though the working area of the worker would be completely safe within the robot's reach, without fences the psychological effect that a high-speed robot creates could at the very least lower work efficiency.

Safety should be kept in mind when programming a robot. Even though a human is not working within the robot's area, the robot's movements should be somewhat predictable. Also movements in extreme positions carry much more kinetic energy, i.e. stopping distances are greater and damage on collision is more severe.

References

[1] Nof, Shimon Y. (editor) (1999). Handbook of Industrial Robotics, 2nd ed. John Wiley & Sons. 1378 pp. ISBN 0-471-17783-0.
[2] Neuronics AG, "Neuronics AG – Katana" Internet site, available: http://www.neuronics.ch/cms_en/web/index.php?id=244&s=katana
[3] Peter Heiligensetzer, Sichere Mensch-Roboter-Kooperation, 33. Sitzung des VDI/VDE-GMAFachausschuß "Steuerung und Regelung von Robotern", 2003
[4] German Aerospace Center (DLR), "DLR – Institute of Robotics and Mechatronics . Light Weight Robots", Internet site, available: http://www.dlr.de/rm-neu/en/desktopdefault.aspx/tabid-3803/
[5] Mensch und Automation - Die Zeitung für Kunden der Pilz GmbH & Co. KG – Ausgabe 4/2006, Pilz GmbH&Co. KG, 2006, Pp 1 – 2
[6] T. Möller, H. Kraft, J. Frey, M. Albrecht and R. Lange, Robust 3D Measurement with PMD Sensors, in Proceedings of the 1st Range Imaging Research Day, 2005.
[7] ISO 10218-1:2006. « Robots for industrial environments -- Safety requirements -- Part 1: Robot » The International Organization for Standardization (ISO). 27 p.

INTEGRATION OF DESIGN AND ASSEMBLY USING AUGMENTED REALITY

Juha Sääski[1], Tapio Salonen[1], Mika Hakkarainen[1], Sanni Siltanen[1], Charles Woodward[1], Juhani Lempiäinen[2]

[1]VTT Technical Research Centre of Finland
[2]Deltatron Ltd

Juhani Lempiäinen
Deltatron Ltd
Soidintie 14, Helsinki
FI-00700 Helsinki, Finland
Tel: +358-9-3452660
Fax: +358-9-3454297
jle@deltatron.fi
www.deltatron.fi

Juhani Lempiäinen
Deltatron Ltd
Soidintie 14, Helsinki
FI-00700 Helsinki, Finland
Tel: +358-9-3452660
Fax: +358-9-3454297
jle@deltatron.fi
www.deltatron.fi

Abstract This paper presents a methodology and a system for augmented reality aided assembly work. We concentrate in particular on the requirements on information processing and data flow for implementing augmented assembly systems in real life production environments. A pilot case with an augmented assembly task at the Finnish tractor company Valtra is described.

Keywords augmented reality, assembly work, assembly instruction, PDM, CAD, design for assembly

1 Introduction

This paper presents augmented reality (AR) system architecture for assisting assembly work by visual information superimposed on the physical assembly parts. Such AR methods are particularly well suited for complex, short manufacturing series or in a customized production factory environment. Each individual product may have a slightly different configuration: the order of assembling parts may vary for different products and/or the number of phases in the assembly line may be large. The traditional approach is to use assembly drawings (blueprints) and possibly instruction manuals with guiding pictures to describe the content of each work task. As the assemblies become even smaller, the need for guiding the worker with all available tools becomes increasingly important. The AR system can also reduce assembly times, accelerate learning of the assembly tasks and provide more quality assurance to the factory floor.

The emphasis in this paper is on content authoring for AR. Our ultimate goal is to automate the augmented assembly content creation pipeline, starting from sales/ordering systems up to product specific assembly instructions displayed on mobile hardware to the workers at the production line. The specific focus in this paper is on integrating design for assembly (DFA) software tools to the design systems (CAD/PDM/PLM). The main objectives are to develop a content creation process from design systems via DFA to AR based assembly instructions, and to show the natural links between DFA work and AR content authoring.

The organization of this paper is as follows. First we present a brief overview of related work on AR assisted assembly research. A brief overview of augmented reality technology is provided next. The main body of the article then discusses 3D CAD and assembly information data flow to content creation and authoring for AR. The two last sections present our first experiments and results, as well as conclusions of the work so far.

2 Related work

Augmented reality based technology is closely linked to different kinds of display systems, especially so-called head-mounted displays (HMD). The HMD display technology has been available for many years (Furness 1969), the traditional application being in military field. Examples of some more recent display devices are shown in Fig. 1.

Fig. 1. Examples of augmented reality display devices: hand held units and see through HMD.

One of the first projects dealing with manufacturing was launched by Boeing in which Claudell and Mizell (Claudell and Mizell 1992) described the challenges in aircraft manufacturing. "Much of the information is derived from engineering designs for parts and processes stored in CAD systems. In many cases, this information comes to the factory floor in the form of assembly guides, templates, drawings, wiring lists, and location markings on sheet metal". "A significant source of expense and delay in aircraft manufacturing comes from the requirement to mirror changes in the engineering design in the guides, templates and so on used to control the manufacturing process. If manufacturing workers were able to directly access digital CAD data when performing manufacturing or assembly operations, several sources of expense and error would be eliminated". Their concept was to provide a "see-thru" display to the factory worker, and use this device to augment the worker's visual field of view with dynamically changing information. In their demonstration system the challenges were to align real and virtual objects with each other and also the capacity of portable computing unit allowing only representing simple graphics in real time. Today we are witnessing huge improvement in mobile computing units and graphics; also the recent HMD development is now led by gaming industry with the so called i-glasses, providing good full screen picture resolution with reasonable pricing.

Recently ARVIKA (www.arvika.de) was a large industrial AR project in Germany 1999-2003, and it consisted of an industrial consortium led by Siemens. The goal was to develop prototypes of portable and fixed location AR systems for development, production and service in the automotive and aircraft industry, both in systems and mechanical engineering. Participants included: automobile manufacturers such as Audi, BMW, DaimlerChrysler, Ford and Volkswagen; aircraft manufacturers such as EADS and Airbus; equipment manufacturers such as MicroVision, Physoptics and Zeiss. ARTESAS (www.artesas.de) (2004-2006) aimed at the exploration and evaluation of augmented reality base technologies for applications in industrial service environments. The project was based on the results of the ARVIKA project. On-going AR research projects for manufacturing industries in Europe are e.g. SmartFactory (www.smartfactory-kl.de), Wearit@Work (www.wearitatwork.com) and Ultra (www.ist-ultra.org). Augmented reality is also explored in many universities and research institutes in U.S.A and Asia.

3 Augmented reality technology

Augmented reality systems combine digital information and real world in a way that a user experiences this as a whole. An important property is especially that virtual objects are located to the right place and position. AR system follows dynamically the user's point of view and keeps virtual objects aligned with real world objects. The basic components in AR applications are a display, a camera and a computer with application software (Azuma et al 2001). Various different kinds of hardware can be used to implement this, e.g. camera phones, PDAs, laptops, HMDs etc.

Typically, a camera is attached to the display device which shows the real time camera view "through" its screen. To determine the relation between real and virtual worlds, computer vision techniques are used to find (track) a marker in the camera image and determine the position and the pose of the camera relative to it. Once the position and the pose of the real camera are known, a digital 3D-model can be exactly overlain on (or near) the marker in the camera image. Thus the user experiences video see-through augmented reality, seeing the real world through the real time video with virtual models. Figure 2 summarizes the tracking and display process.

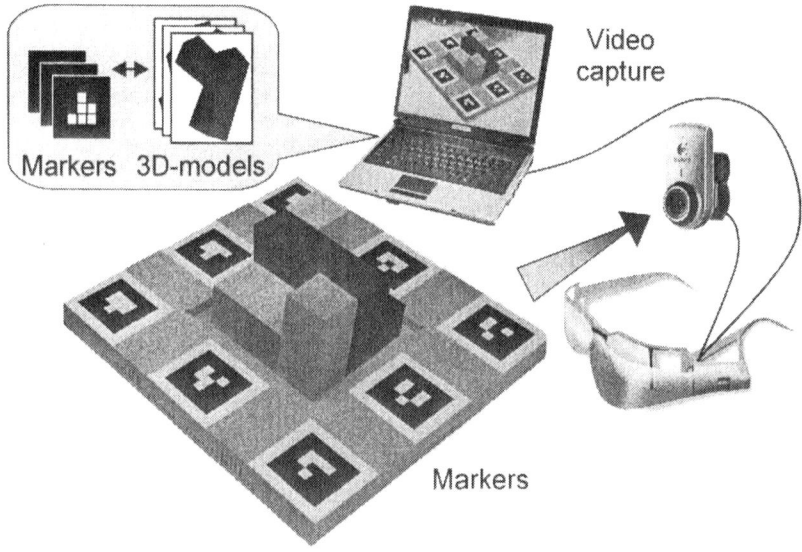

Fig. 2. The augmented reality tracking and display process: the computer-generated graphical augmentation is integrated in the user's view of the real world.

AR has been studied for years, the focus being in hardware technology and not the usability. Thereafter the AR systems have been clumsy, that is heavy and big. However, the rapid development of mobile devices (handhelds) has lead to small units with enough processing capacity and long lasting batteries to enable lightweight mobile AR systems. Recently PDAs (Pasman and Woodward 2003), camera phones (Henrysson et al 2005, Rohs 2006) and miniature PCs (Honkamaa et al 2007) have been successfully used in AR applications. Mobile augmented reality is considered one of the most promising emerging technologies (Jonietz 2007) "most likely to alter industries, fields of research, and even the way we live".

4 Augmented Assembly

One of the biggest challenges for utilisation of AR technology in manufacturing industry is re-use of existing product data. Product data is stored in PDM/PLM systems and these systems include all relevant product data (3D geometry, product structure, simulation results, part fabrication plan, assembly plan etc.). The information should be retrieved from PDM/PLM systems into forms suitable for AR display as automatically as possible; cf. (Matysczock and Ebbesmeyer 2004). Besides finding suitable 3D data representations and conversion methods for AR use, also the assembly related guidance information (annotations, animations etc.) should be considered. Our approach is to exploit ISO 10303 (ISO 10303) definitions for product data representation. However, there is a lack of semantics dealing with representing with AR type of information. Heilala (Heilala et al 2007) have reviewed the latest possibilities integrating information in manufacturing industry.

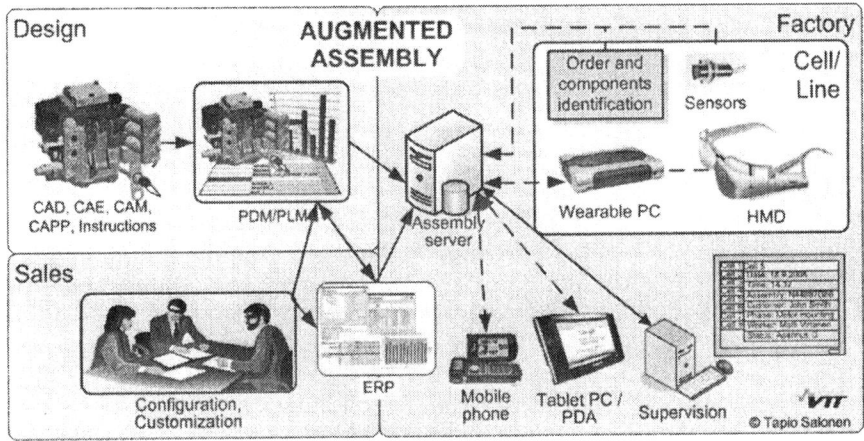

Fig. 3. The proposed augmented reality based information processing architecture.

AR based manufacturing instructions affect different information processing systems of the company in many ways. In order to implement augmented reality based assembly instructions we propose the information processing architecture shown in Fig. 3. The majority of the product data is created in design systems and stored to PDM/PLM system. Sales configure and customise the individual product, for instance. That is also stored to the PDM/PLM system, from which AR based instructions will be created as described below. The ERP system controls production planning, and assembly server, which manages augmented instructions to the worker.

Figure 4 shows the proposed methodology how the AR instructions are created from the product's 3D model. First the CAD model is exported to standard STEP (ISO 10303 1994) format file, that includes the product structure and 3D models of the parts. Because of the designer's preferences, company specific part libraries and features of typical CAD systems the generated product structure usually does not conform to the real parts to be assembled in assembly line. The design structure of the product is not equal to the manufacturing assembly structure of the product. Therefore, the assembly structure (i.e. the definition of the assembled parts) and work phases have to be re-configured.

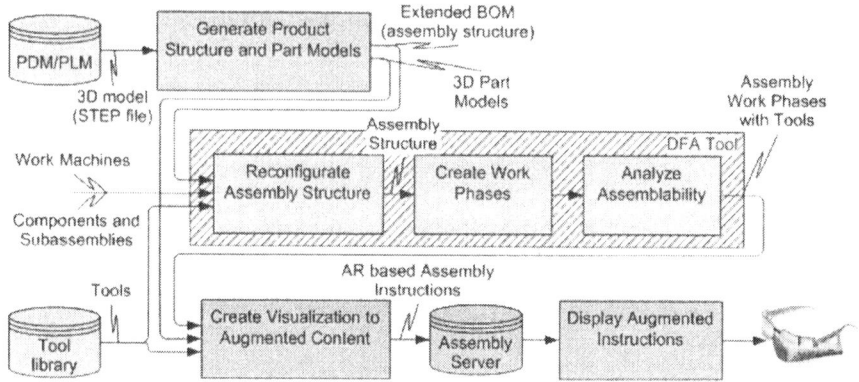

Fig. 4. Creation of AR based assembly instruction using the 3D product model.

As described in (Salmi and Lempiäinen 2005) assemblability analysis can be a very effective tool in authoring the assembly description for the worker. This includes 1) the right and natural component order in a subassembly, 2) use of specific assembly tools for the components, and 3) the special actions needed to be carried out at this point. All these three characteristics are the key elements in the traditional assemblability analysis and suddenly they are equally the main interest for AR authoring in assembly work.

The right assembly order and instructions and hints are generated normally in the late prototype/pre-production phase of the product development. Now with these described AR tools this production documentation can be generated in an early phase and with automated means. This reduces the need for extra written assembly instruction sheets and documented pictures/photos in which assembly sequence these instructions will be applied. The other benefit is the semi-automated assemblability analysis of the product. This analysis will point out any special problematic assembly sequence where some revised constructions can be applied.

In this study the DFA-Tool® assemblability analysis software is used to carry out the assembly analysis task. The extended Bill Of Materials (BOM) that includes design structure, cf. Fig 4, will be imported to DFA-Tool. The analysis will be carried out there with the reconfiguration the assembly structure, the generation of specific actions, tools and right component assembly order. Fig 5 shows a view of DFA-Tool. Finally the assembly work phases and the 3D part models are exported for augmentation, cf. "Create Visualization to Augmentation Content" in Fig 4. The visual instructions and possibly some text notes are generated using 3D models of assembly tools and components.

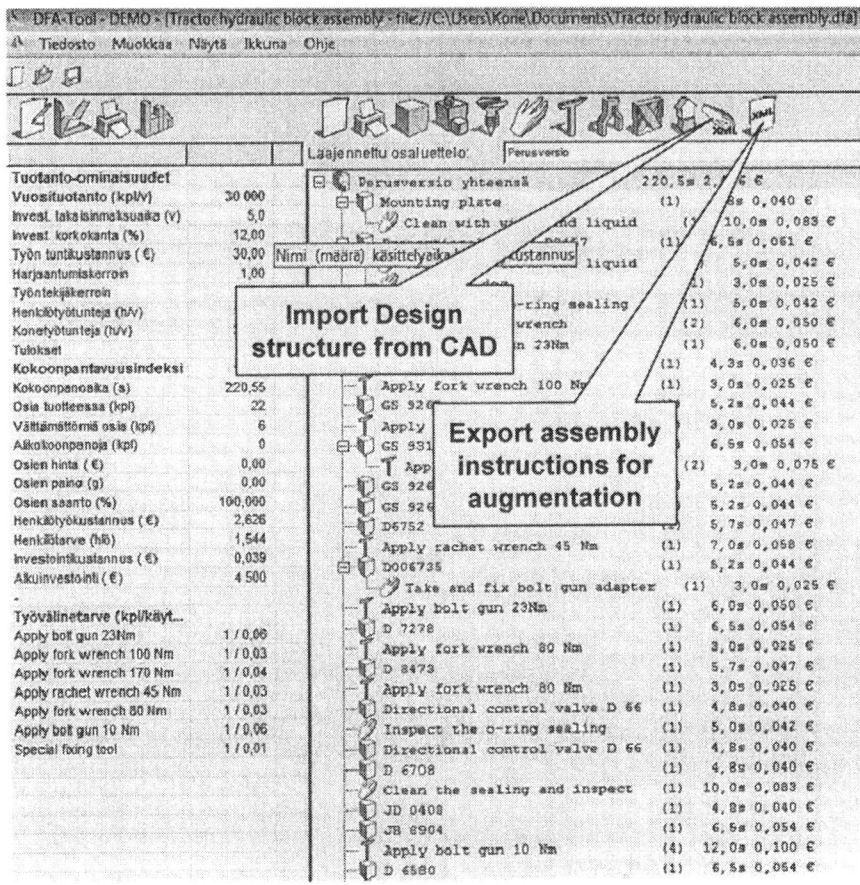

Fig. 5. Augmentation content of a hydraulic block created by DFA analysis tool.

5 Experiments and results

In the preliminary tests we have used a simplified assembly task that simulates real assembly work. The task is to put parts in a 3D puzzle box as shown in Fig. 2. At each stage the system shows to the user which part to pick and animates how to place it in the box. The user follows the instructions and puts the parts piece-by-piece according to augmented instructions at the right place and in the right order. The task is real-enough using the actual devices (HMD, camera, etc) and it also serves for our preliminary testing of the content creation components and data flow. User experiences with this test bed system have been most encouraging (Salonen et al. 2007, Sääski et al. 2007); cf. also similar work by (Pathomaree and Charoenseang 2005).

Our industrial pilot case focuses on the assembly of a tractor accessory's power unit at the Finnish tractor factory Valtra Plc. The CAD model of the finished block, our case study, is shown left in Fig. 6. Markers on the rotating metal plate are used for tracking the physical unit's orientation. Assembled parts are shown to the worker task by task according to the work phase instructions, overlain on the physical unit. We are investigating alternatives for display devices, from static and hand held PCs to HMDs, as well as multiple camera systems for enhanced accuracy; cf. (Sääski et al. 2007). The preliminary system already includes most of the above described components for automatic data flow from CAD systems to DFA and augmentation; however integration of sales and ERP remains for future work.

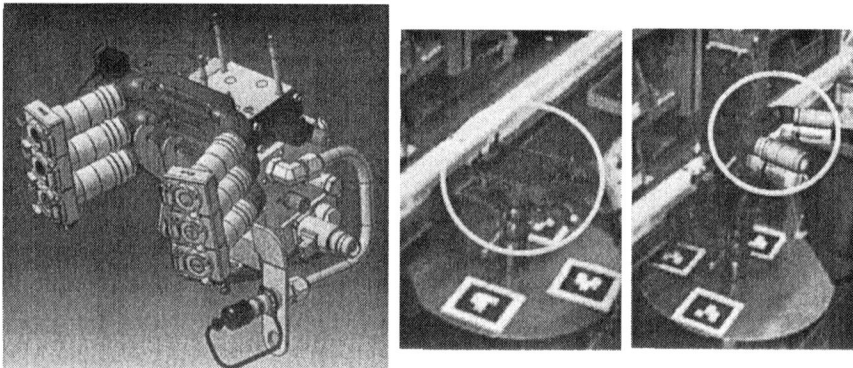

Fig. 6. 3D CAD model of finished hydraulic block assembly (Courtesy of Valtra Plc). On the center the first part is assembled to the fixture. On the right the real parts are assembled, and the worker can see the next component as a virtual part.

6 Conclusions

AR-technology (hardware and software) and price have matured to the level that manufacturing industry is starting to seek applications in real factory environments. There have been large industrial research projects with good and promising results. AR-technology has been studied mostly how to display assembly information, but only a few studies are dealing with integration of AR related information with current PDM systems used in industry. In this paper we show a method to retrieve assembly knowledge to AR-system. Our approach relies on usage of ISO 10303 (known as STEP), so that smooth information flow can be realized. Our preliminary tests with 3D puzzle show that AR is promising technology at least in laboratory environment. In this on-going project the industrial case is used to test our assumptions with the real case in the factory, and with the real assembly tasks and assembly workers.

As assembly constructions in the future will be even smaller in size and have even more functions, it is important that right assembly tools and assembly guidance are immediately available for every assembly operator in synchronization when a new product version is created. We feel these described techniques with the fast development of light weight wearable display devices will make a huge improvement for the quality of both the augmented assembly constructions and their physical counterparts as well.

DFA-Tool® is a registered trade mark of Deltatron Ltd, see www.dfa-tool.eu

Acknowledgements

This contract based research project is funded by Tekes, The Finnish Agency for Technology and Innovation through its manufacturing technology programme SISU 2010. Key industrial companies also cover partially the costs of this work. The research team thanks for the contributions.

References

1. Azuma R, Baillot Y, Behringer R, Feiner S, Julier S, and MacIntyre B, (2001). Recent advances in augmented reality, IEEE Computer Graphics and Applications, 21 (2001), no. 6, 34—47, issn 0272-1716.
2. Caudell TP, Mizell DW, (1992). Augmented reality: an application of heads-up display technology to manual manufacturing processes. Proceedings of the Twenty-Fifth Hawaii International Conference on System Sciences. Volume ii, page(s):659 - 669
3. Furness T, (1969). Helmet-mounted displays and their aerospace applications. National aerospace electronic conference, Dayton, OH, U.S.A.
4. Heilala J, Helaakoski H, Peltomaa I (2007). Smart Assembly - data and model driven. Fourth International, Precision Assembly Seminar, IPAS'2008. Chamonix, France, 10-13 February 2008. Submitted for publication.
5. Henrysson A, Billinghurst M, Ollila M, (2005). Virtual object manipulation using a mobile phone", Proc. 15th International Conference on Artificial Reality and Telexistence (ICAT 2005), Dec 5th 8th, 2005, Christchurch, New Zealand, pp. 164-171.
6. Honkamaa P, Siltanen S, Jäppinen J, Woodward C, Korkalo O, (2007). Interactive outdoor mobile augmentation using markerless tracking and GPS. To appear in Proc. VRIC – Laval Virtual 2007.
7. ISO 10303-203 (1994). Industrial automation systems and integration -- Product data representation and exchange -- Part 203: Application protocol: Configuration controlled 3D designs of mechanical parts and assemblies. Geneva, Switzerland: ISO. 581 p
8. Jonietz E, (2007). Augmented Reality: Special Issue 10 Emerging Technologies 2007, MIT Technology Review, March/April 2007.

9. Matysczock C, Ebbesmeyer P, (2004). Efficient creation of augmented reality content by using an intuitive authoring system, Proc. DETC'04, Salt Lake City, Utah, USA (2004), pp. 53-60.
10. Pahl G, Beitz W, (2003). Engineering Design. A systematic approach. 2nd edition. London: Springer Verlag. 544 p.
11. Pang Y, Nee AYC, Ong SK, Yuan ML and Youcef-Toumi K, (2006). Assembly feature design in an augmented reality environment", Assembly Automation Journal, Vol. 26, No. 1, 34-43 (2006).
12. Pasman W, Woodward C, (2003). Implementation of an augmented reality system on a PDA, Proc. The Second IEEE and ACM International Symposium on Mixed and Augmented Reality (ISMAR 2003), Tokyo, Japan, October 2003, pp. 276-277.
13. Pathomaree N, Charoenseang S, (2005). Augmented reality for skill transfer in assembly task, 2005 IEEE International Workshop on Robotics and Human Interactive Communication, pp. 500-504.
14. Rohs M, (2006). Marker-Based Embodied Interaction for Handheld Augmented Reality Games, Proceedings of the 3rd International Workshop on Pervasive Gaming Applications (PerGames) at PERVASIVE 2006, Dublin, Ireland, May 2006
15. Salmi T, Lempiäinen J (2006). First stemps in Integrating Microassembly Features into Industrially Used DFA Software, Proceedings of the IFIP Third International Precision Assembly Seminar IPAS 2006 , Bad Hofgastein, Austria, 19-21 February, 2006, pp 149-154.
16. Salonen T, Sääski J, Hakkarainen M, Kannetis T, Perakakis M, Siltanen S, Potamianos A, Korkalo O, Woodward C, (2007). Demonstration of Assembly Work Using Augmented Reality, Proceedings of the ACM International Conference on Image and Video Retrieval, 2007, Amsterdam, The Netherlands, 9-11 July 2007, pp. 124-126.
17. Sääski J, Salonen T, Siltanen S, Hakkarainen M, Woodward C, (2007). Augmented Reality Based Technologies For Supporting Assembly Work, Proceedings of 6th Eurosim Congress on Modelling and Simulation. Vol.2, 2007, Ljubljana, Slovenia, 9-13 September 2007.

CONCEPT FOR AN INDUSTRIAL UBIQUITOUS ASSEMBLY ROBOT

Juhani Heilala*, Mikko Sallinen[§]

*VTT, P.O. BOX 1000, 02044 VTT, Finland,
Juhani.Heilala@vtt.fi
[§]VTT, P.O. BOX 1100, 90571 Oulu, Finland
Mikko.Sallinen@vtt.fi

Abstract We present a concept of a ubiquitous industrial robot. It has been defined to consist of technologies of artificial intelligence, ubiquitous computing, sensor network and industrial robots. The advantages compared with current intelligent robots are that they are more autonomous and they have cognitive skills. A ubiquitous robot is interoperable with all sensors, computers and other devices around it. One important factor is the natural interaction with humans. The ubiquitous approach is more common in consumer applications but still new in the industrial environment, even if many research efforts are being made. We also present an example of a ubiquitous robot: the isle of automation.

Keywords ubiquitous industrial robot, concept

1 Introduction

Manufacturing is facing growing challenges, and the global competition business environment is rapidly changing and filled with uncertainties. Typically, production is a high mix and low volume of customised personalised products. Production systems are also getting more complex due to advanced manufacturing technologies. Small lot sizes, versatility of product variations and shorter life cycles all set their own requirements. Production equipment must be fast to install and re-configurable, and production set-up times must be short. In the flexible production of small lot sizes, human workers with their problem-solving abilities and cognitive capabilities are still the single best way to provide the required flexibility, adaptability and reliability. Cost competition and the need for productivity are forcing the trend towards automation. Another incentive for automation is the future likelihood of a lack of qualified personnel. The complexity of products sets further limitations and constraints on both automation and the skills and capabilities of human operators. Especially product miniaturisation has precision requirements that mean that some tasks are beyond the reach of the human worker.

The Preliminary Smart Assembly roadmap defined by the NIST workshop [1] identifies that empowered, knowledgeable people, a multi-disciplined, highly skilled workforce empowered to make the best overall decisions and working collaboratively with automation in a safe, shared environment for all tasks is part of the solution. The Manufuture Strategic Research Agenda 2006 [2] and the Technology Platform on Robotics, EUROP Strategic Research Agenda 2006 [3] both identify adaptive manufacturing as including the field of automation and robotics, robots as assistants of humans, hybrid assembly, and service robots. Adaptive manufacturing includes new automation solutions through the integration of new methods of cognitive information processing, signal processing and production control by high-speed information and communication systems.

In collaborative human centred automation, the main objective is to support human workers with qualified tools to increase productivity. The approach combines human creativity, intelligence, knowledge, flexibility, and skills as well as the advantages of sophisticated technical systems and tools, industrial cognitive robotics and efficient use of ICT (Fig 1). There are both physical and cognitive tasks that can and should be automated, as shown in Fig. 1. One key issue is to define the level of automation as discussed elsewhere [4].

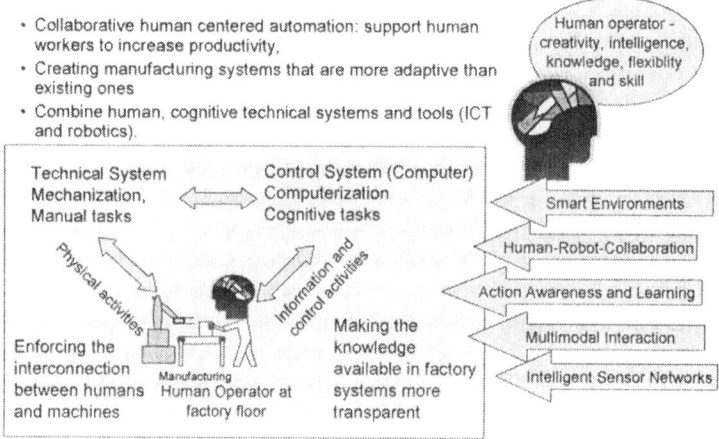

Fig. 1. Collaborative human-centred automation

A clear trend is emerging in the field of autonomous robotics, towards the integration of robotics with ambient intelligence and ubiquitous computing. This paper presents a concept for ubiquitous industrial robotics as well as future research needs. The concept combines an autonomous smart robot, networked smart production equipment, sensor networks and human collaboration in the intelligent assembly space.

2 Concept for ubiquitous industrial robot workcell

In recent years, due to the emergence of ubiquitous computing technology, a new class of networked robots called ubiquitous robots has been introduced. The networked robotic devices in smart environments, the ubiquitous robotics, provide a radically new way to build intelligent robot systems in the service of people [5]. Correspondingly, ubiquitous robot systems require a new way of thinking on the part of the designers of such systems, as well as new tools to build them.

The Ubiquitous Industrial Robot Workcell is our conceptual vision of ubiquitous service robots that provide the services the user needs on the factory floor. This kind of robotic system needs to be interoperable with sensors, sensor networks and devices in its current service environments automatically, rather than statically pre-programmed for its environment. Therefore artificial intelligence is needed [5]. The robot system must be autonomous, capable also of working with a human operator in the same smart or intelligent space (see Figs. 1, 2). In the industrial environment, since the environment is to some degree defined, fixed and known, a lot of autonomy can be built into the robot workcell even if many environment-, process- and task-related inaccuracies or uncertain parameters remain.

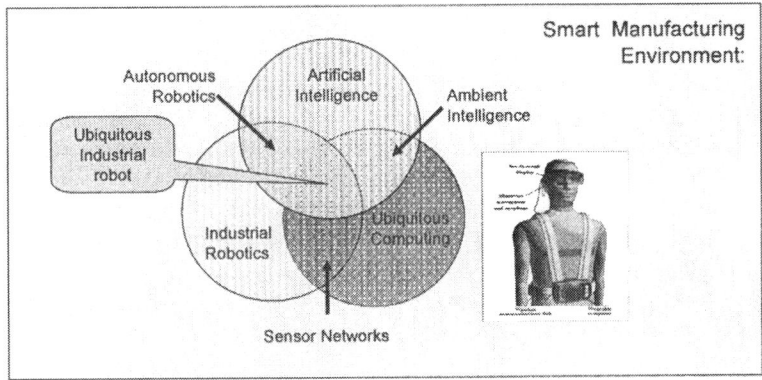

Fig. 2. The ubiquitous robot consists of three areas: Artificial intelligence, ubiquitous computing and industrial robotics, adapted from [5].

Typically, users' main tasks in automated, e.g. robot-based manufacturing systems are: configuring and teaching the system, monitoring and fine-tuning the system, charging and unloading, dealing with unplanned events, and communication with other team personnel. The general experience from the past is that the more advanced, complex and smarter the system is, the more critical is the role of the human operator [7], [8], [9]. Smart systems are vulnerable to failure and, therefore, they are even more dependent on human intervention than less intelligent ones. Usually the situation is the same if off-line programming is used in the system, especially when the lot sizes are small. Smart tools and artefacts should support users' normal behaviour, but more important is also managing an unpredictable situation; for example in the case of human error, the tools should be robust enough while being smart. They should help people manage complex situations, make good decisions, and act in a reasonable way.

A key element in flexible agile manufacturing is the ability to easily configure and program the systems concurrently, taking into account the relevant human factors and considering the change of systems and human activities in order to provide sufficient affordance. Several approaches have been developed to teach robots to perform complex actions. Some promising new approaches are based on learning by demonstration and learning by instruction paradigms, in which users first demonstrate the to-be-learned task or procedure [10, 11, 12, and 13]. Important issues in this approach are: 1) understanding the procedure and the human intention; 2) learning and representing the procedure; and 3) mapping and executing the procedure with the system [13]. Knowledge of human-robot interaction plays a key role in the development of these systems.

2.1 ICT for manufacturing and human operator

One consequence of the deep penetration of ICT (Information and Communication Technology) into daily life is migration from the conventional factory floor to intelligent manufacturing environments built around the AmI (Ambient Intelligence) paradigm. That is, workplaces with emphasis on greater user-friendliness, more efficient service support, user-empowerment, and support for human interaction.

AmI has been defined [6] as "A concept in IST that presents what should come beyond the current "keyboard and screen" interfaces to enable ALL citizens to access IST services wherever they are, whenever they want, and in the form that is most natural for them. It involves new technologies and applications both for the access to, and for the provision of applications and services. It calls for the development of multi-sensorial interfaces which are supported by computing and networking technologies present everywhere and embedded in everyday objects. It also requires new tools and business models for service development and provision and for content creation and delivery." AmI is very similar to ubiquitous computing, since ubiquitous refers to "anywhere any time".

There are many needs for communication, as well as available technologies, in human–robot–smart artefact or smart manufacturing environment collaboration. Firstly, in the communication aspect, there is a two-tier communication environment: body area communication and operator-infrastructure communication. Body area communication can be carried out using RFID type communication, NFC (near field communication) or other short range communication technologies. In operator-infrastructure communication, WLAN, Bluetooth or GPRS communication can be used. It is important use as standard technologies as possible.

Communication can have many purposes. Sensor data can provide context-awareness and thus integration with manufacturing information systems, provide task support for operators, and the system can adapt to human operator needs. Operator hand movements, body gestures other measurements of physical actions can be used for the novel user interface. We can use either optical tracking, i.e. camera systems, or mechanical tracking with wearable motion sensors, force sensors or similar.

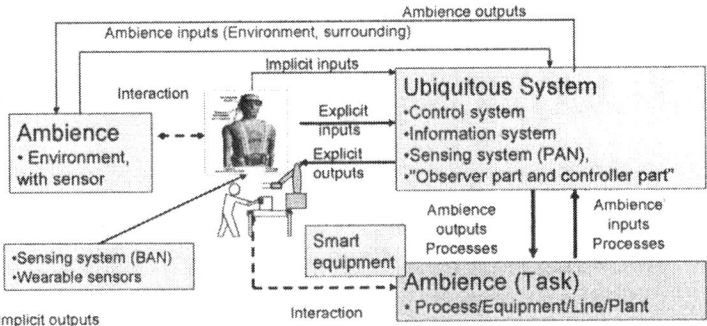

Fig. 3. A wearable, embedded computing system supports a human operating with a ubiquitous robot.

In the most advanced system, human operators carry a wearable, embedded system with sensors and local computing (see Fig. 3). The wireless body area network and sensors with localization and indoor positioning capability provide context and task awareness information even in a sensor-less environment. The industrial robot workcell is a smart environment; there are a lot of sensors for monitoring the manufacturing process and controlling the equipment.

3 State of the art

There are many initiatives for the development of ubiquitous robots, but they are mainly targeted for other areas than production [14, 15], such as service robots for the home and robot companions. Cognitive robotics is also a topic of EU focused research, including manufacturing applications. There are several EU-level large-scale development projects focusing on collaborative industrial robots, SMERobot (www.smerobot.org) and PISA-IP (www.pisa-ip.org), just to mention a few. Industrial AmI is also being developed in several EU projects, but these are not dedicated to robotics.

The paradigm of the Cognitive Factory has been outlined by Zäh [16]. The Cognitive Factory is a factory environment with its machines, robots, storages, planning processes and human workers, that is equipped with a sensor network and an IT structure to allow the resources and processes to perceive what they are doing, enable them to control themselves and plan further actions in cooperation with other machines and human workers.

Acording to Vassos [17], cognitive robotics is a design paradigm that, when applied to robotic agents (in contrast to software agents), involves taking care of issues that lie in several different fields of research and applications, including:

1. The mechanical part of the robot that is responsible for movement and affecting the environment (e.g. mechanical arms, body, motors).
2. The software and hardware part responsible for getting meaningful information from the environment in which the robot is situated (e.g. the hardware and

information processing software for doing feature extraction from visual images and sound).
3. The software part responsible for the representation of the environment and the way that the robot can interact with it (e.g. a logical specification of the properties of the environment as well as how these are affected by the available actions that the robot can perform).
4. The software part responsible for the specification of the task that the robot should do, based on the previous representation.
5. The software and hardware part that makes use of the representation of the environment and the specification of the task that the robot should do (numbers 3 and 4) in order to compute the behaviour of the robot at any given moment.
6. The software and hardware part that provides the interface between the reasoning component (number 5) and the actual sensors and actuators in the environment (numbers 1 and 2).

These guidelines can also be used for ubiquitous industrial robot system design.

4 Example of a ubiquitous robot — the automation island

An autonomous robotic workcell can be called a "production island". This island of automation provides a flexible way of producing short series production [18]. The concept of automation islands is modular and realises highly flexible and controllable robotised automation, and includes software components and hardware components operating synchronously. The basic element of the automation island is an industrial robot equipped with different kinds of sensors and auxiliary devices optimally combining mechanics, sensor technology and software. This gives high-level flexibility in terms of programmability, reusability and price. The production system easily adapts to new products or product variants and to the deviation in work pieces.

The concept is an optimal combination of three technology elements: mechanics, sensor technology and software, as outlined in Fig. 4; the fourth element is the human operator and related interaction technology. Mechanics is utilised with clever design principles by applying low cost solutions whenever possible. The automation island is a part of the material flow process, from order to delivery. It is also part of the process from data file to work program and finally to the finished part. As a part of different processes and information flow, the automation island has appropriate contact points so that it is able to communicate and receive material and information. In addition, data acquisition presents new possibilities when open interfaces are offered up to sensor level.

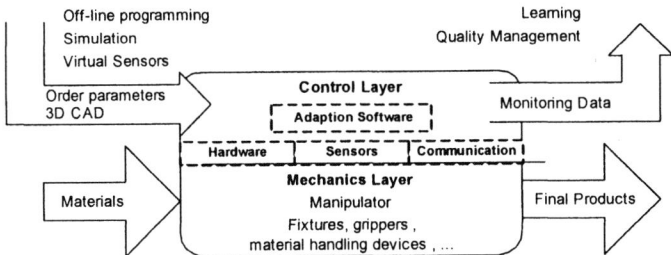

Fig. 4. Modular structure of the island of automation [18]

Interaction between the human operator and the island can be carried out in several ways. One solution is that the isle is making the measurements and provides processed results to the user, who makes the final decision. This approach provides an optimal fast solution to the control of production. Most of the need is required in the programming phase where information for manipulating a new object is fed to the system. In the future, robot systems will also include cognitive skills, which means that they learn from the tasks they are doing. Flexibility can be achieved using a multi-sensor system for observing the target object, advanced programming techniques, and a component-based approach for solving the application-related problems.

5 Development needs

A manufacturing environment where workforces are surrounded by a collection of reconfigurable production components (physical agents) that include mechatronics, control and intelligence (intelligent sensors and data processing units, autonomous, self-tuning and self-repair machines, intuitive multi-modal human machine interfaces, etc.) is an intelligent assembly space. In these circumstances, the challenge is to develop production automation and control systems with autonomy and intelligence or cognitive capabilities for co-operative/collaborative work, agile and fast adaptation to environmental changes, robustness against the occurrence of disturbances, and the easier integration of manufacturing resources and legacy systems. One of the major challenges is also human safety issues, which are not addressed in this article.

6 Conclusions

Recent manufacturing roadmaps are pinpointing the development needs in human-system interaction, beyond the current HMI systems, and use of the keyboard as an input device and the display as the output device. In this paper, we present a novel system concept to be designed to fulfil the requirements for hybrid, knowledge-intensive manufacturing in the future, where humans and robots operate in close cooperation.

Related to the topic, the paradigm of the cognitive factory has been outlined by Zäh [16]. This is a factory environment with its machines, robots, storages, planning processes and human workers, that is equipped with a sensor network and an IT structure to allow the resources and processes to perceive what they are doing, enabling them to control themselves, and plan further actions in cooperation with other machines and human workers. It can be considered that the ubiquitous industrial robot concept introduced in this paper is one part of the cognitive factory.

Acknowledgements

The authors wish to acknowledge the financial support received from VTT. The development is part of the VTT Thema Complex System Design project KNOWMAN - Knowledge and Skill-Driven Manufacturing of High Added-Value Products.

References

1. Caie, J. ARC STRATEGIES. NIST Workshop defines preliminary Roadmaps for Smart Assembly. ARC Advisory Group, February 2007, 16 p.
2. MANUFUTURE - Strategic Research Agenda. Manufuture Technology Platform, September 2006. available from http://www.manufuture.org/strategic.html, accessed 13th July, 2007
3. EUROP - Strategic Research Agenda. Technology Platform on Robotics. May 2006, available from http://www.robotics-platform.eu.com/documents.htm accessed 13th July, 2007
4. Dencker, K., Stahre, J., Grondahl, P., Martensson, L., Lundholm, T., Johansson, C. An Approach to Proactive Assembly Systems: -Towards competitive assembly systems. IEEE International Symposium on Assembly and Manufacturing, ISAM '07. Ann Arbor, Michigan, USA, 22-25 July 2007, pp. 294 – 299.
5. A. Saffiotti and M. Broxvall, "PEIS Ecologies: Ambient intelligence meets autonomous robotics," in *Proc of the Int Conf on Smart Objects and Ambient Intelligence* (sOc-EUSAI), Grenoble, France, 2005, pp. 275–280.
6. http://istresults.cordis.europa.eu/index.cfm?section=overview&tpl=glossary accessed 13th July. 2007
7. Bainbridge, L. 1983. Ironies of automation. *Automatica* 19, 775-779.
8. Rasmussen, J. 1986. Information processing and human-machine interaction: An approach to cognitive engineering. New York: North-Holland. 215 p.
9. Vicente, K. J., & Rasmussen, J. 1992. Ecological interface design: Theoretical foundations. *IEEE Transactions on Systems, Man, and Cybernetics*, 22, 4, p. 589-606.
10. Schraft, R. D., Meyer, C. The Need for an Intuitive Teaching Method for Small and Medium Enterprises. In: VDI-Wissensforum et al.: *ISR 2006 - ROBOTIK 2006 : Proceedings of the Joint Conference on Robotics*, May 15-17, 2006, Munich: Visions are Reality. Düsseldorf, 2006, 10 p. (CD-ROM), Abstract p. 95 (VDI-Berichte 1956). Available from www.smerobot.org - sientific publications.

11. Barna Reskó, Andor Gaudia, Péter Baranyi, Trygve Thomessen. Ubiquitous Sensory Intelligence in Industrial Robot Programming. *5th International Symposium of Hungarian Researchers on Computational Intelligence.* November 11-12, 2004, Budapes
12. Billard, A. & Siegwart, R. 2004. Robot learning from demonstration. *Robotics and Autonomous Systems* 47, 65-67.
13. Dillman, R. 2004. Teaching and learning of robot tasks via observation of human performance. *Robotics and Autonomous Systems* 47, 109-116.
14. Kirchhoff, Uwe; Stokic, Dragan; Sundmaeker, Harald; AmI Technologies Based Business Improvement in Manufacturing SMEs. Paper at t*he eChallenges e-2006 Conference;* 25 - 27 October, Barcelona, Spain.
15. Jong-Hwan Kim, Yong-Duk Kim, and Kang-Hee Lee. The Third Generation of Robotics: Ubiquitous Robot. *2nd International Conference on Autonomous Robots and Agents* December 13-15, 2004 Palmerston North, New Zealand
16. Zäh, M. F.; Lau, C.; Wiesbeck, M.; Ostgathe, M.; Vogl, W. 2007. Towards the Cognitive Factory, in *Proceedings of the 2nd International Conference on Changeable, Agile, Reconfigurable and Virtual Production* (CARV 2007)
17. Stavros Vassos. Cognitive robotics in the industry. available at http://stavros.lostre.org/2007/05/14/cognitive-robotics-in-the-industry/ accessed 2nd October, 2007
18. Sallinen, M.; Salmi, T.; Haataja, K.; Göös, J.; Voho, P. 2006. A Concept for Short Series Production and Manufacturing: Isles of Automation. *Smart Systems 2006 & ICMA 2006, Conference Proceedings (2006).* 6th International Conference on Machine Automation (ICMA2006).Seinäjoki, 7 - 8 June 2006.

AUTHOR INDEX

Adragna, P. A.	23, 189	Jones, C. W.	307
Alvarado, J. R.	121	Kallela, T.	99
Banse, X.	207	Kapoor, S. G.	37
Beckert, E.	139	Koelemeijer Chollet, S.	315, 345
Behera, A. K.	37	Laine, E.	385
Boufercha, N.	149	Lambert, P.	265
Bourgeois, F.	67, 315	Lanz, M.	99
Braun, M.	315	Leach, R. K.	5, 307
Brecher, C.	285, 297	Lempiäinen, J.	395
Burgard, M.	149	Lenders, C.	265
Burisch, A.	337	Luetzelschwab, M.	129
Burkhardt, T.	139	Lutz, P.	235
Cartiaux, O.	207	Malm, T.	385
Ceyssens, F.	75	Malukhin, K.	243
Charvier, L.	67	Marastio, I.	385
De Volder, M.	75	Osuna, R. V.	121
Degen, R.	257	Othman, N.	149
Delchambre, A.	265	Papastathis, T.	177
Denimal, D.	55, 189	Parusel, A.	109
Desaedeleer, M.	265	Paul, L.	207
Desmulliez, M. P. Y.	129	Peltomaa, I.	371
DeVor, R. E.	37	Pillet, M.	23, 189
Dickerhof, M.	109	Plak, R.	251, 325
Docquier, P. L.	207	Porta, M.	223
Dufey, S.	345	Puers, R.	75
Eberhardt, R.	139	Puik, E.	251, 325
Ehmann, K.	243	Pyschny, N.	285
Fantoni, G.	223	Raatz, A.	161, 199, 337
Favreliere, H.	23	Rabenorosoa, K.	235
Freundt, M.	285	Ratchev, S.	5, 171, 277, 359
Garcia, F.	99	Rathmann, S.	161
Gengenbach, U.	87	Raucent, B.	75, 207
Germain, F.	55	Reynaerts, D.	75
Giordano, M.	55	Ronaldo, R.	5, 171, 177
Görtzen, R	251, 325	Ryll, M.	359
Haag, M.	353	Sääski, J.	395
Haddab, Y.	235	Sägebarthl, J.	149
Hakkarainen, M.	395	Sallinen, M.	405
Härer, S.	353	Salmi, T.	385
Heilala, J.	371, 405	Salonen, T.	395
Helaakoski, H.	371	Samper, S.	23, 189
Hesselbach, H.	161, 199, 337	Sandmaier, H.	149
Hoch, A.	353	Schäfer, W.	149
Jacot, J.	67, 315, 345	Scharnowell, R.	87

Schlenker, D.	149
Schöttler, K.	199
Sieber, I.	87
Siltanen, S.	395
Simons, F.	353
Smal, O.	75
Staiger, A.	257
Tanase, D.	13
Tichem, M.	13
Tietje, C.	5, 177
Tsiklos, G.	171
Tünnermann, A.	139
Tuokko, R.	99, 121
Turitto, M.	5, 277
Valsamis, J. B.	265
Wehrli, F.	345
Weiland, D.	129
Weinzierl, M.	285
Wenzel, C.	285
Woodward, C.	395
Yang, H.	171, 177

Printed in the United States of America